Palgrave Studies in
and F

Series Editor
Christopher B. Barrett, Cornell University, Ithaca, NY, USA

Agricultural and food policy lies at the heart of many pressing societal issues today and economic analysis occupies a privileged place in contemporary policy debates. The global food price crises of 2008 and 2010 underscored the mounting challenge of meeting rapidly increasing food demand in the face of increasingly scarce land and water resources. The twin scourges of poverty and hunger quickly resurfaced as high-level policy concerns, partly because of food price riots and mounting insurgencies fomented by contestation over rural resources. Meanwhile, agriculture's heavy footprint on natural resources motivates heated environmental debates about climate change, water and land use, biodiversity conservation and chemical pollution. Agricultural technological change, especially associated with the introduction of genetically modified organisms, also introduces unprecedented questions surrounding intellectual property rights and consumer preferences regarding credence (i.e., unobservable by consumers) characteristics. Similar new agricultural commodity consumer behavior issues have emerged around issues such as local foods, organic agriculture and fair trade, even motivating broader social movements. Public health issues related to obesity, food safety, and zoonotic diseases such as avian or swine flu also have roots deep in agricultural and food policy. And agriculture has become inextricably linked to energy policy through biofuels production. Meanwhile, the agricultural and food economy is changing rapidly throughout the world, marked by continued consolidation at both farm production and retail distribution levels, elongating value chains, expanding international trade, and growing reliance on immigrant labor and information and communications technologies. In summary, a vast range of topics of widespread popular and scholarly interest revolve around agricultural and food policy and economics. The extensive list of prospective authors, titles and topics offers a partial, illustrative listing. Thus a series of topical volumes, featuring cutting-edge economic analysis by leading scholars has considerable prospect for both attracting attention and garnering sales. This series will feature leading global experts writing accessible summaries of the best current economics and related research on topics of widespread interest to both scholarly and lay audiences.

Christophe Béné · Stephen Devereux
Editors

# Resilience and Food Security in a Food Systems Context

palgrave
macmillan

*Editors*
Christophe Béné
International Center for Tropical Agriculture
Cali, Colombia

Stephen Devereux
Institute of Development Studies
University of Sussex
Brighton, UK

Open access of this publication was made possible through funding from South Africa's National Research Foundation (grant number: 98411) and from the CGIAR-funded Initiative "Sustainable Healthy Diets through Food Systems Transformation" (SHiFT).

ISSN 2662-3889　　　　　　　　ISSN 2662-3897　(electronic)
Palgrave Studies in Agricultural Economics and Food Policy
ISBN 978-3-031-23537-5　　　ISBN 978-3-031-23535-1　(eBook)
https://doi.org/10.1007/978-3-031-23535-1

© The Editor(s) (if applicable) and The Author(s) 2023, corrected publication 2023. This book is an open access publication.

**Open Access** This book is licensed under the terms of the Creative Commons Attribution 4.0 International License (http://creativecommons.org/licenses/by/4.0/), which permits use, sharing, adaptation, distribution and reproduction in any medium or format, as long as you give appropriate credit to the original author(s) and the source, provide a link to the Creative Commons license and indicate if changes were made.

The images or other third party material in this book are included in the book's Creative Commons license, unless indicated otherwise in a credit line to the material. If material is not included in the book's Creative Commons license and your intended use is not permitted by statutory regulation or exceeds the permitted use, you will need to obtain permission directly from the copyright holder.

The use of general descriptive names, registered names, trademarks, service marks, etc. in this publication does not imply, even in the absence of a specific statement, that such names are exempt from the relevant protective laws and regulations and therefore free for general use.

The publisher, the authors, and the editors are safe to assume that the advice and information in this book are believed to be true and accurate at the date of publication. Neither the publisher nor the authors or the editors give a warranty, expressed or implied, with respect to the material contained herein or for any errors or omissions that may have been made. The publisher remains neutral with regard to jurisdictional claims in published maps and institutional affiliations.

Cover credit: SONY-ILCE-7M3/Unsplash Images

This Palgrave Macmillan imprint is published by the registered company Springer Nature Switzerland AG
The registered company address is: Gewerbestrasse 11, 6330 Cham, Switzerland

The original version of this book was revised: Acknowledgement of open access funder has been included. The correction to this book is available at https://doi.org/10.1007/978-3-031-23535-1_13

# Contents

1 Resilience, Food Security and Food Systems: Setting the Scene  1
Christophe Béné and Stephen Devereux

2 Achieving Food Security Through a Food Systems Lens  31
Jessica Fanzo

3 The Global Food System is Not Broken but Its Resilience is Threatened  53
Patrick Caron, Ellie Daguet, and Sandrine Dury

4 Food Security and the Fractured Consensus on Food Resilience: An Analysis of Development Agency Narratives  81
Karl-Axel Lindgren and Tim Lang

5 Food Security and Resilience: The Potential for Coherence and the Reality of Fragmented Applications in Policy and Research  147
Mark A. Constas

6 Food Systems, Resilience, and Their Implications for Public Action  185
John Hoddinott

| 7 | Food Security Under a Changing Climate: Exploring the Integration of Resilience in Research and Practice<br>Alessandro De Pinto, Md Mofakkarul Islam, and Pamela Katic | 207 |
|---|---|---|
| 8 | Gender, Resilience, and Food Systems<br>Elizabeth Bryan, Claudia Ringler, and Ruth Meinzen-Dick | 239 |
| 9 | COVID-19, Household Resilience, and Rural Food Systems: Evidence from Southern and Eastern Africa<br>Joanna Upton, Elizabeth Tennant, Kathryn J. Fiorella, and Christopher B. Barrett | 281 |
| 10 | Place-Based Approaches to Food System Resilience: Emerging Trends and Lessons from South Africa<br>Bruno Losch and Julian May | 321 |
| 11 | Urban Food Security and Resilience<br>Gareth Haysom and Jane Battersby | 355 |
| 12 | Reflections and Conclusions<br>Stephen Devereux and Christophe Béné | 389 |

Correction to: Resilience and Food Security in a Food Systems Context   C1
Christophe Béné and Stephen Devereux

Index   407

# Editors and Contributors

## About the Editors

**Christophe Béné** is Senior Policy Advisor at the International Center for Tropical Agriculture (CIAT). He is currently affiliated to the CGIAR Systems Transformation Science Group and seconded to the Economic Research Group (WEcR), at Wageningen University in the Netherlands. He has 20+ years of experience conducting inter-disciplinary research and advisory work in different parts of the world (Africa, Asia, Pacific), focusing on poverty alleviation and food security in low- and middle-income countries. In his career, he worked on a wide range of topics, including natural resource management, policy analysis, science-policy interface, resilience (measurement), and more recently, food system (sustainability).

**Stephen Devereux** is co-Director of the Centre for Social Protection at the Institute of Development Studies, University of Sussex, UK. He holds a research chair in Social Protection for Food Security at the Institute for Social Development, University of the Western Cape, South Africa, supported by the National Research Foundation (NRF) and the Newton Fund, and affiliated to the DSI-NRF Centre of Excellence in Food Security. He currently holds a Mercator Fellowship at the SOCIUM Research Centre on Inequality and Social Policy, University of Bremen, Germany.

He has (co-)authored and (co-)edited books on famine, food security, seasonality, and social protection.

## Contributors

**Christopher B. Barrett** is an Agricultural and Development Economist on Faculty at Cornell University. He is co-editor-in-chief of the journal *Food Policy*, edits the Palgrave Macmillan book series Agricultural Economics and Food Policy, co-edited the Elsevier *Handbook of Agricultural Economics*, volumes 5 and 6, and previously was editor of the *American Journal of Agricultural Economics*. He is an elected Member of the National Academy of Sciences and an elected Fellow of the American Association for the Advancement of Science, the Agricultural and Applied Economics Association, and the African Association of Agricultural Economists.

**Jane Battersby** is a Senior Lecturer in the Department of Environmental and Geographical Science at the University of Cape Town. She has worked on urban food security, urban food systems and their governance in Southern and Eastern Africa for over a decade. Her research interests lie in the relationships between food environments, urban systems, and social systems, with a particular focus on the dual burden of malnutrition. She has worked in consultancy and advisory roles for a number of UN Agencies and local and provincial governments. She holds a D.Phil. in Geography from Oxford University.

**Elizabeth Bryan** is a Senior Scientist in IFPRI's Environment and Production Technology Division, where she conducts policy-relevant research on gender, climate change, resilience, sustainable agricultural production, natural resource management, and small-scale irrigation, using mixed-methods. She has published numerous articles based on her work and regularly presents research results to diverse audiences. She holds a doctorate in agricultural sciences from Hohenheim University, an M.A. in development economics from American University, and a B.A. in international affairs from Wagner College.

**Patrick Caron** is a geographer, focusing on territorial dynamics. His works relate to the analysis of food systems in rural transformations. He is Director of the Montpellier Advanced Knowledge Institute for Transitions (MAK'IT) and Vice President for International Affairs of the

University of Montpellier. He joined CIRAD in 1988 and was Director for research from 2010 to 2016. He was the Chair of the High Level Panel of Experts (HLPE) of the UN Committee on World Food Security from 2015 to 2019. He is Board member of CGIAR and member of the Scientific Group of the UN Food Systems Summit.

**Mark A. Constas** is an Associate Professor of Applied Economics and Policy in the Charles H. Dyson School of Applied Economics and Management at Cornell University. His research is focused on the development of assessment tools and evaluation designs to measure the ways in which households and communities situated in shock-prone contexts can achieve and maintain well-being. His work has appeared in a selection of competitive peer-reviewed journal and in other publication outlets intended for policy audiences. He regularly participates in expert panels on resilience measurement and has served as an advisor to several United Nations organizations.

**Ellie Daguet** is a French engineer in agronomy and life sciences, specialized in human nutrition. After working for the French government for 9 years on the European regulation on nutrition and health claims made on foods, as the representative for France, she joined the CIRAD—French agricultural research center for international development—to work on the sustainability of food systems and more particularly on the UN Food Systems Summit 2021.

**Alessandro De Pinto** is Professor of Climate Change and Food Security at the Natural Resources Institute, University of Greenwich. He is an environmental and natural resource economist with some 30 years of experience. Throughout his career, he has worked on providing support for the design of policies promoting sustainable use of natural resources. He has conducted research in Latin America, South Asia, and sub-Saharan Africa with a focus on climate smart agriculture, climate resilience, and low emissions development strategies. He also worked extensively on developing modelling techniques connecting local issues with global forces such as climate change and global markets.

**Sandrine Dury** is an agricultural economist at the center of agricultural research for international development (CIRAD). In the last years, she has been working on the linkages between food and nutrition security and farm production with several African partners and at global level. She managed the Relax project (2017–2021) which explored the linkages between agricultural, wild resources, and food consumption diversity in order to understand why higher crop production does not always translate into better nutrition for farm household members. In 2019, she co-edited the CIRAD-FAO-EU report food systems at risk: New trends and challenges.

**Jessica Fanzo,** Ph.D. is the Bloomberg Distinguished Professor of Global Food Policy and Ethics at Johns Hopkins University. She holds appointments in the Nitze School of Advanced International Studies (SAIS), Berman Institute of Bioethics, and the Bloomberg School of Public Health. She serves as the Director of Hopkins' Global Food Policy & Ethics Program, and as Director of Food & Nutrition Security at Hopkins' Alliance for a Healthier World. Before Hopkins, she held positions at Columbia University, the Food and Agriculture Organization, the World Food Programme, Bioversity International, and the Doris Duke Charitable Foundation. She holds a Ph.D. in nutrition from the University of Arizona.

**Katie Fiorella** is Assistant Professor in the Department of Public and Ecosystem Health at Cornell University. She is environmental scientist and epidemiologist. Her research aims to understand the interactions among environmental change and livelihood, food, and nutrition security. Her work is focused on global fisheries and the households that are reliant on the environment to access food and income. She uses interdisciplinary methods to foster a deeper understanding of how ecological and social systems interact, the ways communities and households adapt to and mitigate environmental change, and the links between human well-being and ecological sustainability.

**Gareth Haysom** is an Urban Food systems researcher at the African Centre for Cities, University of Cape Town. His work uses food as a lens to understand the rapidly changing nature of African and Southern cities, working across countries and city types. Motivating this work is the premise that the future of African and Southern countries will be urban, and a narrow window exists in which to respond to the development and specific food system needs. His research spans the food and urban systems, across scales and governance domains. He holds a Ph.D. from the University of Cape Town.

**John Hoddinott** is the H.E. Babcock Professor of Food and Nutrition Economics and Policy, Cornell University and a Non-Resident Senior Fellow, International Food Policy Research Institute (IFPRI). Previously, he was a Deputy Division Director at IFPRI and University Lecturer in Economics and Fellow and Tutor, Lady Margaret Hall, University of Oxford. His research focuses on issues surrounding poverty, food insecurity, and undernutrition. Born in Canada, he has a B.A. in Economics from the University of Toronto and a doctorate in economics from the University of Oxford.

**Md Mofakkarul Islam** is a Senior Fellow at the Natural Resources Institute (NRI) of the University of Greenwich. Previously, he held teaching and research positions at Nottingham Trent University, Scotland's Rural College, the Open University, Massey University, and Bangladesh Agricultural University. He is an interdisciplinary social scientist, with over 20 years of professional experience, working on sustainable development issues in agriculture, food, and environment. His recent works have focused on indicators of sustainable agricultural intensification; smallholders' climate change adaptation and vulnerability; food security in a conflict context; and justice in global climate finance.

**Pamela Katic** is a Senior Social Scientist at the Natural Resources Institute, University of Greenwich. She has extensive experience in environmental economic modeling and ecosystem services valuation, including water economics and policy, trade-off analysis in agroforestry systems, Indigenous Peoples' food systems, biodiversity economics, aquaculture diversification, and mediation analysis of the pathways between natural resource management and health/nutrition outcomes. She is Vice-chair of Future Earth's Water-Food-Energy Knowledge-Action Network and a member of the UK Future Earth National Committee. She is also an active member of FAO's Global Hub on Indigenous Peoples' Food Systems and the Water Justice Hub.

**Tim Lang** is Professor Emeritus of Food Policy at City University of London's Centre for Food Policy which he founded in 1994 and directed until 2017. Hill farming in Lancashire in the 1970s formed his interest in the relationship between food, health, environment, and culture which he's worked on ever since. He was policy lead on the EAT-Lancet Commission's 2019 Food in the Anthropocene report and is Senior Advisor to the Food Research Collaboration linking academia with civil society on food matters. His books include *Feeding Britain* (Pelican, 2021), *Sustainable Diets* (Routledge, 2017), and *Food Wars* (2015).

**Karl-Axel Lindgren** is a Postdoctoral Researcher at the Centre for Food Policy, City University of London. His thesis on India's 2013 National

Food Security Act critically analyzed the narratives around urban food security among policy-makers, exploring the extent that urban food security was considered and understood in the formulation of the Act. He has contributed one chapter in the book 'Food Poverty and Insecurity: International Food Inequalities' on the Right to Food in India, analyzing the shift of entitlements from government responsibility to obligation, and published an article in the journal Food Security on Indian food security.

**Bruno Losch** is a Political Economist at CIRAD (*UMR ArtDev* and University of Montpellier, France) currently posted at the *University of the Western* Cape, South Africa, where he is program investigator at the *Centre of Excellence in Food Security* (food governance and policy program). Before settling in South Africa, he was senior economist at the World Bank, visiting fellow at the University of California, Berkeley, and in charge of CIRAD's Family Agriculture Program. He has published extensively in the field of territorial policies, rural studies, family agriculture, and the political economy of development, with a particular focus on Africa, Central America, and the Pacific.

**Julian May** is the Director of the DSI-NRF Centre of Excellence in Food Security at the University of the Western Cape and holds the UNESCO Chair in African Food Systems. He is an economist and has worked on options for poverty reduction including public nutrition, land reform, social grants, information and communication technologies, and food system governance. He has published extensively and has served on international and national expert groups, including the South African Statistics Council, the Council of the Academy of Science of South Africa, and currently, the third National Planning Commission.

**Ruth Meinzen-Dick** is a Senior Research Fellow in the Environment and Production Technology Division at IFPRI where her research explores how institutions and policies affect the way people manage natural resources, especially land and water. She also studies gender issues in agriculture, with a particular focus on gender differences in control over assets, and the impact of agricultural research on poverty. She is the author of over 150 peer-reviewed publications based on this research. She received her Ph.D. and M.Sc. degrees in development sociology from Cornell University and her bachelor's degree in anthropology from Washington University.

**Claudia Ringler** is Deputy Division Director of IFPRI's Environment and Production Technology Division where she leads the natural resource management theme. Her research focuses on water resources management and agricultural and natural resource policy. Her recent research also focuses on the impacts of global warming for developing country agriculture and on appropriate adaptation and mitigation options. She has more than 100 journal articles on these research areas. She received her Ph.D. in Agricultural Economics from the Center for Development Research, Bonn University, Germany, and her M.A. in International and Development Economics from Yale University.

**Elizabeth Tennant** is a Research Associate in the Department of Economics at Cornell University. Her research uses integrated earth- and human-systems approaches to answer questions related to disaster risk reduction, climate adaptation, resilience and sustainable development. In recent work, she has studied how institutions, inequalities, and other socioeconomic patterns shape our ability to absorb and recover from extreme events.

**Joanna Upton** is a Research Associate in the Charles H. Dyson School of Applied Economics and Management at Cornell. She focuses with her research and outreach on food security, food policy, and resilience in sub-Saharan Africa and has a passion in particular for working on questions with policy-relevant applications. She works closely with a wide range of collaborators, including academics from diverse disciplines, CGIAR institutions, UN organizations, government agencies, think tanks, and non-profit organizations implementing programs in developing countries. She also thrives on working with students and mentees, both in the classroom and in the field.

# List of Figures

| | | |
|---|---|---|
| Fig. 2.1 | Global changes in cereal production, yield, and land use, 1961–2018 (*Source* Our World in Data [2021]) | 34 |
| Fig. 2.2 | Food insecurity pathways to multiple forms of malnutrition (*Source* FAO et al. [2018]) | 35 |
| Fig. 2.3 | A food systems framework (*Source* Fanzo, Rudie et al. [2021]) | 40 |
| Fig. 2.4 | Change in number of people who could be nourished without trade (*Source* Wood et al. [2018]) | 42 |
| Fig. 3.1 | Long-term evolution of real agricultural prices (OECD, 2021) (*Source* OECD/FAO [2020], "OECD-FAO Agricultural Outlook", OECD Agriculture statistics [database], http://dx.doi.org/10.1787/agr-outl-data-en) | 55 |
| Fig. 3.2 | Global population, food production and agricultural land use in the long run (OECD, 2021) http://dx.doi.org/10.1787/agr-outl-data-en | 56 |
| Fig. 3.3 | Share of consumer expenditure spent on food worldwide in 2015 (*Data source* United States Department for Agriculture, Economic Research Service, USDA, ERS, 2017) | 58 |
| Fig. 3.4 | Undernourishment, overweight and obesity, 2000–2016 (OECD, 2021) | 59 |
| Fig. 5.1 | Integrated food security and resilience model (*Source* Author) | 154 |
| Fig. 5.2 | Appearance of "*resilience*": trends in SOFI, 1999–2020 (*Source* Author) | 161 |

| Fig. 5.3 | Appearance of *"resilience"*: trends in SOFI (2004–2020) (*Source* Author) | 162 |
| Fig. 5.4 | Appearance of *"resilience"* in the research literature (*Source* Author) | 163 |
| Fig. 5.5 | Relative prevalence of "resilience" in the research literature: Food Policy and Global Food Security (*Source* Author) | 164 |
| Fig. 7.1 | Incidence of the words 'climate change and resilience'; 'food security and resilience'; 'climate change, food security and resilience' in selected academic literature, years 1996–2020 (*Source* Authors) | 211 |
| Fig. 7.2 | The generic resilience intervention framework/compiled from the case study projects (*Source* Authors) | 221 |
| Fig. 8.1 | The Gender, Climate Change, and Nutrition Integration Initiative (GCAN) framework (*Source* Adapted by the authors from Bryan et al. [2017] and Theis et al. [2019]) | 243 |
| Fig. 9.1 | Conceptual framework of covariate shocks, mechanisms, and food security resilience (*Source* Authors) | 288 |
| Fig. 9.2 | COVID-19 control measures' stringency varied over time in Malawi, Madagascar and Kenya—original figured designed from Oxford COVID-19 Government Response Tracker data (Johns Hopkins University, 2021; Hale et al., 2021) | 289 |
| Fig. 9.3 | Mean Household Hunger Scale, over time, Malawi (*Note* Range is 0–6 with higher worse; >=1 is considered 'stressed,' and >=3 'crisis') | 297 |
| Fig. 9.4 | Mean Food Consumption Score, over time, Madagascar. Dashed horizontal lines represent standard thresholds for "Food Secure" (35) and "Borderline" (21). Years run April–March. A higher FCS is better | 298 |
| Fig. 9.5 | Food Security Score (HFIAS), March 2020–March 2021, Kenya | 299 |
| Fig. 9.6 | Reported worry about impacts of (a) COVID-19 and (b) COVID-19 control measures, Kenya, June 2020–March 2021 | 302 |
| Fig. 9.7 | Reported impacts of the pandemic, May–September 2020, Malawi | 303 |
| Fig. 9.8 | Reported impacts of the pandemic, April 2020–March 2021, Madagascar | 304 |

| | | |
|---|---|---|
| Fig. 9.9 | Reports of reasons for lost income among those reporting losses, May 2020–September 2020, Malawi. Of responding households, 87–89% reported lost income in every period, the vast majority (77–80%) a male income earner | 306 |
| Fig. 9.10 | Household reports of working or earning less to avoid COVID-19 infection, June 2020–March 2021, Kenya | 308 |
| Fig. 9.11 | Month-to-month changes in imported rice price, comparing years, Madagascar | 310 |
| Fig. 11.1 | Urban food insecurity within a wider set of intersecting urban processes and vulnerabilities despite absent governance action (*Source* Authors' own representation) | 369 |
| Fig. 11.2 | Household food types by HDDS food groups across all income groups in Cape Town, South Africa (High: $n = 200$; Middle: $n = 504$; Low: $n = 1800$) (*Source* Haysom et al. [2017]—from 2013 data) | 373 |
| Fig. 11.3 | Food access profiles in Cape Town, South Africa ($n = 2{,}503$) (*Source* Haysom et al. [2017]) | 374 |
| Fig. 11.4 | Combined Lived Poverty Index results for key indicators in three African cities in response to the question "how often in the past year did you go without …" (Epworth, Zimbabwe $n = 483$; Kitwe, Zambia $n = 871$; Kisumu, Kenya $n = 841$) (*Source* Haysom and Fuseini [2019, p. 17]) | 375 |
| Fig. 11.5 | Kisumu food trader costs after stock purchase costs (multiple response option; $n = 1{,}839$) (*Source* Opiyo et al. [2018]) | 377 |

# List of Tables

| | | |
|---|---|---|
| Table 4.1 | Types of resilience and examples of agencies who address it | 87 |
| Table 4.2 | Emerging themes and sub-themes around food resilience | 89 |
| Table 4.3 | Bilateral agencies—narratives on food security & resilience | 90 |
| Table 4.4 | Regional development banks—narratives of food security & resilience | 96 |
| Table 4.5 | Multilateral agencies—narratives of food security & resilience | 102 |
| Table 5.1 | Logic of the review | 158 |
| Table 5.2 | SOFI—resilience, food security, and food systems | 166 |
| Table 5.3 | *Global Food Security* articles examined for content analysis | 168 |
| Table 5.4 | Summary of findings of the content analysis of *Global Food Security* | 171 |
| Table 5.5 | SOFI titles/themes 1999–2020 | 176 |
| Table 7.1 | Case study projects | 219 |
| Table 9.1 | Data collection start dates, locations, and sample sizes | 290 |
| Table 9.2 | Food security indicators collected, by location | 290 |
| Table 11.1 | Food security and dietary diversity scores for selected Southern cities—periods 2015–2018 | 372 |

CHAPTER 1

# Resilience, Food Security and Food Systems: Setting the Scene

*Christophe Béné and Stephen Devereux*

## PREAMBLE: WHAT THIS BOOK IS ABOUT

The key questions which underpin the writing of this collective volume revolve around the concept of resilience and the contribution that this concept plays in the international development agenda, specifically in relation to the issue of food security. Put simply: what does resilience bring that is relevant, useful and different from other previous or contemporary concepts that have shaped the food security agenda, such as entitlements (Sen, 1981), vulnerability (Chambers, 1989) or the sustainable livelihoods framework (Scoones, 1998)? How does the concept of

---

C. Béné (✉)
International Center for Tropical Agriculture, Cali, Colombia
e-mail: c.bene@cgiar.org

Wageningen Economic Research group, Wageningen, The Netherlands

S. Devereux
Institute of Development Studies, University of Sussex, Brighton, UK
e-mail: s.devereux@ids.ac.uk

---

© The Author(s) 2023
C. Béné and S. Devereux (eds.), *Resilience and Food Security in a Food Systems Context*, Palgrave Studies in Agricultural Economics and Food Policy, https://doi.org/10.1007/978-3-031-23535-1_1

resilience help in improving our general understanding of the development process, in particular the question of food security, and how is it influencing the way development interventions around food security are now programmed and implemented?

To answer these questions, this edited volume compiles a series of chapters written by a group of international experts recognized for their contributions to a diverse range of development questions. This volume is not, therefore, just advocating for 'more resilience'. Instead, it proposes to step back, take stock of what has been learned and what is still being debated, and assess rigorously and critically the contribution of this concept in advancing our understanding and ability to design and implement development interventions in relation to food security.

In doing so, and in a resolute departure from the narrow and beaten tracks of agriculture and trade that have influenced the mainstream debate on food security for nearly 60 years (cf. FAO, 1980), the book also proposes to adopt a wider, more holistic and integrative perspective, framed around food systems. Our premise is that in the current post-globalization era, the food and nutritional security of the world's population at household, community, municipality or even country level no longer depends just on the performance of the agricultural sector and national or international policies on trade, but rather on the capacity of the entire system to produce, process, transport and distribute safe, affordable and nutritious food in ways that remain environmentally sustainable and socially acceptable. In that context, we posit that adopting a food systems' perspective provides a more appropriate frame for the questions on food security, as it incites us to broaden our conventional thinking and to acknowledge the systemic and interactive nature of the different processes and actors involved (Ericksen, 2008; HLPE, 2017).

The volume comprises 12 chapters that offer a carefully pondered combination of conceptual discussions, historical reviews and empirical analyses. The ambition is that the main questions that drive the reflections around the three concepts central to this volume and their interactions are addressed in a balanced and comprehensive manner, making the discussion relevant to a large audience including researchers, policymakers and practitioners, but also members of UN or bi/multi-lateral development agencies working on food security or related development policy, planning and programming. The book is not designed, therefore, to remain a scholastic exercise or a textbook for academics; neither is it intended to become a practical handbook directed at NGOs looking for ways to implement resilience activities on the ground. Nonetheless, the relevance of this volume and its timeliness are indisputable, and the impact of the COVID-19 pandemic on food systems' resilience and on people's food security has confirmed this strikingly.

In addition to chapters exploring in greater depth, some of the conceptual aspects of this multi-faceted relationship between resilience, food security and food systems in the context of low- and middle-income countries (LMICs), the book also provides empirical analyses of resilience and food security related to gender, cities, climate change, locality and COVID-19. This choice of specific topics reflects what the co-editors consider as important themes in the current discourse. Of course, this list of themes and topics is not encyclopaedic or all-inclusive. This field of research is growing rapidly, and many other issues could also have been selected, such as nutrition, biodiversity, political economy or comparative analyses.

Beyond this issue of which themes are critical when analysing resilience, food security and food systems, it will soon become evident when advancing through the different chapters of this book that many other questions around those concepts and their interactions remain unsettled and contested. The question of the most effective level at which resilience should be considered is a good illustration of this open discussion. On this issue, a growing body of literature insists that local food systems are preferable over global ones (see, e.g., Colet et al., 2009; FAO, 2020; La Trobe & Acott, 2000). Bruno Losch and Julian May (Chapter 10 in this volume) follow this view and posit that adopting a local approach based on community-led responses to food system management (what they refer to as a "place-based" approach) in a time of crisis is key to build food systems' resilience. At the other end of the spectrum, another group of authors claims on the contrary that the more open and connected our food systems, the better. This view builds conceptually on the idea that multiplicity and flexibility are important priorities to build resilience (Ebata et al., 2020; Reardon & Swinnen, 2020). John Hoddinott, for instance (Chapter 6 in this volume), adopts this line of thinking when he reminds us that inward looking systems (such as Ethiopia in the 1980s or North Korea in the 1990s) are more vulnerable to food system collapse than more integrated systems based on strong trade markets. Yet, as COVID-19 demonstrated, too much connectivity is also likely to expose people to "concatenated shocks" (Biggs et al., 2011). In sum, no consensus has yet emerged on whether privileging local or global solutions is the way forward. Or perhaps should we consider both simultaneously? This is for instance what Patrick Caron, Ellie Daguet and Sandrine Dury propose when they remark (Chapter 3 in this volume) "Rather than opposing them, articulating local, national and international processes, arrangements and frameworks, and organising inter-dependencies and regulations among scales are key to promote the diversity and the coexistence of context-specific pathways".

## Clarifying the Concepts

The next step in this introductory chapter is to start clarifying the three concepts that are at its core. The reader will soon observe, however, that this volume and its contributors do not adopt a unique view or definition of "resilience", "food systems" or even "food security". On the contrary, one aspiration of this endeavour was to embrace and promote a pluralist interpretation where several different and sometimes divergent interpretations of these concepts co-exist across the different chapters, reflecting the rich but still unsettled and contested nature of these concepts.

### *Food Security*

Among the three central concepts of this book, many would consider food security as the concept that is most firmly anchored in the development literature. Indeed, as Mark Constas reminds us (Chapter 5 in this volume), food security understood in a broad sense is a relatively old concept that can be traced back to Malthus and his *Essay on the Principle of Population* (1798) or perhaps even earlier, to the work of Giovani Botero (1588) which highlighted the importance of the balance between the capacity of a city to produce food and the size of its population. Thinking around food security remained under the grip of Malthusian productivist interpretations for another two centuries, until Amartya Sen (1981) revisited and challenged this reductionist framing. Sen's reanalysis of several famines, seen not primarily as the consequence of food shortage but quintessentially as the consequence of the inability of certain groups of people to access or purchase food, was instrumental in expanding the understanding of food security beyond its initial productivist roots and in forging what was to become until very recently the mainstream understanding of food security, represented through its four pillars of availability, access, utilization and stability (FAO, 2006).

This interpretation of food security and its four pillars is what most academics and practitioners would follow today, either implicitly or explicitly. As we shall discuss later in this chapter, building food security around those four pillars is also useful as it will allow us to draw some clear conceptual links with both resilience and food systems. In the meantime, Jess Fanzo (Chapter 2 in this volume) reminds us that even though this concept is well established internationally, it is not cast in stone. "Food security and its framing in international development has historically evolved to adapt to the times". She explains that this adaptation process has become more nuanced because of our deepening understanding of the social, political, environmental and biological causes and consequences of food insecurity. As recently as 2020, following an extensive international consultation, the Committee on Word Food Security (CFS) proposed to expand the current definition of food security by adding two more pillars: agency and sustainability (HLPE, 2020).

The inclusion of agency was justified by the fact that "agency implies the capacity of individuals or groups to make their own decisions about what foods they eat, what foods they produce, how that food is produced, processed and distributed within food systems, and their ability to engage in processes that shape food system policies and governance" (HLPE, 2020, p. 8). In sum, while agency is widely accepted as a key aspect of the development process itself (Kabeer, 1999; Sen, 1985; World Bank, 2005), the CFS saw the necessity to make its link to food security more prominent and explicit.

As for sustainability, its inclusion as a new pillar was proposed on the basis that "Sustainability as a dimension of food security implies food system practices that respect and protect ecosystems [...] over the long term, in their complex interaction with economic and social systems required for providing food security and nutrition" (HLPE, 2020, p. 9). Sustainability was therefore seen as a complement to the stability pillar, which had been added to the original three pillars (availability, access and utilization) in 2006, following the recognition that short-term shocks, such as conflict, natural disasters and market turmoil, can rapidly undermine food security (FAO, 2006). This time, with the inclusion of sustainability, the CFS intended to highlight the longer-term dimension of food security and in particular, the importance of "the linkages between the natural resource base, livelihoods and society to continually maintain systems that support food security and ensure that the needs of future generations are taken into account" (HLPE, 2020,

p. 9). The introduction of sustainability to the concept of food security is a logical extension of the recent emergence of sustainability more generally in the development discourse, as reflected in the Sustainable Development Goals (SDGs) in particular (UN, 2015).

## *Resilience*

Even today, it is not uncommon when one mentions "resilience" in a conversation with academics or development practitioners, to see the "R-word" received with signs of exasperation or even annoyance. This reality certainly has something to do with the fact that, for several years in its early 'career' in the development community, resilience was severely misused and abused, often presented as the panacea to every ailment, and instrumentalized as a way to sell old wine in new bottles (Béné et al., 2012; Cannon & Muller-Mahn, 2010). The term was so ubiquitous that some saw it as the new buzzword of the development industry (Béné et al., 2014; Hussain, 2013), generating legitimate scepticism and suspicion. Ten years later, the term has not totally managed to shake off this reputation, and it is still often used as a 'hook' word that is included in the title of project proposals, academic papers or even politicians' discourses, without necessarily any associated substance. Alex De Pinto, Mofa Islam and Pamela Katic (Chapter 7 in this volume) show, for instance, how in many resilience projects funded under the UK Adaptation Fund, the concept is still presented as an 'antidote' to vulnerability and, as such, proposed as an end-goal of the projects themselves. Yet, as De Pinto et al. point out, the monitoring and evaluation approach generally used by practitioners to evaluate project impacts hardly ever includes any measurements of resilience.

Reiterated rigour and scrutiny are therefore still required when it comes to resilience, if one wants to avoid conceptual confusion and contribute to rehabilitating its reputation. That said, resilience is now widely recognized as both a suitable conceptual lens and an operational concept through which to rethink the complex relationships between society and environment and, within this (and more specifically), between food security and development. While the wider relationship between society and environment has been abundantly debated for more than three decades, essentially with resilience conceptualized from a socio-ecological perspective (see, e.g., Berkes & Folke, 1998), the interest in the second, more specific and more recent, exploration between resilience and development is growing. In this regard, there is little doubt that our understanding of what resilience is and how to use it in relation to development has improved very steeply in the last decade (Ansah et al., 2019; Béné et al., 2014; Constas et al., 2014). As a result, many now see resilience as part of the discourse currently used by bi- and multi-lateral development agencies. Karl-Axel Lindgren and Tim Lang (Chapter 4 in this volume) even compare resilience to a "'scaffolding' used by these agencies to support their development frameworks".

This positive trend does not mean, however, that the road has not been bumpy and winding. As is often the case with the appropriation of a new concept by a particular community, the first years have been characterized by an 'explosion' of definitions and interpretations. In that regard, Joanna Upton and her co-authors (Chapter 9 in this volume) refer to it as an "amorphous concept". Historical accounts of science tell us that this unsettled discussion is part of a normal process of appropriation and is set to continue until some form of paradigm or consensus emerges. In the case of resilience, this consensus is yet to materialize even though several contributions towards it have been made, starting with DFID and USAID both publishing a series of seminal position papers in the early 2010s (DFID, 2011; USAID, 2012). Soon after, the Resilience Measurement Technical Working Group was set up by FAO and WFP in 2013 as an attempt to provide clarity and a normative element to the debate.[1] Internal politics and conflicting individual and institutional ambitions, however, led the group to an early 'retirement'. Its legacy is now reduced to a series of working papers. As a result, no consensus exists and several different views of how resilience should be defined in relation to development still prevail among academics, development agencies and NGOs. For instance, from their review of how resilience has been integrated in the development discourse in relation to climate change, De Pinto and his co-authors (Chapter 7) conclude: "Even though the concept of resilience is better understood among practitioners than in the past, there is a lack of consensus regarding its definition and metrics". This conclusion is echoed by Lindgren and Lang, who refer to a "fractured consensus" to describe a situation where "the malleability and lack of strong consensus on how to concretely define [and measure] resilience" is still the reality today.

---

[1] See https://www.fsinplatform.org/resilience-measurement.

Although this fractured consensus is probably one of the main reasons why many are still able to misuse and abuse the term, we believe that this situation does not necessarily refute the idea that some form of common understanding can emerge from this discordant debate. For us, across all these contested and disputed definitions is the proposition that resilience is simply and broadly about the capacities of individuals, households and communities to deal with adverse events (shocks, stressors) in a way that does not affect negatively their long-term well-being, and in particular their food security.

### *Food Systems*

As with resilience, 'food system' still lacks a universal definition. Yet, like resilience, this lack of a generic definition does not mean there is not some degree of common understanding. In that regard, the work of the High Level Panel of Experts (HLPE) on Food Security and Nutrition of the Committee on World Food Security is often cited as a good entry point. In their reports, the HLPE describes a food system as "all the elements (environment, people, inputs, processes, infrastructures, institutions, etc.) and activities that relate to the production, processing, distribution, preparation and consumption of food, and the outputs of these activities, including socio-economic and environmental outcomes" (HLPE, 2017, p. 11). This description has clear similarities with one of the earliest definitions proposed by Polly Ericksen in her seminal paper (2008), where she defined a food system as being made up of four elements: (i) the interactions between and within biogeophysical and human environments; (ii) a set of interconnected and interdependent activities ranging from production all the way through to consumption; (iii) the outcomes of those activities, including contributions to food security, environmental security and social welfare; and (iv) some other determinants of food security stemming in part from the interactions mentioned in point (i).

Thus, although complex because of their interactions and interdependence (Béné et al., 2019; Fanzo et al., 2021), many recognize that the elements that constitute food systems are also relatively clearly defined (HLPE, 2017). What is missing, perhaps, in all these definitions is a more explicit mention of some of the most important attributes of food systems and of their actors. Two of those attributes are particularly relevant in the context of this book: informality and vulnerability.

***Informality***: One of the critical characteristics structuring food systems, especially in LMICs, is the fact that for the most part the daily and seasonal labourers and self-employed men and women who are engaged in various income-generating activities in food systems typically operate in the informal sector (Resnick, 2017; Roever & Skinner, 2016; Young, 2018). This informality starts at the production level, with entire groups of smallholder farmers, pastoralists or fisherfolks generally working informally (Lowder et al., 2014; McCullough et al., 2010). If they do not commercialize their production themselves on local markets, these producers sell it to middlemen/women (aggregators, wholesalers and brokers) the vast majority of whom are also working in the informal sector (Porter et al., 2007; Veldhuizen et al., 2020). This would involve, among others, women who smoke and process fish directly at home (Akintola & Fakoya, 2017), young men who transport chickens and vegetables every morning on their bicycles to the nearby city or men who collect mangoes produced by their neighbours and transport them with their old pickups to town. Further along the supply chains, the retailing segment is also often dominated by a high degree of informality, both in its structures (open markets, street vending and corner stores) and in its transactional arrangements (informal contracts) (Cadilhon et al., 2006; Kawarazuka et al., 2018; Roever & Skinner, 2016; Smit, 2016). Overall, all these small-scale, informal or semi-formal businesses are sources of revenue and income for a very large number of poor but economically active people for whom these activities often represent a last resort livelihood activity.

***Vulnerability***: With informality comes invisibility in official statistics, exclusion from public programmes, and vulnerability, be it in small-scale fisheries for instance (Kolding et al., 2014) or in urban fruit and vegetable retailing (Steenhuijsen Piters et al., 2021a). In LMICs, those small-scale producers and food suppliers typically operate under extremely difficult conditions. They usually face poor or obsolete infrastructure including inadequate roads, power supply, irrigation, market facilities, etc. (Maloney, 2004), insufficient access to financial services (in particular, credit and insurance) (Oviedo et al., 2009) and high dependence on weather conditions (Harvey et al., 2014). In addition, the informal nature of their activities, combined with insufficient cash flow, economic marginalization and even in some cases discrimination and harassment (Kawarazuka et al., 2018), are exacerbated by the absence of labour protection and laws preventing exploitation, forced and child labour

(Marschke & Vandergeest, 2016), leading to extreme physical, economic and social vulnerability.

In these circumstances, any unexpected shock can have catastrophic implications. The COVID-19 pandemic provides a vivid illustration. In Ethiopia, for instance, Hirvonen et al. (2020) document how the imposition of mobility restriction and lockdowns led to disruptions in informal traders' business practices, including increased costs of transport, decrease in downstream demand and subsequent loss of income. Similar disruptions in the activities of traders, processors and other food system actors were observed in many other countries in 2020, including India (Varshney et al., 2020), Bangladesh (Termeer et al., 2020), Senegal (Tounkara, 2020) and Colombia (Burkart et al., 2020). In some cases, the hardship that followed the imposition of lockdowns led to serious social unrest, like protests which erupted in March and April 2020 in South Africa and Malawi, where informal traders took to the streets, brandishing banners with slogans such as: "Lockdown more poisonous than corona" and "We'd rather die of corona than of hunger" (Aljazeera, 2020), demonstrating that ignoring the economic function played by food systems for millions of informal actors in LMICs is politically risky.

In their empirical analysis conducted in eastern and southern Africa, Joanna Upton and her colleagues (Chapter 9) also observe important disruptions induced by the pandemic on food system actors, but these authors provide a slightly different interpretation than what is usually stated in the literature. For them, the pandemic should still be seen as "a major new shock, but [one that] impacted households (…) through recurring processes of structural deprivation activated by the myriad shocks and stressors faced in low-income, rural communities", thus suggesting that adverse events do not occur in isolation but often as part of a continuous and incessant "multi-stressor multi-shock environment". Upton and her co-authors also make the important point that in food systems, "shocks often have indirect impacts that far exceed their direct impacts"—a remark that resonates well with some principles of resilience analysis, precisely because of the "multiple mechanisms that link individuals to features of the social-ecological systems in which they reside, and endogenous, multi-scalar behavioural responses by many people and organizations". With this remark, Upton and her co-authors reinforce the idea that in a food system the ultimate impacts of shocks may have less to do with their frequency or severity than with the types of responses (coping strategies, adaptation, transformation) put in place by the different actors to mitigate them, a concept that is referred to in the resilience literature as a "ripple effect" (Ivanov et al., 2019).

What we see emerging here, therefore, is a series of empirical observations that highlight the significance of some lessons and principles that had initially been discussed while conceptualizing resilience at the household and community levels—such as the importance of better understanding the actors' various sources of vulnerability, or documenting the types of responses that those actors put in place when they are hit by a shock and the impacts that these responses have on other actors—and how some of those principles become even more relevant when the scale of analysis is expanded from the household or community to the whole food system level. These points will be revisited in greater depth in the next section.

## Linking Resilience, Food Security and Food Systems—Some Initial Remarks

### Food Security, Shocks and Resilience...

Since the 1980s, a growing body of evidence has pointed to the debilitating impacts that seasonal or unexpected shocks can have on the livelihoods and food security of poor people in low-income countries (Dercon & Krishnan, 2000; Morduch, 1995; Yamano et al., 2003). Small events such as individual illness and delay in monsoon rainfall, or more severe idiosyncratic or covariant shocks such as disability or two consecutive harvest failures, can have severe impact on people's lives, affecting their income and food security, sometimes with long-term irreversible effects. We know, for instance, that women who are pregnant during a hunger gap give birth to smaller babies (Rayco-Solon et al., 2002) and that poor nutrition during the first 1000 days of life can have irremediable effects on the physical and mental development of those children. Longitudinal studies have shown for instance that height gain among young children displays seasonal variations that are closely linked to the annual hunger season (Maleta et al., 2003). In parallel, the detrimental impact of local or regional armed conflicts on food security and nutrition is increasingly recognized (Breisinger et al., 2014; Quak, 2018). Data show, for instance, that people living in conflict-affected areas are up to three times more likely to be food insecure than those who are living in more stable developing countries (Holleman et al., 2017). Globally, about 60% of the 800+ million undernourished individuals in the world live in regions affected by violent conflict (Martin-Shields & Stojetz, 2018).

The importance of being able to avoid or reduce the detrimental impacts of different types of shocks and stressors on the food security and nutrition of people is therefore critical. It is at this interface between shocks and food security that the relevance of the concept of resilience is probably the strongest. If, as mentioned earlier, resilience is basically the idea that individuals, households and communities are able to deal with adverse events in ways that reduce the long-term negative consequences of those shocks on their well-being and their food security, then supporting those individuals, households and communities to become more resilient should surely be the main objective of any development agency interested in food security. This observation is undoubtedly one of the reasons why FAO, WFP, USAID, DFID and several other development agencies decided, in the early 2010s, to engage in the 'resilience journey' (see Lindgren and Lang in this volume).

Food security is indeed, by its very definition and in particular its fourth pillar (*stability*), very closely linked to household's resilience. Put in simple terms, a household, which is not able to *stabilize*, protect and buffer its own food security pillars (access, availability and quality of food) against the impact of shocks and stressors, will not be food secure. In essence, this means that the experts who suggested adding stability as the fourth pillar to the original three were already embracing the idea that resilience is instrumental to food security. Likewise, many of the empirical studies in the 1980s and 1990s that documented the effects of idiosyncratic or covariant shocks on the food security and/or nutrition of households (see Dercon & Krishnan, 2000; Morduch, 1995; Rayco-Solon et al., 2002; Yamano, et al., 2003; etc.) were already implicitly demonstrating the relevance of the concept of resilience in this discussion of food security and nutrition, even if they did not use the term explicitly at that time. Instead, the discussion was framed around the concepts of risk and risk aversion, in line with the dominant neoclassical narrative of the 1990s where poor farmers were presented as both victims and culpable of causing their own predicament. They were victims, of course, because of the undisputed negative impact that these different adverse events can have on their livelihoods. But they were also 'guilty' because they were adopting 'risk-averse' strategies—a behaviour which from a theoretical economic standpoint is flawed as it is traditionally presented as one of the reasons why poor people don't invest in the 'right' innovations—those innovations which may be riskier in the short run but are presented as the long-run solution to their problem. In this neoclassical understanding of the world where risk is conceptualized essentially as a statistical concept, everything is reduced to probabilities.

A resilience version of this probabilistic vision of poverty has been developed. In that interpretation, resilience is defined in terms of a conditional expectation function or, less technically, as the statistical probability of remaining above the poverty line in the near future. John Hoddinott, in line with others (e.g., Barrett & Constas, 2014; d'Errico et al., 2018; Knippenberg et al., 2019), follows this probabilistic path when he proposes to define resilient food production as the "reciprocal of the probability of total crop failure" (Chapter 6 in this volume). As he concedes, however, applying a statistical frame to resilience in relation to food production or, as a matter of fact, to any more complex issues such as food security is limited as it "does not attempt to disaggregate or disentangle how the outcome has come about". As such, this probabilistic approach may leave many frustrated, as resilience has often been praised for the analytical or conceptual insights that it can offer around these questions of food security in the context of shocks (Ansah et al., 2019; Béné et al., 2016; Constas et al., 2014).

Importantly, conceptualizing resilience as an analytical tool rather than just the "probability to avoid poverty over time" (as in Barrett & Constas, 2014) also allows us to move away from the poor and risk-averse farmers conveyed by the 1990s interpretation of the problem. Resilience analysis understood as an analytical tool, indeed, puts emphasis on the absorptive, adaptive and transformative *capacities* of people (Béné et al., 2012; Vaughan & Frankenberger, 2018). Applied to food security, this more positive and realistic interpretation stresses the *active* choice that people make when faced by shocks and stressors and links this to the observed long-term outcomes, through a clear impact pathway that allows for deconstructing the process and identifying specific entry points for interventions. A particularly concrete example of this more interactive/agency-led interpretation of resilience is the work of Elizabeth Bryan, Claudia Ringler and Ruth Meinzen-Dick (Chapter 8 in this volume). These authors expand a recent gender and resilience framework (Bryan et al., 2017; Theis et al., 2019) with a food systems lens and show how such an approach can be used to identify key entry points to strengthen women's and men's food security and nutrition outcomes in the face of climate change.

Importantly, this interpretation of *people* as agents of their own resilience and the subsequent use of clear impact pathways associated with it allows analysts to go beyond the econometric black box of the relation {shock => change in food security} and start explaining the degradation in food security as a result of, not just the severity of the shocks or its duration—or even the level of assets or endowments characterizing the households—but as the outcome of the decisions or responses that people adopt in the face of those shocks. As such, resilience analysis shows that, yes, people make decisions in the aftermath of disasters which in the long-run may have negative consequences—like when they decide for instance to sell their productive assets or to reduce their food expenditures—and, yes, those decisions generally end up having detrimental long-term repercussions on their food security or the nutrition status of their children. But it also highlights that those "bad decisions" have little to do with risk aversion and instead are more often than not the result of a no-choice situation. People don't *choose* to reduce their food consumption. They do so because they don't have other choices. This point echoes what Haysom and Battersby refer to as "negotiated resilience" to describe the choices that food system actors have to make, recognizing that these choices often "involve having to make challenging trade-offs between immediate needs and the consequences of those choices, a form of negotiated resilience" (Chapter 11 in this volume).

Applying this 'agency-based' or 'people-centred' resilience approach (Bohle et al., 2009) also allows us to identify which interventions could help poor people reduce the risk of adopting those detrimental coping strategies and which interventions could, on the contrary, help them to engage in more positive adaptive or transformative strategies—so that the next time they are hit by a similar shock, they *have choices* and can better protect their food security and the nutrition of their children while, at the same time, recover from that shock. At this point, the justification for the introduction of human agency as one of the new pillars in the HLPE definition of food security comes back to mind. As explained by the HLPE report on Food Security and Nutrition, "agency [applied to food systems] implies the capacity of individuals or groups to make their own decisions about what foods they eat, what foods they produce, (…), their ability to engage in processes that shape food system policies and governance"(HLPE, 2020, p. 8); building on the discussion on resilience just above, we suggest adding "and the ability to make more informed and empowered choices when hit by a shock".

To this point, this section has shown how two specific pillars of food security (stability and agency) can help make more explicit the ways that resilience improves our understanding of food security in the context of shocks and stressors. Does this mean that the integration of the two concepts is now strong and well established? Certainly not, and several chapters in this volume make this reality crystal clear. Karl-Axel Lindgren and Tim Lang, for instance, in their review of development agencies' narratives, compare the two concepts to "totem poles around which policy lobbies dance, calling for consensus", but their analysis leaves little doubt about the weak integration of resilience in the food security agendas of different agencies. The authors conclude: "There seems to be a continuous delay in integrating the concept of resilience wholly into food security" (Chapter 4). Mark Constas seems to reach the same conclusion (in Chapter 5) when he salutes "the potential for coherence" but regrets "the reality of fragmented applications in policy and research" that characterizes resilience.

What remains to be discussed in this first chapter is the extent to which a food system approach expands (or perhaps modifies) the perception and the understanding that academics and experts have developed of the relation between resilience and food security.

### *Food Security, Resilience and Food Systems*

"Food systems have become the predominant theme among food actors and scholars to frame, understand and adequately address food security" claims Jess Fanzo in Chapter 2. She adds: "A food systems approach is a departure from traditional, historical approaches, which tend to be sectoral, technical, and short-term with a narrowly defined focus and scope of food security. Instead, a food systems approach uses a holistic, comprehensive view of the entire system". This statement is in line with the more general understanding of food systems as now widely adopted in the literature (see, e.g., HLPE, 2017).

One of the main benefits of adopting this more holistic view is certainly around the question of scale. In theory, both food security and resilience are acknowledged to be relevant within and across a wide range of scales or levels—from individuals, households and communities to countries or even possibly a continent (referring for instance to "the level of food security of Africa"). In ecology, resilience has also often been described as a multiscale concept (see Cummings et al., 2015). Yet, a quick review of the literature reveals that the majority of discussions to date on resilience in the context of food security and humanitarian interventions have taken place essentially at the household or community levels (see Béné et al., 2011; Brück et al., 2018; Cutter et al., 2008; Smith & Frankenberger, 2018). In a recent analysis of food systems in the urban context, Battersby and Watson made the same observation:

> previous work on food security has conventionally focused at either the household scale or at aggregate food production, with far less focus on the food system itself and its intersection with cities." (Battersby & Watson, 2018; p. 3)

A probable explanation for this observation is the fact that the most frequent levels at which practitioners (NGOs, development and humanitarian agencies) design and implement their interventions are at household and community levels. In fact, although not explicitly acknowledged, the level at which food security is conceptualized and defined in most UN and other official documents is often at the individual or household. As evidence we would point to the fact that most food security indicators (food consumption score; coping strategy index, household food insecurity access scale, etc.) are designed to measure and report status of food (in)security at the household level.

Yet we know that, even at household level, food security does not depend solely on capacity, ability or other socio-economic characteristics such as social capital, education, assets, income or wealth. To make this point more tangible, imagine a scenario where a household is well endowed in all these characteristics to the point that, in theory, it would be able to secure a satisfactory level of availability, access, utilization and stability of food. Yet if the local food system on which it depends for its food collapses, then its food security will be directly threatened. A good illustration of this scenario would be the case where an armed conflict severely disrupts the functioning of a local food system (for instance by interrupting the transportation of food through armed attacks and roadblocks) to the point that no supplies reach the shops and markets in a certain small town for several months. A wealthy household living in this town may have at its disposal all the attributes necessary to ensure, in theory, the food security of its members, yet these attributes will not save its members' food security from being threatened.

In sum, just as Sen taught us that food security does not depend only on food availability at the household level but also on financial and physical accessibility, we argue here that at a higher level, household food security also depends on the resilience of local food systems to deal with shocks and stressors (Béné, 2020). Very little empirical evidence is available, however, to substantiate this statement, which remains therefore a hypothesis for two main reasons. First is the fact that, as mentioned above, the quantitative information and the analyses or assessments that are available on food security are generally conducted at household level. This is the case, for instance, for the very rich literature that documents the impact of armed conflict on food security: all the analyses looking at potential correlation between conflict and food security (or nutrition) use indicators and variables monitored at household, community (or individual) levels (see Brück et al., 2018; Martin-Shields & Stojetz, 2018; Tusiime et al., 2013, etc.). Second, there is at present no clear methodology to measure (local) food systems' resilience (Béné, 2020). It is therefore very difficult to demonstrate *empirically* any form of correlation between household food security and local food system resilience, even if theoretically or conceptually many have tried—see, e.g., Meyer (2020), de Steenhuijsen Piters et al. (2021b) or John Hoddinott (in Chapter 6).

In this context, one would expect that the integration of sustainability as a new pillar in the definition of food security would be useful. It is indeed correct that the addition of this new pillar has already been accompanied by a more frequent use of the term 'food system' in the general discourse on food security; for instance when the HLPE report remarks "Sustainability refers to the long-term ability of *food systems* to provide food security and nutrition today in such a way that does not compromise the environmental, economic, and social bases that generate food security and nutrition for future generations" (HLPE, 2020, p. 9, our emphasis). Jess Fanzo (Chapter 2) also builds a convincing argument for why a food system framing is necessary to better understand and adequately address food security issues—even if, relying on the situation observed in her own country, she reminds us that although "food systems are well functioning [in the US], food insecurity can still occur". This is echoed by Losch and May (Chapter 10) at the subnational level. Looking at the Western Cape province in South Africa, these authors note: "Despite the Western Cape's prosperity when compared to other provinces and its well-established food system, […] the prevalence of the indicators of malnutrition among its population […] are similar to national trends". Contrasting with those assessments, Caron and his co-authors (Chapter 3) try to depict a more positive vision: "Contrarily to what is usually stated, the food system is not broken". Instead, they argue, "the food system has become more and more resilient at the global level […] [as it was] able to adapt during the twentieth century to what can be considered a huge shock on a long-term basis: an unprecedented increase in population".

In sum, our food systems are not broken and they may even be resilient. Yet they still result in food insecurity and inequity. While highlighting the links between food security, resilience and food systems is therefore totally relevant (as we hopefully demonstrated in this introductory chapter), what remains to be built is a much clearer conceptual and empirical connection between them. This is the object of the rest of this volume.

## Outline of the Volume

Following this introductory chapter, the contributions to this volume have been clustered into two parts, with a logic attached to the sequencing of the whole endeavour. Part I "From concepts to policy and narratives" offers a series of critical and sometimes provocative perspectives on resilience, looking at various conceptual and discursive aspects of its recent 'institutionalization' in the academic and development communities in relation to food security and food systems. In Part II "Specific issues and empirical analyses", the objective is to complement or ponder this initial series of theoretical reflections with some more empirical analyses and concrete case studies around issues that include gender, cities, climate change, locality and COVID-19. A final conclusion is then presented which reflects on the conceptual, empirical and policy-related contributions that this book has made.

### Part I: From Concepts to Policy and Narratives

Chapter 2: *Achieving Food Security through a Food Systems Lens* (Jess Fanzo). This chapter examines the history of how food security has been framed and addressed in international development discourse, and why it is important to adopt a food systems approach in tackling food security. In reviewing these questions, Fanzo also shows how food security has become more complex in the modern, challenged world, and makes the important point that functional food systems do not necessarily equate to improved food security.

Chapter 3: *The Global Food System is not Broken but Its Resilience is Threatened* (Patrick Caron, Ellie Daguet and Sandrine Dury). In sharp contrast to the conclusion of Chapter 2, Patrick Caron and his co-authors demonstrate why they consider that "the food system is not broken". They, however, also insist that even if the system is not broken, a great transformation is still needed; and they warn us that the road will be bumpy. Obstacles and barriers, including conflicts of interest, and even possibly the resilience of the system itself, may make that great transformation difficult.

Chapter 4: *Food Security and the Fractured Consensus on Food Resilience* (Karl-Axel Lindgren and Tim Lang). In this chapter, Lindgren and Lang conduct an insightful analysis of the different (and often contrasted) narratives that have been adopted on food resilience by major development agencies. The authors expose that, although there is little doubt that resilience has now graduated to become a core concept in the development industry, it is still characterized by a "fractured consensus". Agencies use different—and sometimes antagonistic—definitions, approaches and measurements to build their own discourses around food resilience. They warn us that unless more inter-disciplinary attention is paid to how food resilience is measured, there is a risk that the benefits of adopting this concept will eventually become "diluted" as it becomes more ubiquitous.

Chapter 5: *Food Security and Resilience: The Potential for Coherence and the Reality of Fragmented Applications in Policy and Research* (Mark Constas). The parallel between the conclusion of Lindgren and Lang and their 'fractured consensus' in Chapter 4 and Mark Constas' 'fragmented application' in this chapter is, to say the least, striking. Using a combination of conceptual reflections and reviews of selected policy and research documents, Constas shows that beyond some rhetoric, attempts to integrate food security and resilience have so far been limited, inconsistent, and largely superficial. Not surprisingly, he concludes on the urgent need for more coherent integration at the intersection of food security and resilience.

Chapter 6: *Food Systems, Resilience, and Their Implications for Public Action* (John Hoddinott). In the last chapter of Part I, John Hoddinott proposes a mix of theoretical and rhetorical reflections about the potential links between food systems and resilience. Starting with a stylized model of food production resilience, he then moves to a general discussion in which he shares a series of perspectives, including on the potential importance of market openness for food systems to remain resilient. He concludes, however, by acknowledging important knowledge gaps, leading him to remark that "it would be unwise to make strong statements regarding (...) applying a resilience lens to food systems for the purposes of contributing to improved food security interventions and policy".

## Part II. Specific Issues and Empirical Analyses

Chapter 7: *Food Security Under a Changing Climate: Exploring the Integration of Resilience in Research and Practice* (Alex de Pinto, Md Mofakkarul Islam, Pamela Katic). After a series of five chapters focusing on the conceptual and discursive elements of the interactions between food security, resilience and food systems, Alex de Pinto and his co-authors provide us in this seventh chapter with the first of a series of analyses refocusing the discussion on empirical case studies. In the present case, these authors propose to explore the extent to which resilience is (or is not) appropriately integrated in the climate change literature, using a sample of projects implemented through the Adaptation Fund. Their conclusion is unambiguous: while the concept of resilience may have favoured a transition towards more integrated approaches and interventions in work related to climate change and food security, the pathways through which actions translate into resilience and then into food security remain unclear.

Chapter 8: *Gender, Resilience, and Food Systems* (Elizabeth Bryan, Claudia Ringler, and Ruth Meinzen-Dick). In this chapter, Elizabeth Bryan and her co-authors develop a new framework to better analyse the articulation between gender and resilience in the context of food systems. Building on this new framework, Bryan and her IFPRI colleagues were able to deliver a comprehensive review of the literature on the topic, reviewing and discussing more than 170 documents. The chapter is likely to become a 'must-read' for whoever looks for empirical evidence around the question of gender and resilience analysed from a food system perspective.

Chapter 9: *COVID-19, Household Resilience, and Rural Food Systems: Evidence from Southern and Eastern Africa* (Joanna Upton, Elizabeth Tennant, Kathryn Fiorella and Christopher Barrett). In this empirical chapter, Joanna Upton and her colleagues from Cornell analyse the effects of COVID-19 and the various policy responses triggered by the pandemic, on food system actors in three rural areas of Malawi, Madagascar and Kenya. For this, they developed and then applied a conceptual framework that helped them explore the multiple paths through which observed shocks interact with systemic mechanisms to influence resilience. Among many key findings, the analysis demonstrates that, in some settings, the direct health effects—in this case severe illness and mortality—have impacted fewer people than the indirect impacts that arise as behaviours, markets and policies adjusted to the first wave of the pandemic.

Chapter 10: *Place-based Approaches to Food System Resilience: Emerging Trends and Lessons from South Africa* (Bruno Losch and Julian May). Fully aligned with the ambition of this second part of the book to provide empirical case studies, Bruno Losch and Julian May test further the concept of resilience in the context of local (food) governance. Drawing on the experience of the Western Cape Province in South Africa during the COVID-19 pandemic, they illustrate how a place-based approach can facilitate food system resilience, through the identification of opportunities for community-led adaptation initiatives and through the design of locally specific risk management strategies to deal with external shocks.

Chapter 11: *Urban Food Security and Resilience* (Gareth Haysom and Jane Battersby). In this last chapter of Part II, Gareth Haysom and Jane Battersby conclude our series of empirical analyses by diving further into the question of the relevance of resilience, this time in the context of urban food systems. Drawing on their own 'on-the-ground' experience and using findings from different cities in five African countries, they argue convincingly for the re-framing of urban food system resilience into a more inclusive planning tool, where local factors that shape the form and the function of food systems are better acknowledged and included. They insist in particular that the agency of urban food system users is a key component that needs to be explicitly accounted for and better incorporated if we want to improve our abilities to strengthen urban dwellers' food systems resilience.

### *Concluding Chapter*

In the *Reflections and Conclusion* chapter, the two editors of this book synthesize the main contributions that the different authors have made on a wide range of issues—conceptual, empirical and policy-related. They stress some of the paradoxes that emerged throughout the chapters, such as the coexistence of high levels of undernutrition and overnutrition in countries where food systems appear to be performing efficiently. They also point to unsettled debates or unresolved issues, such as the question of whether food systems are or are not resilient. At times, their conclusions sound as if resilience, food system and even food security are likely to remain elusive and contested concepts for ever. But their final words are resolutely assertive: as the world is becoming increasingly complex and unpredictable, achieving food security for all will not be possible without a new international consensus where the resilience of food systems is seen as a major priority at global, national and local levels.

## References

Akintola, S. L., & Fakoya, K. A. (2017). Small-scale fisheries in the context of traditional post-harvest practice and the quest for food and nutritional security in Nigeria. *Agriculture & Food Security, 6*, 34. https://doi.org/10.1186/s40066-017-0110-z

Aljazeera. (2020). Informal vendors rally against coronavirus lockdown in Malawi, 16 April 2020 https://www.aljazeera.com/economy/2020/4/16/informal vendors-rally-against-coronavirus-lockdown-in-malawi, last accessed July, 3, 2020.

Ansah, I. G. K., Gardebroek, C., & Ihle, R. (2019). Resilience and household food security: A review of concepts, methodological approaches and empirical evidence. *Food Security, 11*, 1187–1203. https://doi.org/10.1007/s12571-019-00968-1

Barrett, C. B., & Constas, M. A. (2014). Toward a theory of resilience for international development applications. *Proceedings of the National Academy of Sciences of the United States of America, 111*(40), 14625–14630.

Battersby, J., & Watson, V. (2018). *Urban food systems governance and poverty in African cities* (290 p.). Taylor & Francis.

Béné, C., Evans, L., Mills, D., Ovie, S. et al., (2011). Testing resilience thinking in a poverty context: Experience from the Niger river basin. *Global Environmental Change, 21*(4), 1173–1184. https://doi.org/10.1016/j.gloenvcha.2011.07.002

Béné, C. (2020). Resilience of local food systems and links to food security—A review of some important concepts in the context of COVID-19 and other shocks. *Food Security ,12*(4), 805–822.

Béné, C., Godfrey-Wood, R., Newsham, A., & Davies, M. (2012). Resilience: New utopia or new tyranny? —Reflection about the potentials and limits of the concept of resilience in relation to vulnerability reduction programmes. *IDS working Paper 405*. Institute of Development Studies, p. 61

Béné, C., Newsham, A., Davies, M, Ulrichs M., & Godfrey-Wood, R. (2014). Review article: Resilience, poverty and development. *Journal of International Development*, 26, 598–623.

Béné, C., Headey, D., Haddad, L., & von Grebmer, K. (2016). Is resilience a useful concept in the context of food security and nutrition programmes? Some conceptual and practical considerations. *Food Security*, 8(1), 123–138.

Béné, C., Prager, S. D., Achicanoy, H. A., Toro, P. A., Lamotte, L., Cedrez, C. B., & Mapes, B. R. (2019). Understanding food systems drivers: A critical review of the literature. *Global Food Security*, 23, 149–159.

Berkes, F., & Folke, C. (Eds.). (1998). *Linking social and ecological systems management practices and social mechanisms for building resilience*. Cambridge University Press.

Biggs, D., Biggs, R., Dakos, V., Scholes, R. J., & Schoon, M. (2011). Are we entering an era of concatenated global crises? *Ecology and Society*, 16(2), 27.

Bohle, H. G., Etzold, B., & Keck, M. (2009). Resilience as agency. International Human Dimensions Programme on Global Environmental Change, IHDP Update 2.2009, 6 p.

Botero, G. (1588) [2012]. *On the causes of the greatness and magnificence of cities*. Geoffrey Symcox (Trans and intr.). University of Toronto Press (Original work published in 1588).

Breisinger, C., Ecker, O., & Trinh Tan, J. F. (2014). *Chapter 7: Conflict and food insecurity—How do we break the links?* In 2014–2015 Global Food Policy Report, International Food Policy Research Institute.

Brück, T., d'Errico, M., & Pietrelli, R. (2018). The effects of violent conflict on household resilience and food security: Evidence from the 2014 Gaza conflict. *World Development*, 119, 203–223.

Bryan, E., Theis, S., Choufani, J., De Pinto, A., Meinzen-Dick, R., & Ringler, C. (2017). Gender-sensitive, climate-smart agriculture for improved nutrition in Africa South of the Sahara. In A. De Pinto & J. M. Ulimwengu (Eds.), *ReSAKSS annual trends and outlook report: The contribution of climate-smart agriculture to malabo and sustainable development goals* (pp. 113–145). International Food Policy Research Institute.

Burkart, S., Díaz, M. F., Enciso-Valencia, K., Urrea-Benítez, J. L., Charry-Camacho, A., & Triana-Ángel, N. (2020). *COVID-19 y el sector ganadero bovino en Colombia: Desarrollos actuales y potenciales, impactos y opciones de mitigación*. Cali (Colombia), Centro Internacional de Agricultura Tropical (CIAT). 36 p.

Cadilhon, J. J., Moustier, P., Poole, N. D., Tam, P. T. G., & Fearne, A. P. (2006). Traditional vs. modern food systems? Insights from vegetable supply chains to Ho Chi Minh City (Vietnam). *Development Policy Review, 24*(1), 31–49.

Cannon, T., & Muller-Mahn, D. (2010). Vulnerability, resilience and development discourses in context of climate change. *Natural Hazards, 55*(3), 621–635.

Chambers, R. (1989). Editorial introduction: Vulnerability, coping and policy. *IDS Bulletin, 20*(2), 1–7. https://doi.org/10.1111/j.1759-5436.1989.mp20002001.x

Coley, D., Howard, M., & Winter, M. (2009). Local food, food miles and carbon emissions: A comparison of farm shop and mass distribution approaches. *Food Policy, 34*, 150–155.

Constas, M., Frankenberger, T. R., Hoddinott, J., Mock, N., Romano, D., Béné, C., & Maxwell, D. (2014). A common analytical model for resilience measurement—Causal framework and methodological options. Resilience Measurement Technical Working Group, FSiN Technical Series Paper No. 2, World Food Program and Food and Agriculture Organization, 52 p.

Cumming, G. S., Allen, C. R., Ban, N. C., Biggs, D., Biggs, H. C., Cumming, D. H., De Vos, A., Epstein, G., Etienne, M., Maciejewski, K., Mathevet, R., Moore, C., Nenadovic, M., & Schoon, M. (2015). Understanding protected area resilience: A multi-scale, social-ecological approach. *Ecological Applications, 25*(2), 299–319. https://doi.org/10.1890/13-2113.1

Cutter, S., Barnes, L., Berry, M., Burton, C., Evans, E., Tate, E., & Webb, J. (2008). A place-based model for understanding community resilience to natural disasters. *Global Environmental Change, 18*(4), 598–606.

d'Errico, M., Romano, D., & Pietrelli, R. (2018). Household resilience to food insecurity: Evidence from Tanzania and Uganda. *Food Security, 10*(4), 1033–1054.

Department for International Development (DFID). (2011). *Defining Disaster Resilience: A DFID Approach Paper*. London: Department for International Development.

Dercon, S., & Krishnan, P. (2000). Vulnerability, poverty and seasonality in Ethiopia. *Journal of Development Studies, 36*(6), 25–53.

de Steenhuijsen Piters, B., Dijkxhoorn, Y., Hengsdijk, H., Guo, X., Brouwer, I., Eunice, L., Tichar, T., Carrico, C., Conijn, S., Oostewechel, R., & de Boef, W. (2021a). Global scoping study on fruits and vegetables: Results from literature and data analysis. Report No. 2021a-092, Wageningen Economic Research, 158 p.

de Steenhuijsen Piters, B., Termeer, E., Bakker, D., Fonteijn, H., & Brouwer, H. (2021b). Food system resilience. Towards a joint understanding and implications for policy. Wageningen Economic Research | Policy paper, https://edepot.wur.nl/549244

Ebata, A., Nisbett, N., & Gillespie, S. (2020). *Food systems and building back better* (p. 20). Institute of Development Studies, University of Sussex.

Ericksen, P. J. (2008). Conceptualizing food systems for global environmental change research. *Global Environmental Change, 18*, 234–245.

FAO. (2020). *Cities and local governments at the forefront in building inclusive and resilient food systems: Key results from the FAO survey "Urban Food Systems and COVID-19."* Food and Agricultural Organization.

Fanzo, J., Haddad, L., Schneider, K. R., Béné, C., Covic, N. M., Guarin, A., Herforth, A. W., Herrero, M., Sumaila, U. R., Aburto, N. J., Amuyunzu-Nyamongo, M., Barquera, S., Battersby, J., Beal, T., Bizzotto Molina, P., Brusset, E., Cafiero, C., Campeau, C., Caron, P., & Rosero Moncayo, J. (2021). Viewpoint: Rigorous monitoring is necessary to guide food system transformation in the countdown to the 2030 global goals. *Food Policy, 104*, 102163.

FAO. (1980). *Agriculture, trade and food security: issues and options in the WTO negotiations from the perspective of developing countries.* Report and papers of an FAO Symposium held at Geneva on 23–24 September 1999, Food and Agriculture Organization of the United Nations, Rome.

FAO. (2006). *Food Security* (Issue 2). UN Food and Agriculture Organization. https://www.fao.org/fileadmin/templates/faoitaly/documents/pdf/pdf_Food_Security_Cocept_Note.pdf

Harvey, C. A., Rakotobe, Z. L., Rao, N. S., Dave, R., Razafimahatratra, H., Rabarijohn, R. H., Rajaofara, H., & MacKinnon, J. L. (2014). Extreme vulnerability of smallholder farmers to agri-cultural risks and climate change in Madagascar. *Philosophical Transactions of the Royal Society B, 369*(1639), 20130089.

Hirvonen, K., Mohammed, K., Minten, B., & Tamru, S. (2020). *Food marketing margins during the COVID-19 pandemic—Evidence from vegetables in Ethiopia.* Strategy Support Program Working Paper 150, International Food Policy Research Institute, 16 p. http://ebrary.ifpri.org/utils/getfile/collection/p15738coll2/id/133931/filename/134143.pdf

Hussain, M. (2013). *Resilience: meaningless jargon or development solution?* The Guardian, 5 March 2013, retrieved from https://www.theguardian.com/global-development-professionals-network/2013/mar/05/resilience-development-buzzwords

HLPE. (2017). *Nutrition and food systems.* A report by the High Level Panel of Experts on Food Security and Nutrition of the Committee on World Food Security, Rome.

HLPE. (2020). *Food security and nutrition: building a global narrative towards 2030. A report by the High Level Panel of Experts on Food Security and Nutrition of the Committee on World Food Security.*

Holleman, C., Jackson, J., Sánchez, M. V., & Vos, R. (2017). *Sowing the seeds of peace for food security—Disentangling the nexus between conflict, food security and peace*. FAO Agricultural Development Economics Technical Study 2. Rome, FAO

Ivanov, D., Dolgui, A., & Sokolov, B. (2019). Ripple effect in the supply chain: Definitions, frameworks and future research perspectives. In D. Ivanov, A. Dolgui, & B. Sokolov (Eds.), *Handbook of Ripple Effects in the Supply Chain* (pp. 1–33). Springer.

Kabeer, N. (1999). Resources, agency, achievements: Reflections on the measurement of women's empowerment. *Development and Change, 30*(3), 435–464.

Kawarazuka, N., Béné, C., & Prain, G. (2018). Adapting to a new urbanizing environment: Gendered strategies of Hanoi's street food vendors. *Urbanization and Environment, 30*(1), 233–248.

Knippenberg, E., Jensen, N., & Constas, C. (2019). Quantifying household resilience with high frequency data: Temporal dynamics and methodological options. *World Development, 121*, 1–15. https://doi.org/10.1016/j.worlddev.2019.04.010

Kolding, J., Béné, C., & Bavinck, M. (2014). Small-scale fisheries—Importance, vulnerability and deficient knowledge In S. Garcia, J. Rice, & A. Charles (Eds.). *Governance for marine fisheries and biodiversity* (pp. 317–331). Wiley-Blackwell.

Lowder, S. K., Skoet, J., & Singh, S. (2014). *What do we really know about the number and distribution of farms and family farms worldwide?* Background paper for The State of Food and Agriculture 2014. ESA Working Paper No. 14-02. FAO. http://www.fao.org/3/i3729e/i3729e.pdf

La Trobe, H. L., & Acott, T. G. (2000). Localising the global food system. *International Journal of Sustainable Development & World Ecology, 7*(4), 309–320.

Maleta, K. V., & S. M., Espo M., & Ashorn P. (2003). Seasonality of growth and the relationship between weight and height gain in children under three years of age in rural Malawi. *Acta Paediatrica, 92*, 491–497.

Maloney, W. F. (2004). Informality revisited. *World Development, 32*(7), 1159–1178.

Malthus, T. (1798). An essay on the principle of population as it affects the future improvement of society, with remarks on the speculations of Mr. Godwin, M. Condorcet, and other writers. London: J. Johnson in St. Paul's Churchyard, www.esp.org/books/malthus/population/malthus.pdf

Marschke, M., & Vandergeest, P. (2016). Slavery scandals: Unpacking labour challenges and policy responses within the off-shore fisheries sector. *Marine Policy, 68*, 39–46.

Martin-Shields, C., & Stojetz, W. (2018). *Food security and conflict. Empirical challenges and future opportunities for research and policy making on food security and conflict*. FAO Agricultural Development Economics Working Paper 18-04. Rome, FAO.

McCullough, E. B., Pingali, P. L., Stamoulis, K. G., & K.G. (2010). Small farms and the transformation of food systems: An overview. In R. Haas, M. Canavari, B. Slee, C. Tong, & B. Anurugsa (Eds.), *Looking east, looking west Organic and quality food marketing in Asia and Europe* (pp. 47–83). Wageningen Academic Publisher.

Meyer, M. A. (2020). The role of resilience in food system studies in low- and middle-income countries. *Global Food Security, 24*, 100356. https://doi.org/10.1016/j.gfs.2020.100356

Morduch, J. (1995). Income smoothing and consumption smoothing. *Journal of Economic Perspectives, 9*(3), 103–114.

Porter, G., Lyon, F., & Potts, D. (2007). Market institutions and urban food supplying West and Southern Africa: A review. *Progress in Development Studies, 7*(2), 115–134.

Oviedo, A.M., Thomas M., & Karakurum-Ozdemir, K. (2009). *Economic informality: Causes, costs, and policies—A literature survey*. World Bank Working Paper No. 167, World Bank, Washington, DC.

Quak, E. (2018). *Food systems in protracted crises: Strengthening resilience against shocks and conflicts*. K4D Helpdesk Report 447. Institute of Development Studies.

Reardon, T., & Swinnen J. (2020). COVID-19 and resilience innovations in food supply chains. In J. Swinnen & J. McDermott (Eds.) (2020). *COVID-19 and global food security* (pp. 132–136). International Food Policy Research Institute.

Resnick, Danielle. (2017). "Informal Food Markets in Africa's Cities." in *Global Food Policy Report*. Washington, DC: International Food Policy Research Institute.

Rayco-Solon, P., Fulford, A. J., & Prentice, A. M. (2002). Differential effects of seasonality on preterm birth and intrauterine growth restriction in rural Africans. *The American Journal of Clinical Nutrition, 81*(1), 134–139.

Roever, S., & Skinner, C. (2016). Street vendors and cities. *Environment and Urbanization, 28*(2), 359–374.

Scoones, I. (1998). Sustainable rural livelihoods: A framework for analysis, *IDS Working Paper 72*. IDS.

Sen, A. (1981). *Poverty and famine, an essay on entitlement and deprivation* (p. 266). Oxford University Press.

Sen, A. (1985). Well-being, agency and freedom: The Dewey lectures 1984. *The Journal of Philosophy, 82*(4), 169–221.

Smit, W. (2016). Urban governance and urban food systems in Africa: Examining the linkages. *Cities, 58*, 80–86. https://doi.org/10.1016/j.cities.2016.05.001

Smith, L. C., & Frankenberger, T. R. (2018). Does resilience capacity reduce the negative impact of shocks on household food security? Evidence from the 2014 floods in northern Bangladesh. *World Development, 102*, 358–376. https://doi.org/10.1016/j.worlddev.2017.07.003

Termeer, E., Brouwer, I., & de Boef W. (2020). *Rapid country assessment: Bangladesh. The impact of COVID-19 on the food system*. Wageningen University & Research, Global Alliance for Improved Nutrition and CGIAR Research Program on Agriculture for Nutrition and Health, 11 p.

Theis, S., Bryan, E. & Ringler, C. (2019). Addressing Gender and Social Dynamics to Strengthen Resilience for All. In *2019 Annual Trends and Outlook Report (ATOR): Gender Equality in Rural Africa. From Commitments to Outcomes*. Eds. A. Quisumbing, R. Meinzen-Dick, and J. Njuki. International Food Policy Research Institute (IFPRI).

Tounkara. (2020). *La COVID-19 et la Chaîne de Valeur Mangue au Sénégal: Effets, Stratégies d'adaptation et Recommandations*. Initiative Prospective Agricole et Rurale, 58 p. https://media.africaportal.org/documents/le_covid_19_et_la_chaine_de_valeur_mangue_au_senegal.pdf

Tusiime, H. A., Renard, R., & Smets, L. (2013). Food Aid and Household Food Security in a Conflict Situation: Empirical Evidence from Northern Uganda. *Food Policy, 43*, 14–22. https://doi.org/10.1016/j.foodpol.2013.07.005

United Nation. (2015). *Transforming our world: The 2030 agenda for sustainable development*. United Nations.

USAID. (2012). *Building resilience to recurrent crisis USAID policy and program guidance*. US Agency for International Development, 32 p.

Varshney, D., Roy, D., & Meenakshi, J. V. (2020). Impact of COVID-19 on agricultural markets: Assessing the roles of commodity characteristics, disease caseload and market reforms. *Indian Economic Review, 55*, 83–103.

Vaughan, E. and Frankenberger, T. (2018). *Resilience Measurement Practical Guidance Note Series 3: Resilience Capacity Measurement*. Produced by Mercy Corps as part of the Resilience Evaluation, Analysis and Learning (REAL) Associate Award, 28 p.

Veldhuizen, L., Giller, K. E., Oosterveer, P., Brouwer, I. D., Janssen, S., van Zanten, H. H. E., & Slingerland, M. A. (2020). The Missing Middle: Connected action on agriculture and nutrition across global, national and local levels to achieve Sustainable Development Goal 2. *Global Food Security, 24*. https://doi.org/10.1016/j.gfs.2019.10033

World Bank. (2005). *World Development Report 2006: Equity and Development*. Washington, DC.

Young, G. (2018). De-democratisation and the rights of street vendors in Kampala, Uganda. *International Journal of Human Rights, 22*, 1007–1029.

Yamano, T., Alderman, H., & Christiaensen, L. (2003). *Child growth, shocks and food aid in rural Ethiopia*. World Bank.

**Open Access** This chapter is licensed under the terms of the Creative Commons Attribution 4.0 International License (http://creativecommons.org/licenses/by/4.0/), which permits use, sharing, adaptation, distribution and reproduction in any medium or format, as long as you give appropriate credit to the original author(s) and the source, provide a link to the Creative Commons license and indicate if changes were made.

The images or other third party material in this chapter are included in the chapter's Creative Commons license, unless indicated otherwise in a credit line to the material. If material is not included in the chapter's Creative Commons license and your intended use is not permitted by statutory regulation or exceeds the permitted use, you will need to obtain permission directly from the copyright holder.

CHAPTER 2

# Achieving Food Security Through a Food Systems Lens

*Jessica Fanzo*

## INTRODUCTION

The twentieth and early twenty-first centuries have been marked with many international commitments, starting with the first United Nations declaration of a goal of zero hunger in 1943 (Byerlee & Fanzo, 2019). This commitment has been periodically renewed in international fora, most recently, with the Sustainable Development Goals (SDGs) and the specific goal of SDG2 to end hunger by 2030 and the Food Systems Summit in 2021 (Covic et al., 2021). Despite noble intentions, the world can claim only mixed success in eliminating food insecurity (and with it, hunger and other forms of malnutrition) (Fanzo, 2019).

Undernourishment has increased for the fourth straight year in a row since 2016, with approximately 828 million people considered undernourished in 2020 (FAO et al., 2022). Roughly 20% of the world's

J. Fanzo (✉)
Bloomberg School of Public Health, Johns Hopkins University, Baltimore, MD, USA
e-mail: jfanzo1@jhu.edu

© The Author(s) 2023
C. Béné and S. Devereux (eds.), *Resilience and Food Security in a Food Systems Context*, Palgrave Studies in Agricultural Economics and Food Policy, https://doi.org/10.1007/978-3-031-23535-1_2

young children are chronically undernourished or stunted, and overweight is increasing in that same demographic with 39 million children under the age of five considered overweight (Micha et al., 2020). The growing pandemic of obesity now tops 2 billion adults struggling with overweight and obesity, with significant risk of diet-related noncommunicable diseases (Micha et al., 2020). Food systems have a role to play in ending or perpetuating food insecurity and malnutrition in all its forms.

This chapter has three objectives. First, it describes the evolution of the concept of food security and its historical framing. Second, it presents three mainstream approaches to address food security that have failed to deliver promised outcomes and improve overall food security. Third, it articulates why a food systems perspective is now necessary, but not always sufficient, to guide effective food security improvements.

## The Evolution of Food Security and Its Framing

Food security and its framing in international development has historically evolved to adapt to the times. This adaptation process has become more nuanced because of our further understanding of the social, political, environmental, and biological causes and consequences of food insecurity.

The evolution began in 1943 when the Hot Springs conference—the first of a series of conferences on the post-war architecture of the proposed United Nations—set the goal of "freedom from want of food, suitable and adequate for the health and strength of all peoples" and agreed that "the most fundamental of necessities is adequate food which should be placed within the reach of all men in all lands within the shortest possible time" (Department of State, 1943). This goal was equivalent in many respects to SDG2 to end hunger but without a firm end date and nebulous targets (Barona, 2008).The conference also urged countries to "maintain optimum levels of productivity consistent with ensuring the preservation of basic resources"—in other words, a call for integrating more sustainable practices with agriculture. There were also explicit calls for the state to take a more proactive role in promoting better nutrition and coordinating efforts to improve food security through various ministries (Barona, 2008).

Starting in the 1960s, the UN held several high-profile world food summits that reaffirmed global commitments to ending hunger. Notably, these conferences emphasized the supply of calories through increased production of food staples but largely ignored the broader dimensions of malnutrition. At the World Food Congress in 1963, the President of the United States, John F. Kennedy, declared "we have the capacity to eliminate hunger in our lifetime, we only need the will" (Shaw, 2007b). Another food summit was held in 1974 during the 1973–1975 world food crisis that declared that "every man, woman, and child have the inalienable right to be free from hunger and malnutrition" (Shaw, 2007a). The conference called on "all governments to accept the goal that within a decade no child will go to bed hungry, that no family will fear for its next day's bread, and no human being's future capacity will be stunted by malnutrition" (Shaw, 2007c).

At the next World Food Summit in 1996, food security was further articulated. The definition is still widely used and says: "Food security means that all people, at all times, have physical, social, and economic access to sufficient, safe, and nutritious food that meets their food preferences and dietary needs for an active and healthy life" (FAO, 1996). This definition thus evolved from when the term was first used at the World Food Conference in 1974, where food security was defined as: "Availability at all times of adequate world food supplies of basic foodstuffs to sustain a steady expansion of food consumption and to offset fluctuations in production and prices" (FAO, 1974).

The major focus for many of these summits was on agriculture—to produce enough calories to feed a growing population and stave off famines. What was considered successes, such as the Green Revolution, became the paradigm of how agriculture was designed, managed, and governed since then, resulting in a significant increase in yields and starchy calories coming largely from cereal grains with minimal extensification into land (Fig. 2.1). However, cracks became apparent with large-scale trade-offs for environmental sustainability, nutrition, and some livelihoods (Pingali, 2012, 2015).

To further articulate food security, four distinct but connected pillars have been defined to bring clarity and to enable measurement across discrete areas of food security (FAO, 2008). It has also created silos of thought and political action in some regards. The first pillar is *food availability*, which refers to ensuring sufficient quantity and diversity of food is available for consumption from the farm, the marketplace, or elsewhere. Such food can be supplied through household production, other domestic output, commercial imports, or food assistance. Second, *food access* refers

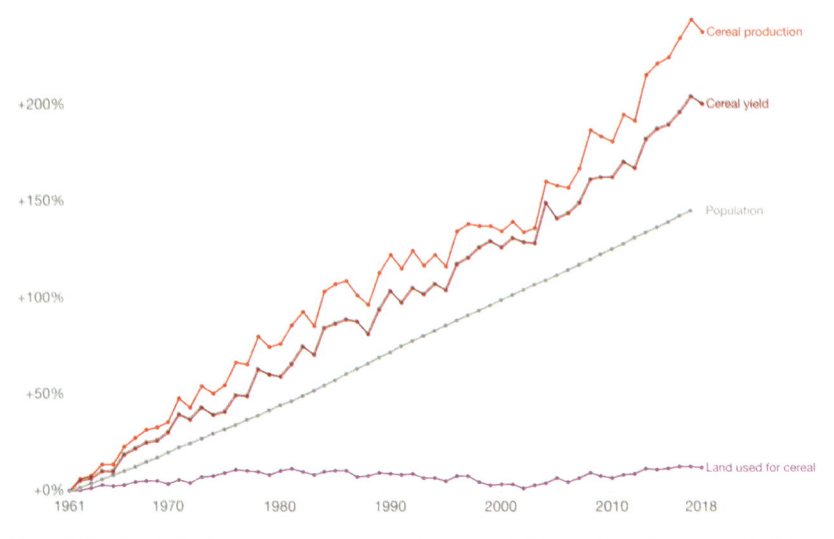

**Fig. 2.1** Global changes in cereal production, yield, and land use, 1961–2018 (*Source* Our World in Data [2021])

to households having the physical and financial resources required to obtain appropriate foods for a nutritious diet. Access depends on income available to the household, on the distribution of income within the household, distance to markets, and on the price of food. Third, *food utilization* implies the capacity and resources necessary to use food appropriately to support healthy diets including sufficient energy and essential nutrients, potable water, and adequate sanitation. Utilization often refers to an individual's ability to absorb nutrients, based on their health status. Effective food utilization depends, in large measure, on knowledge within the household of food storage and processing techniques, basic principles of nutrition and proper childcare, and illness management. *Stability* is considered a fourth pillar, which mainly refers to the stability of the food supply/access but can also refer to stability in availability and quality. Stability is reliant on food imports and domestic production and can be negatively impacted by disruptions in the food supply such as price volatility, seasonality, and conflicts (FAO, 2006). Instability can significantly impact low-income households, especially those in low-income countries, such as in South and Southeast Asia, Africa and Latin America who spend a large share of their income on food (Ivanic & Martin,

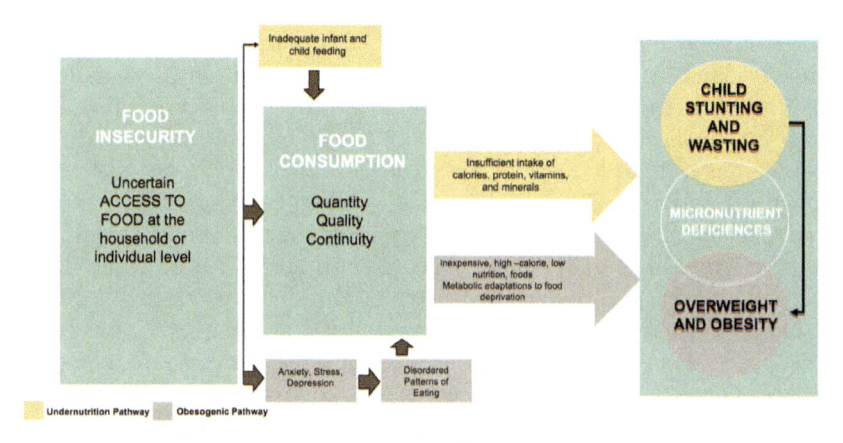

**Fig. 2.2** Food insecurity pathways to multiple forms of malnutrition (*Source* FAO et al. [2018])

2008; Martin & Ivanic, 2016; Raghunathan et al., 2021; Vellakkal et al., 2015). Often, what is purchased can be of poor nutritional quality, made up mainly of grain staple crops (Bouis & Saltzman, 2017). Since 1996, the definition of food security has even further evolved to not only consider the four pillars of food security but to account for agency and sustainability as well (Clapp et al., 2021; HLPE, 2020).

Fast forward to 2021, after two global goal setting agendas—the Millennium Development Goals (MDGs) and the Sustainable Development Goals (SDGs)—which were met with limited success in zeroing hunger, the UN Food Systems Summit (UNFSS) in September of 2021 adopted a food systems approach to address food security (Covic et al., 2021). As a consequence, food security is now recognized as a highly complex outcome that is influenced by hunger, as well as poverty, conflict, and climate change, among other causal factors (Fanzo, 2018; FAO et al., 2018). Resulting food insecurity can contribute to multiple forms of malnutrition, as shown in Fig. 2.2.

## Achieving Food Security Has Become More Complex in the Modern World

While our understanding of food security as a concept has progressed in how it fits within a food systems lens, we still grapple with how to solve massive food insecurity and address its root causes. At the same time, much related to food systems and security, malnutrition in all its forms is increasing; the risk in zoonotic and other infectious diseases related to food production is rising (Rohr et al., 2019); climate change and environmental and biodiversity degradation associated with food systems are volatile (Fanzo, Haddad et al., 2021; Willett et al., 2019); and rural poverty and urban poverty are increasing along with social unrest and conflict (Breisinger et al., 2014; Makita et al., 2019; Micha et al., 2020; Mozaffarian et al., 2018; Rivera-Ferre et al., 2021; Webb et al., 2020). Layered upon these trends are larger, complex political and social forces that are making it challenging to transform food systems (Béné, Fanzo, Haddad et al., 2020). Vexing issues beyond food system components and outcomes such as nationalism, geopolitics, and conflicts plague progress to address food security and transform food systems in positive directions that benefit human health, well-being, and planetary resilience (Brands & Gavin, 2020).

Adding to this complexity has been the approach taken by global food system actors and policies to address food security. Three mainstream approaches have dominated the global food policy agenda and architecture. The first has been the *vertically sectoral approach*. There have been many attempts to improve food security historically, but often these have been siloed, with a singular focus through one sector's lens, a vertically driven approach (Jeppsson & Okuonzi, 2000). One such example is the Green Revolution. This revolution averted social and economic upheaval and large-scale famines in the 1960s in Asia due to the development and widespread adoption of genetically improved high-yielding varieties of cereal crops (maize, rice, and wheat) that were responsive to the application of advanced agronomic practices, including most importantly, fertilizers and improved irrigation (Swaminathan, 2006). Between 1966 and 2005, food production in South Asia increased by almost 250% (FAOStat, n.d.).

While the Green Revolution had a tremendous impact on food production and socio-economic conditions in Asia, it had its fair share of trade-offs including insufficient improvements in nutrition outcomes, increases in environmental stress such as overuse of water, and minimal focus on women's empowerment issues (Negin et al., 2009; Pingali, 2012). To ensure improvements in food security for individuals, households, or communities, the literature suggests that instead, multi-sectoral approaches are essential (Fanzo, 2014; Garrett & Natalicchio, 2010).

At minimum, there are three key sectors that need to engage, collaborate, and contribute to nutrition improvements: agriculture, health, and water sectors (Garrett & Natalicchio, 2010; Lamstein et al., 2016; Pelletier et al., 2016, 2018). These sectors have the capability of injecting nutrition across functioning and effective food, health, and water and sanitation systems (Fanzo, 2014). While the multi-sectoral approach is usually adopted with good intentions, effectively engaging across diverse sectors and distinct systems has proven to be complex (Pelletier et al., 2016).

The second is the *technological treatment approach*. This approach promotes interventions and policy responses through a technocratic lens—focusing on addressing the symptoms more than the root causes. As such, the technocratic solution often attempts to find the "low hanging fruit" or "quick wins" to address what are usually more complex challenges that are entrenched in systemic issues of poverty, marginalization, and food system constraints. Instead, this approach aims to treat the consequences through a technology-driven solution. Some aspects of the Green Revolution fall under this type of approach, but addressing vitamin A deficiency and genetically modified organisms could be seen as other examples.

Vitamin A deficiency is the cause of what is known as preventable blindness in children and severe visual impairment as well as increased susceptibility to succumbing from measles, extreme diarrhea, or respiratory infection (Black et al., 2013). There is much debate and politics around vitamin A deficiency and how to treat it. In the 1970s and 1980s, nutrition planners had three options for dealing with widespread vitamin A deficiency: (1) provide all young children with megadose capsules of vitamin A semi-annually; (2) fortify commonly eaten foods with vitamin A; or (3) improve dietary diversity of foods and ensure people get access to vitamin A-rich sources of foods. The dominating intervention has been through supplementation, and now, most countries are now distributing vitamin A capsules twice a year to children under five years of age quite successfully. Some have argued that a singular, short-term focus on supplements has diverted necessary funds away from improving the diversity of the food supply to ensure that foods rich in micronutrients (such as vitamin A) are available, accessible, and utilized (Chambers et al., 2017; Mason et al., 2015).

Another example of such a technology-driven approach that has polarized the food security world is genetically modified organisms (GMOs).

Some have questioned whether GMOs are necessary when the existing pool of genetic diversity of crops could potentially address concerns of drought, flood, pest, wind, and saline tolerance and resistance, nutrient content, or high-yielding traits (Klümper & Qaim, 2014; Zilberman et al., 2018). GMOs have garnered a lack of consensus on their benefits, risks, and potential impacts on the environment and on human health (Glass & Fanzo, 2017; Sarkar et al., 2021). At the center of the debate is the idea that GMOs can increase crop yields and are thus necessary or at least part of the solution to feeding the world's population and staving off hunger (Klümper & Qaim, 2014; Kovak et al., 2021). Others argue that investing in conservation and use of agricultural biodiversity is a better approach (Jacobsen et al., 2013). There are also scholars that contend GMOs prohibit rights of smallholders and indigenous peoples and their traditional knowledge and values (Calabrò & Vieri, 2014; Koutouki & Marin, 2013). Some argue that GMOs perpetuate "agrarian dispossession," farmers losing the control over seeds and other inputs necessary for food production that are owned by agricultural input and chemical manufacturers and companies (Kloppenburg, 2014). The challenge is to make the best possible use of innovation and technologies to meet the needs of a growing population while also preserving natural resources, biodiversity, ecosystem health, and livelihoods of the most vulnerable. While fair and sustainable use of innovations and technologies has great potential, there are also significant risks and inequities that need to be considered at the same time (Glass & Fanzo, 2017).

The third is the *short-view approach*, which has been disproportionately prioritized by donors in recent decades. This approach often does not elicit lasting change and can have unintended consequences. One such example is food aid. Though food aid is crucial in times of crisis, it is by no means a sustainable, long-term solution to addressing the root causes of hunger (Garrett, 2008). While the cost of emergency food assistance is small compared to the cost of hunger, it is a comparatively expensive measure. One study estimated that food aid costs $812 to deliver one ton of maize as United States food assistance to a distribution point in Africa, whereas it costs only $135 to give local farmers the seed, fertilizer, and technical support to grow an extra ton of maize themselves (Sanchez, 2009). Furthermore, purchasing a ton of maize locally to use as food assistance, rather than maize donated from the United States, was only $320—a much cheaper alternative to international food aid. One potential way to improve the cost effectiveness of food aid would be either by improving the nutritional quality of the staple commodities (flour, sugar, and oil) provided or by diversifying the basket. By improving its nutritional content, billions of dollars can be saved that would have to be later spent on saving lives from nutrition-related illnesses (Rosenberg et al., 2012; Webb et al., 2017).

## Food Systems Lens to Tackling Food Security

Food systems have become the predominant theme among food actors and scholars to frame, understand, and adequately address food security. A "food systems approach" is a departure from traditional, historical approaches, which (as we saw above) tend to be sectoral, technical, and short term with a narrowly defined focus and scope of food security. Instead, a food systems approach uses a holistic, comprehensive view of the entire system. This approach includes the actors within the food supply chain and the governance mechanisms that shape their roles. A food systems approach requires "food systems thinking," which identifies and describes the influences, or "drivers," and relationships in the systems. Food systems thinking also considers how these influences intersect with each other in both positive and negative ways (Hawkes & Fanzo, 2019).

Figure 2.3 shows the components of the entire food system, including food supply chains, food environments, individual factors, consumer behaviors, and diets; the outcomes, including nutrition and health, environment, economic (livelihoods and wages), and social equity; and the drivers, many of which are exogenous to food systems but "push" or "pull" systems in various directions (Béné, Oosterveer et al., 2019; Béné, Prager et al., 2019; Fanzo, Rudie et al., 2021; HLPE, 2017).

Food supply chains are the components that link food production, food storage, loss and distribution, processing and packing, and retail. These links in the chain are influenced by the decisions of many food-specific and indirect actors from small to multi-national scales. The types of foods generated by supply chains avail foods to food environments, which are the places where people buy and order food (Constantinides et al., 2021; Swinburn et al., 2015; Turner et al., 2018, 2019). These environments vary from informal, wild environments (e.g., forests) to highly formalized environments (e.g., supermarkets) (Downs et al., 2020). The architecture of these environments is influenced by the types of food on offer, their affordability, their properties, and their promotion and advertising. Consumer decision-making and behavior are shaped

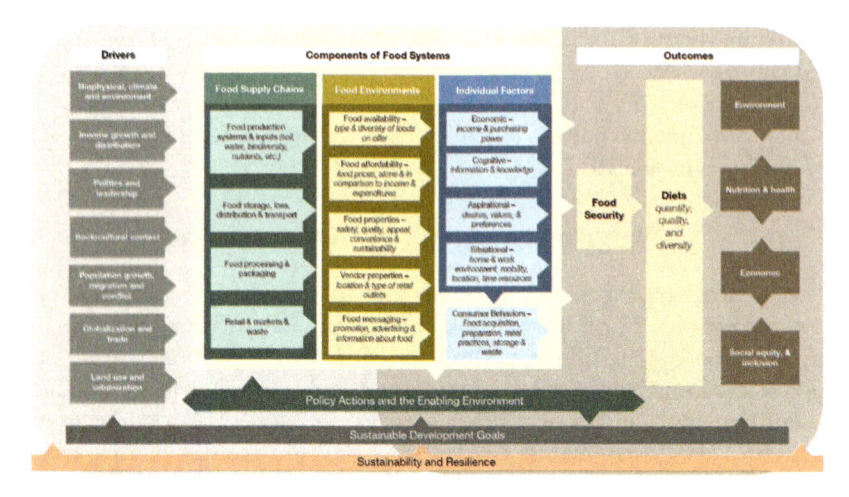

**Fig. 2.3** A food systems framework (*Source* Fanzo, Rudie et al. [2021])

by individual's purchasing power, knowledge, aspirations, resources, and culture. These factors subsequently influence diets (Bell et al., 2021; Mancino et al., 2018).

All of these components affect many outcomes that include health, nutrition, wages, income, the environment, climate, cultures, and systemic societal equity (Ericksen, 2008; Fanzo, Rudie et al., 2021; Webb et al., 2020). Drivers such as climate change, urbanization, population pressure, policies and politics, and migration, to name just a few, can influence the directionality and dynamism of food systems in both positive and negative ways (Béné, Fanzo, Prager et al., 2020). Political, programmatic, and institutional actions can both influence and be influenced by food system components, outcomes, and drivers, all of which affect progress on the SDGs. The pillars of food security intersect with food systems (HLPE, 2020), and risks, shocks, and vulnerabilities consistently threaten resilience and various outcomes of food systems (Barrett, 2020; Gaupp et al., 2019).

With the examples above on the Green Revolution, vitamin A supplementation, GMOs, and food aid, a food systems approach is a way of considering food systems in their totality, which includes all the elements, their relationships, and related impacts. It goes beyond one element (e.g., a value chain, a food environment) and considers the intricate web and networks of activities, actors, and feedback loops of the different directions food systems can take. It also does not focus on one single sector. It considers many sectors that interact with food and considers the many impacts that food system can bring, including health, nutrition, environment, livelihoods, and equity.

Beyond just definitions, food system solutions to ensure food security require integrated actions across multiple scales (from global to local systems; from long to short supply chains), actors, and sectors (e.g., agriculture, trade, policy, health, environment, education, transport, and infrastructure). Food systems are highly interconnected—any intervention or policy that addresses one part of the system will affect other parts. Health, politics, society, the economy, and environmental systems all intersect with food systems. As a result of this interconnectedness, any action can lead to unintended consequences. The global COVID-19 pandemic further highlights the need to take a systems approach—a shock to the health system had ramifications on every other system with the pandemic (Brands & Gavin, 2020; Fanzo, 2021).

Food systems result in many trade-offs due to decisions made within food systems and the many drivers that result in a diverse set of interactions (Béné, Oosterveer et al., 2019), and positive and negative feedback loops which can mitigate harmful outcomes or highlight trade-offs (Béné, Fanzo, Prager et al., 2020). One such example is trade. Trade is of critical importance in moving a diversity of food and the nutrients that food contains around the world. Figure 2.4 shows the number of extra people, in billions, who could be nourished if nutrients in excess of current global needs were evenly distributed. Without trade, there would be increased deficiencies in protein, zinc, and iron for example (Wood et al., 2018) (Fig. 2.4). At the same time, trade moves nutrient-poor, highly processed, packaged foods to the far reaches of the world and that trend is increasing (Garton et al., 2020). These foods have been associated with obesity and non-communicable diseases (Baker et al., 2020; Elizabeth et al., 2020). Both have implications for food security, when taken in its broader view (c.f. Figure 2.2).

## Functional Food Systems Do Not Always Equate to Food Security

If food systems don't function well (see, however, Caron et al., Chapter 3 in this volume), it will be difficult to achieve food security; yet even if

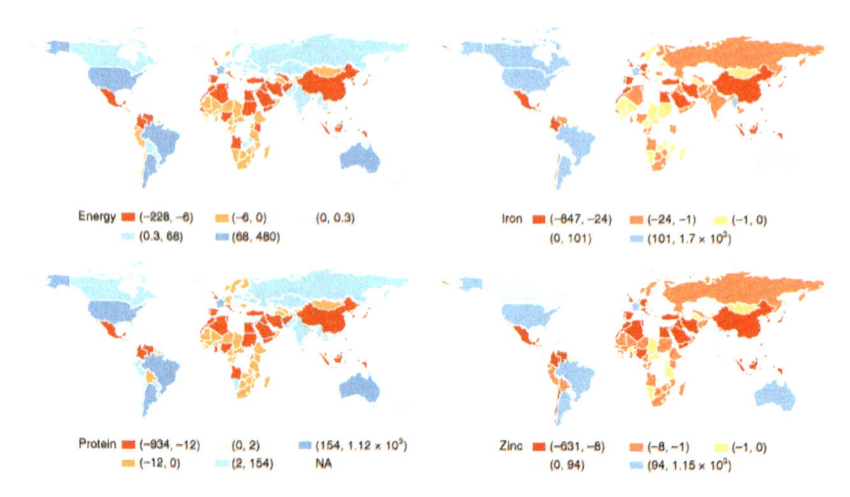

**Fig. 2.4** Change in number of people who could be nourished without trade. For each country, the number of people (in millions) who could be nourished under current (average of 2007–2011) scenarios was subtracted from the number of people who could be potentially nourished under a no-trade scenario. Map breaks correspond to minimum, first quantile, medium, third quantile, and maximum for each nutrient (*Source* Wood et al. [2018])

food systems are well functioning, food insecurity can still occur. Food systems are involved in an intimate societal interdependence with many other systems. These systems influence food security as well. Food security also requires functioning health systems, education systems, water and sanitation, transports, energy, etc. The 1996 World Food Summit definition of food security stated earlier alludes to the necessity of a whole systems approach in that the physical and economic access alone indicates that built environments, urban and rural development and infrastructure, economics, livelihoods, and fair wages and equality of access to resources—all outside the technicalities of food system components—influence food security outcomes.

One illustration is the COVID-19 pandemic. There is growing evidence to suggest that while food supplies have been largely protected during the pandemic, food insecurity still increased for various reasons including loss of income and the global economic slowdown (Béné, Bakker, Chavarro et al., 2021; Béné, Bakker, Rodriguez et al., 2021). The longer-term implications of this loss of income and employment are profound. Food prices have been rising and are 3–25% higher in some parts of the world one year later than they were in July 2020 (FAOStat, n.d.). In addition, models suggest that by 2022, COVID-19 related disruptions may increase the number of children who are undernourished with an additional 9.3 million children wasted and 2.6 million children stunted—essentially unraveling the progress made over the last two decades to reduce these devastating growth outcomes (Osendarp et al., 2021). Thus, it is not just food systems alone that can tackle food security—other systems and their functioning effectively also impact food security and nutrition outcomes.

Another example is the United States (U.S.) and its prevalence of food insecurity—which sits at 10.5% (USDA, n.d.). At the crude level, one can argue that the U.S. food system works effectively well at ensuring food is available and accessible and is abundant in calories and diversity. However, food insecurity still exists. The question is why? Is it a food system problem per se, or a systemic societal problem of injustices and disadvantage that plague populations and their ability to access and afford a healthy diet? Evidence suggests the latter (Cooksey Stowers et al., 2020; Myers & Painter, 2017). The U.S. is plagued with issues of systemic racism that have impacted communities' ability to physically, economically, and socially access healthy foods (Bowen et al., 2021; Odoms-Young & Bruce, 2018). Much of this inaccessibility has to do with redlining—a historic, systematic denial of various services to residents of specific, often racially associated, neighborhoods or communities, either explicitly or through the selective raising of prices (Zhang & Ghosh, 2016). Of the foods that are often available to poor, marginalized populations, are processed, packaged foods that are cheap, convenient, and unhealthy make up a large proportion of the American diet, and are highly traded across the world (Baker et al., 2016; Development Initiatives, 2018; Thow, 2009). Thus, while the U.S. food system is one that has brought about incredible technological advances and abundance, not everyone benefits from the system.

## Conclusion

With only 8 years remaining to achieve the SDGs, the UNFSS and the Convention of Parties (COP26) climate change meetings of 2021 were important moments for States to commit once again to ending food insecurity through a food systems approach or mitigating climate change, respectively. However, history has taught the world what has worked, and what has not, and simplification of definitions and singular or overly technocratic fixes have not been completely effective in driving down, in aggregate, food insecurity. Taking a food systems approach to the problem is a step forward in that it would allow for a more holistic approach to address multiple problems and their root causes at the same time. It also allows for solutions that may serve to benefit multiple outcomes. Moreover, ignoring the importance of other systems—such as health, water, and economic systems—and how they influence food systems is also a dangerous path to take. To effectively address food security, especially for the most marginalized and vulnerable, actions must be seen through a wider, multi-system lens.

## REFERENCES

Baker, P., Friel, S., Schram, A., & Labonte, R. (2016). Trade and investment liberalization, food systems change and highly processed food consumption: A natural experiment contrasting the soft-drink markets of Peru and Bolivia. *Globalization and Health, 12*(1), 24.

Baker, P., Machado, P., Santos, T., Sievert, K., Backholer, K., Hadjikakou, M., Russell, C., Huse, O., Bell, C., Scrinis, G., Worsley, A., Friel, S., & Lawrence, M. (2020). Ultra-processed foods and the nutrition transition: Global, regional and national trends, food systems transformations and political economy drivers. *Obesity Reviews, 21,* 497.

Barona, J. L. (2008). Nutrition and health: The international context during the inter-war crisis. *Social History of Medicine, 21*(1), 87–105.

Barrett, C. B. (2020). Actions now can curb food systems fallout from COVID-19. *Nature Food.* https://doi.org/10.1038/s43016-020-0085-y

Bell, W., Coates, J., Fanzo, J., Wilson, N. L. W., & Masters, W. A. (2021). Beyond price and income: Preferences and food values in peri-urban Viet Nam. *Appetite, 166,* 105439. https://doi.org/10.1016/j.appet.2021.105439

Béné, C., Bakker, D., Chavarro, M. J., Even, B., Melo, J., & Sonneveld, A. (2021). Global assessment of the impacts of COVID-19 on food security. *Global Food Security, 31,* 100575.

Béné, C., Bakker, D., Rodriguez, M. C., Even, B., Melo, J., & Sonneveld, A. (2021). *Impacts of COVID-19 on people's food security: Foundations for a more resilient food system* (Report prepared for the CGIAR COVID-19 Hub Working Group 4, CGIAR, 90 p.). https://doi.org/10.2499/p15738coll2.134298

Béné, C., Fanzo, J., Haddad, L., Hawkes, C., Caron, P., Vermeulen, S., Herrero, M., & Oosterveer, P. (2020). Five priorities to operationalize the EAT–Lancet Commission report. *Nature Food, 1*(8), 457–459.

Béné, C., Fanzo, J., Prager, S. D., Achicanoy, H. A., Mapes, B. R., Toro, P. A., & Cedrez, C. B. (2020). Global drivers of food system (un)sustainability: A multi-country correlation analysis. *PLoS ONE, 15*(4), e0231071. https://doi.org/10.1371/journal.pone.0231071

Béné, C., Oosterveer, P., Lamotte, L., Brouwer, I. D., de Haan, S., Prager, S. D., Talsma, E. F., & Khoury, C. K. (2019). When food systems meet sustainability—Current narratives and implications for actions. *World Development, 113*, 116–130.

Béné, C., Prager, S. D., Achicanoy, H. A., Toro, P. A., Lamotte, L., Cedrez, C. B., & Mapes, B. R. (2019). Understanding food systems drivers: A critical review of the literature. *Global Food Security, 23*, 149–159.

Black, R. E., Victora, C. G., Walker, S. P., Bhutta, Z. A., Christian, P., de Onis, M., Ezzati, M., Grantham-McGregor, S., Katz, J., Martorell, R., Uauy, R., & Maternal and Child Nutrition Study Group. (2013). Maternal and child undernutrition and overweight in low-income and middle-income countries. *The Lancet, 382*(9890), 427–451.

Bouis, H. E., & Saltzman, A. (2017). Improving nutrition through biofortification: A review of evidence from HarvestPlus, 2003 through 2016. *Global Food Security, 12*, 49–58.

Bowen, S., Elliott, S., & Hardison-Moody, A. (2021). The structural roots of food insecurity: How racism is a fundamental cause of food insecurity. *Sociology Compass, 15*(7). https://doi.org/10.1111/soc4.12846

Brands, H., & Gavin, F. J. (2020). *COVID-19 and world order: The future of conflict, competition, and cooperation.* JHU Press.

Breisinger, C., Ecker, O., Maystadt, J.-F., Trinh Tan, J.-F., Al-Riffai, P., Bouzar, K., Sma, A., & Abdelgadir, M. (2014). *How to build resilience to conflict: The role of food security.* International Food Policy Research Institute.

Byerlee, D., & Fanzo, J. (2019). The SDG of zero hunger 75 years on: Turning full circle on agriculture and nutrition. *Global Food Security, 21*, 52–59.

Calabrò, G., & Vieri, S. (2014). The use of GMOs and consumers' rights in the European Union. *International Journal of Environment and Health, 7*(2), 128.

Chambers, J. D., Anderson, J. E., Salem, M. N., Bügel, S. G., Fenech, M., Mason, J. B., Weber, P., West, K. P., Jr, Wilde, P., Eggersdorfer, M., & Booth, S. L. (2017). The decline in vitamin research funding: A missed opportunity? *Current Developments in Nutrition, 1*(8), e000430.

Clapp, J., Moseley, W. G., Burlingame, B., & Termine, P. (2021). The case for a six-dimensional food security framework. *Food Policy, 106*, 102164.

Constantinides, S. V., Turner, C., Frongillo, E. A., Bhandari, S., Reyes, L. I., & Blake, C. E. (2021). Using a global food environment framework to understand relationships with food choice in diverse low- and middle-income countries. *Global Food Security, 29,* 100511.

Cooksey Stowers, K., Marfo, N. Y. A., Gurganus, E. A., Gans, K. M., Kumanyika, S. K., & Schwartz, M. B. (2020). The hunger-obesity paradox: Exploring food banking system characteristics and obesity inequities among food-insecure pantry clients. *PLoS ONE, 15*(10), e0239778.

Covic, N., Dobermann, A., Fanzo, J., Henson, S., Herrero, M., Pingali, P., & Staal, S. (2021). All hat and no cattle: Accountability following the UN food systems summit. *Global Food Security, 30*(100569), 100569.

Department of State. (1943, May 18–June 3). *United Nations Conference on Food and Agriculture: Hot Springs, Virginia*. United States Government Printing Office.

Development Initiatives. (2018). *The global nutrition report 2017*. Development Initiatives.

Downs, S. M., Ahmed, S., Fanzo, J., & Herforth, A. (2020). Food environment typology: Advancing an expanded definition, framework, and methodological approach for improved characterization of wild, cultivated, and built food environments toward sustainable diets. *Foods (Basel, Switzerland), 9*(4). https://doi.org/10.3390/foods9040532

Elizabeth, L., Machado, P., Zinöcker, M., Baker, P., & Lawrence, M. (2020). Ultra-processed foods and health outcomes: A narrative review. *Nutrients, 12*(7). https://doi.org/10.3390/nu12071955

Ericksen, P. J. (2008). Conceptualizing food systems for global environmental change research. *Global Environmental Change: Human and Policy Dimensions, 18*(1), 234–245.

Fanzo, J. (2014). Strengthening the engagement of food and health systems to improve nutrition security: Synthesis and overview of approaches to address malnutrition. *Global Food Security*. https://www.sciencedirect.com/science/article/pii/S2211912414000352

Fanzo, J. (2018). Does global goal setting matter for nutrition and health? *AMA Journal of Ethics, 20*(10), E979–986.

Fanzo, J. (2019). Healthy and sustainable diets and food systems: The key to achieving sustainable development goal 2? *Food Ethics*. https://doi.org/10.1007/s41055-019-00052-6

Fanzo, J. (2021). Achieving equitable diets for all: The long and winding road. *One Earth*. https://doi.org/10.1016/j.oneear.2021.03.007

Fanzo, J., Haddad, L., Schneider, K. R., Béné, C., Covic, N. M., Guarin, A., Herforth, A. W., Herrero, M., Sumaila, U. R., Aburto, N. J., Amuyunzu-Nyamongo, M., Barquera, S., Battersby, J., Beal, T., Bizzotto Molina, P., Brusset, E., Cafiero, C., Campeau, C., Caron, P., ... Rosero Moncayo, J.

(2021). Viewpoint: Rigorous monitoring is necessary to guide food system transformation in the countdown to the 2030 global goals. *Food Policy, 104,* 102163.

Fanzo, J., Rudie, C., Sigman, I., Grinspoon, S., Benton, T. G., Brown, M. E., Covic, N., Fitch, K., Golden, C. D., Grace, D., Hivert, M.-F., Huybers, P., Jaacks, L. M., Masters, W. A., Nisbett, N., Richardson, R. A., Singleton, C. R., Webb, P., & Willett, W. C. (2021). Sustainable food systems and nutrition in the 21st century: A report from the 22nd Annual Harvard Nutrition Obesity Symposium. *The American Journal of Clinical Nutrition.* https://doi.org/10.1093/ajcn/nqab315

FAO. (1974, December 9). *World Food Conference, Rome, 5 to 16 November 1974.* Communication from the Commission to the Council. SEC (74) 4955 final.

FAO. (1996). *Report of the World Food Summit.* http://www.fao.org/3/w35 48e/w3548e00.htm

FAO. (2006). *Food security* (Issue 2). UN Food and Agriculture Organization. https://www.fao.org/fileadmin/templates/faoitaly/documents/pdf/pdf_Food_Security_Cocept_Note.pdf

FAO. (2008). *An introduction to the basic concepts of food security.* https://www.fao.org/3/al936e/al936e.pdf

FAO, IFAD, UNICEF, WFP and WHO. (2018). *The state of food insecurity in the world 2018: Building climate resilience for food security and nutrition.* Rome, FAO. http://www.fao.org/docrep/018/i3434e/i3434e.pdf

FAO, IFAD, UNICEF, WFP and WHO. (2022). *The state of food security and nutrition in the world 2020.* Rome, FAO.

FAOStat. (n.d.). FAOStat. Retrieved September 11, 2021, from http://www.fao.org/faostat/en/

Garrett, J. L., & Natalicchio, M. (2010). *Working multisectorally in nutrition: Principles, practices, and case studies.* International Food Policy Research Institute.

Garrett, L. (2008). *Food failures and futures.* Council on Foreign Relations.

Garton, K., Thow, A. M., & Swinburn, B. (2020). International trade and investment agreements as barriers to food environment regulation for public health nutrition: A realist review. *International Journal of Health Policy and Management.* https://doi.org/10.34172/ijhpm.2020.189

Gaupp, F., Hall, J., Hochrainer-Stigler, S., & Dadson, S. (2019). Changing risks of simultaneous global breadbasket failure. *Nature Climate Change, 10*(1), 54–57.

Glass, S., & Fanzo, J. (2017). Genetic modification technology for nutrition and improving diets: An ethical perspective. *Current Opinion in Biotechnology.* https://www.sciencedirect.com/science/article/pii/S0958166916302488

Hawkes, C., & Fanzo, J. (2019). Health. In *Healthy and sustainable food systems* (pp. 11–22). Routledge.

HLPE. (2017). *Food systems and nutrition* (A report by the High Level Panel of Experts on Food Security and Nutrition of the Committee on World Food Security).

HLPE. (2020). *Food security and nutrition: Building a global narrative towards 2030* (A report by the High Level Panel of Experts on Food Security and Nutrition of the Committee on World Food Security).

Ivanic, M., & Martin, W. (2008). Implications of higher global food prices for poverty in low-income countries. *Agricultural Economics, 39*, 405–416.

Jacobsen, S.-E., Sørensen, M., Pedersen, S. M., & Weiner, J. (2013). Feeding the world: Genetically modified crops versus agricultural biodiversity. *Agronomy for Sustainable Development, 33*(4), 651–662.

Jeppsson, A., & Okuonzi, S. A. (2000). Vertical or holistic decentralization of the health sector? Experiences from Zambia and Uganda. *The International Journal of Health Planning and Management, 15*(4), 273–289.

Kloppenburg, J. (2014). Re-purposing the master's tools: The open source seed initiative and the struggle for seed sovereignty. *The Journal of Peasant Studies, 41*(6), 1225–1246.

Klümper, W., & Qaim, M. (2014). A meta-analysis of the impacts of genetically modified crops. *PLoS ONE, 9*(11), e111629.

Koutouki, K., & Marin, P. H. (2013). The use of GMOs in Chile and the protection of indigenous culture. In M.-C. Cordonier Segger, F. Perron-Welch, & C. Frison (Eds.), *Legal aspects of implementing the cartagena protocol on biosafety* (pp. 433–444). Cambridge University Press.

Kovak, E., Qaim, M., & Blaustein-Rejto, D. (2021). *The climate benefits of yield increases in genetically engineered crops.* BioRxiv. https://doi.org/10.1101/2021.02.10.430488

Lamstein, S., Pomeroy-Stevens, A., Webb, P., & Kennedy, E. (2016). Optimizing the multisectoral nutrition policy cycle. *Food and Nutrition Bulletin, 37*(4), S107–S114. https://doi.org/10.1177/0379572116675994

Makita, K., de Haan, N., Nguyen-Viet, H., & Grace, D. (2019). Assessing food safety risks in low and middle-income countries. In P. Ferranti, E. M. Berry, & J. R. Anderson (Eds.), *Encyclopedia of food security and sustainability* (pp. 448–453). Elsevier.

Mancino, L., Guthrie, J., & Just, D. R. (2018). Overview: Exploring ways to encourage healthier food purchases by low-income consumers—Lessons from behavioral economics and marketing. *Food Policy, 79*, 297–299.

Martin, W., & Ivanic, M. (2016). Food price changes, price insulation, and their impacts on global and domestic poverty. In *Food price volatility and its implications for food security and policy* (pp. 101–113). Springer International Publishing.

Mason, J., Greiner, T., Shrimpton, R., Sanders, D., & Yukich, J. (2015). Vitamin A policies need rethinking. *International Journal of Epidemiology, 44*(1), 283–292.

Micha, R., Mannar, V., Afshin, A., Allemandi, L., Baker, P., Battersby, J., Bhutta, Z., Chen, K., Corvalan, C., Di Cesare, M., Dolan, C., Fonseca, J., Hayashi, C., Rosenzweig, C., Schofield, D., & Grummer-Strawn, L. (2020). *2020 Global nutrition report: Action on equity to end malnutrition* (N. Behrman, Ed.). Development Initiatives, Bristol, UK.

Mozaffarian, D., Angell, S. Y., Lang, T., & Rivera, J. A. (2018). Role of government policy in nutrition—Barriers to and opportunities for healthier eating. *BMJ, 361.* https://doi.org/10.1136/bmj.k2426

Myers, A. M., & Painter, M. A., II. (2017). Food insecurity in the United States of America: An examination of race/ethnicity and nativity. *Food Security, 9*(6), 1419–1432.

Negin, J., Remans, R., Karuti, S., & Fanzo, J. C. (2009). Integrating a broader notion of food security and gender empowerment into the African green revolution. *Food Security.* https://doi.org/10.1007/s12571-009-0025-z

Odoms-Young, A., & Bruce, M. A. (2018). Examining the impact of structural racism on food insecurity: Implications for addressing racial/ethnic disparities. *Family and Community Health, 41*(2 Food Insecurity and Obesity), S3–S6.

Osendarp, S., Akuoku, J. K., Black, R. E., Headey, D., Ruel, M., Scott, N., Shekar, M., Walker, N., Flory, A., Haddad, L., Laborde, D., Stegmuller, A., Thomas, M., & Heidkamp, R. (2021). The COVID-19 crisis will exacerbate maternal and child undernutrition and child mortality in low- and middle-income countries. *Nature Food, 2*(7), 476–484. https://doi.org/10.1038/s43016-021-00319-4

Pelletier, D., Gervais, S., Hafeez-ur-Rehman, H., Sanou, D., & Tumwine, J. (2016). Multisectoral nutrition: Feasible or fantasy? *The FASEB Journal, 30*(1), 669.9.

Pelletier, D., Gervais, S., Hafeez-ur-Rehman, H., Sanou, D., & Tumwine, J. (2018). Boundary-spanning actors in complex adaptive governance systems: The case of multisectoral nutrition. *The International Journal of Health Planning and Management, 33*(1), e293–e319.

Pingali, P. (2015). Agricultural policy and nutrition outcomes—Getting beyond the preoccupation with staple grains. *Food Security, 7*(3), 583–591. https://doi.org/10.1007/s12571-015-0461-x

Pingali, P. L. (2012). Green revolution: Impacts, limits, and the path ahead. *Proceedings of the National Academy of Sciences of the United States of America, 109*(31), 12302–12308.

Raghunathan, K., Headey, D., & Herforth, A. (2021). Affordability of nutritious diets in rural India. *Food Policy, 99,* 101982.

Rivera-Ferre, M. G., López-i-Gelats, F., Ravera, F., Oteros-Rozas, E., di Masso, M., Binimelis, R., & El Bilali, H. (2021). The relation of food systems with the COVID-19 pandemic: Causes and consequences. *Agricultural Systems, 191*, 103134.

Rohr, J. R., Barrett, C. B., Civitello, D. J., Craft, M. E., Delius, B., DeLeo, G. A., Hudson, P. J., Jouanard, N., Nguyen, K. H., Ostfeld, R. S., Remais, J. V., Riveau, G., Sokolow, S. H., & Tilman, D. (2019). Emerging human infectious diseases and the links to global food production. *Nature Sustainability, 2*(6), 445–456.

Rosenberg, I., Rogers, B., Webb, P., & Schlossman, N. (2012). Enhancements in food aid quality need to be seen as a process, not as a one-off event. *The Journal of Nutrition, 142*(9), 1781.

Sanchez, P. A. (2009). A smarter way to combat hunger. *Nature, 458*(7235), 148.

Sarkar, B., Bakshi, U. G., Sayeed, C., & Goswami, S. (2021). Impact of genetically modified organisms on environment and health. In *Multidimensional approaches to impacts of changing environment on human health* (pp. 263–273). CRC Press.

Shaw, D. J. (2007a). Food surpluses: Historical background. In *World Food Security* (pp. 12–14). Palgrave Macmillan UK.

Shaw, D. J. (2007b). World food crisis. In *World Food Security* (pp. 115–120). Palgrave Macmillan UK.

Shaw, D. J. (2007c). World Food Summit, 1996. In *World Food Security* (pp. 347–360). Palgrave Macmillan UK.

Swaminathan, M. S. (2006). An evergreen revolution. *Crop Science, 46*(5), 2293–2303.

Swinburn, B., Kraak, V., Rutter, H., Vandevijvere, S., Lobstein, T., Sacks, G., Gomes, F., Marsh, T., & Magnusson, R. (2015). Strengthening of accountability systems to create healthy food environments and reduce global obesity. *The Lancet, 385*(9986), 2534–2545.

Thow, A. M. (2009). Trade liberalisation and the nutrition transition: Mapping the pathways for public health nutritionists. *Public Health Nutrition, 12*(11), 2150–2158.

Turner, C., Aggarwal, A., Walls, H., Herforth, A., Drewnowski, A., Coates, J., Kalamatianou, S., & Kadiyala, S. (2018). Concepts and critical perspectives for food environment research: A global framework with implications for action in low- and middle-income countries. *Global Food Security, 18*, 93–101. https://doi.org/10.1016/j.gfs.2018.08.003

Turner, C., Kalamatianou, S., Drewnowski, A., Kulkarni, B., Kinra, S., & Kadiyala, S. (2019). Food environment research in low- and middle-income countries: A systematic scoping review. *Advances in Nutrition*. https://doi.org/10.1093/advances/nmz031

USDA. (n.d.). *Trends in U.S. food security*. USDA ERS. Retrieved September 11, 2021, from https://www.ers.usda.gov/topics/food-nutrition-assistance/food-security-in-the-us/interactive-charts-and-highlights/#trends

Vellakkal, S., Fledderjohann, J., Basu, S., Agrawal, S., Ebrahim, S., Campbell, O., Doyle, P., & Stuckler, D. (2015). Food price spikes are associated with increased malnutrition among children in Andhra Pradesh, India. *The Journal of Nutrition, 145*(8), 1942–1949.

Webb, P., Benton, T. G., Beddington, J., Flynn, D., Kelly, N. M., & Thomas, S. M. (2020). The urgency of food system transformation is now irrefutable. *Nature Food, 1*(10), 584–585.

Webb, P., Caiafa, K., Walton, S., & Food Aid Quality Review Group. (2017). Making food aid fit-for-purpose in the 21st century: A review of recent initiatives improving the nutritional quality of foods used in emergency and development programming. *Food and Nutrition Bulletin, 38*(4), 574–584.

Willett, W., Rockström, J., Loken, B., Springmann, M., Lang, T., Vermeulen, S., Garnett, T., Tilman, D., DeClerck, F., Wood, A., Jonell, M., Clark, M., Gordon, L. J., Fanzo, J., Hawkes, C., Zurayk, R., Rivera, J. A., De Vries, W., Majele Sibanda, L., … Murray, C. J. L. (2019). Food in the Anthropocene: the EAT-Lancet Commission on healthy diets from sustainable food systems. *The Lancet, 393*(10170), 447–492.

Wood, S. A., Smith, M. R., Fanzo, J., & Remans, R. (2018). Trade and the equitability of global food nutrient distribution. *Nature, 1*(1), 34–37.

Zhang, M., & Ghosh, D. (2016). Spatial supermarket redlining and neighborhood vulnerability: A case study of Hartford, Connecticut. In *Transactions in GIS, 20*(1), 79–100. https://doi.org/10.1111/tgis.12142

Zilberman, D., Holland, T., & Trilnick, I. (2018). Agricultural GMOs—What we know and where scientists disagree. *Sustainability, 10*(5), 1514.

**Open Access** This chapter is licensed under the terms of the Creative Commons Attribution 4.0 International License (http://creativecommons.org/licenses/by/4.0/), which permits use, sharing, adaptation, distribution and reproduction in any medium or format, as long as you give appropriate credit to the original author(s) and the source, provide a link to the Creative Commons license and indicate if changes were made.

The images or other third party material in this chapter are included in the chapter's Creative Commons license, unless indicated otherwise in a credit line to the material. If material is not included in the chapter's Creative Commons license and your intended use is not permitted by statutory regulation or exceeds the permitted use, you will need to obtain permission directly from the copyright holder.

CHAPTER 3

# The Global Food System is Not Broken but Its Resilience is Threatened

*Patrick Caron, Ellie Daguet, and Sandrine Dury*

INTRODUCTION

Bringing together food systems' transformation and resilience raises a series of questions in two directions: what does resilience of food systems

P. Caron (✉)
CIRAD—Centre de Coopération Internationale en Recherche Agronomique pour le Développement and ART-Dev: Acteurs, Ressources et Territoires dans le Développement, University of Montpellier, Montpellier, France
e-mail: patrick.caron@cirad.fr

E. Daguet
CIRAD—Centre de Coopération Internationale en Recherche Agronomique pour le Développement, University of Montpellier, Montpellier, France
e-mail: ellie.daguet@cirad.fr

S. Dury
CIRAD—Centre de Coopération Internationale en Recherche Agronomique pour le Développement, MoISA—Montpellier Interdisciplinary Center on Sustainable Agri-food Systems, University of Montpellier, Montpellier, France
e-mail: sandrine.dury@cirad.fr

© The Author(s) 2023
C. Béné and S. Devereux (eds.), *Resilience and Food Security in a Food Systems Context*, Palgrave Studies in Agricultural Economics and Food Policy, https://doi.org/10.1007/978-3-031-23535-1_3

means as their adaptation has, on the one hand, managed to avoid any global food shortage (even though it might have occured locally), yet, on the other hand, led to massive negative multidimensional impacts? How can change take place as it is unanimously expected and in the same time so difficult to orchestrate because of the huge diversity of contexts and actors?

To address these questions, we will first look at the way food systems have been able to evolve in the past under huge and numerous constraints. This first section will thus focus on the incredible changes that have taken place since the Second World War and that have successfully prevented a massive global food shortage. Challenging many statements that rightly point out the current deficiencies of food systems, this highlights the success of past transformation. We will then consider the reasons and challenges for future adaptation and finally formulate questions regarding the pathways and conditions to undertake future transformation.

## The Global Food System is Resilient in Terms of Food Supply

Contrarily to what is often stated (Schmidt-Traub et al., 2019), the global food system is not broken and is more and more resilient in terms of food supply!

In terms of development, the global food system has proven to be resilient in the last decades if we consider specific outcomes and metrics such as food production and more specifically global caloric availability (Porkka et al., 2013 or Roser & Ritchie, 2021). According to these authors, the percentage of population living in countries with sufficient food supply (>2500 kcal/cap/day) has almost doubled from 33% in 1965 to 61% in 2005; the population living with critically low food supply (<2000 kcal/cap/day) has dropped from 52% to just 3%. Similarly, a long-run downward trend of international food prices has been observed up to the mid-2000s. Between 1961 and 2006, the World Bank's international index of food grain prices fell by more than 30% according to Baldos and Hertel (2016). OECD recent report (2021) clearly illustrates this past trend (Fig. 3.1) for important food commodities.

Global food production and trade systems have made possible the delivery of staple food all over the world and all year-round, and to recover after stock shortage crises such as in 2008. Today, the 155 million

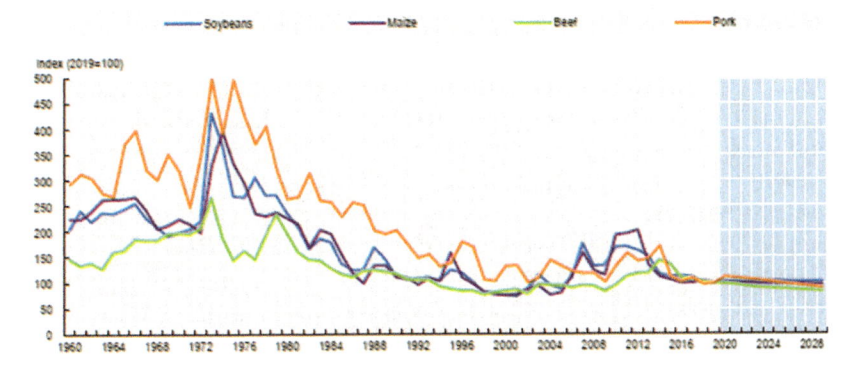

**Fig. 3.1** Long-term evolution of real agricultural prices (OECD, 2021) (*Note* Historical data for soybeans, maize and beef from World Bank, "World Commodity Price Data" [1960–1989]. Historical data for pork from USDA QuickStats [1960–1989]. *Source* OECD/FAO [2020], "OECD-FAO Agricultural Outlook", OECD Agriculture statistics [database], http://dx.doi.org/10.1787/agr-outl-data-en

acutely food insecure people in need of urgent assistance are suffering more from persistent conflict or insecurity, economic shocks and weather extremes (the Global Network Against Food Crisis, 2021) than from a lack of global food supply.

The systemic "hunger riots" crisis in 2008 did not translate into either long-lasting skyrocketing prices or serious food shortages in the global food market. Similarly, during the COVID-19 crisis in 2020, the production and trade systems did not collapse and international supply chains continued to function (Béné et al., 2021).

The global situation of food availability as defined by the 1974 food security definition has improved in recent decades thanks to the so-called modernization of the food system (Burchi & De Muro, 2016), including the Green Revolution. This built in particular on food and agricultural research and innovation systems and on strong national and international agricultural and public trade policies. Thanks to strong progress in agricultural productivity, the food supply more than doubled (2.5 times) between 1960 and 2000, increasing even faster than the doubling of the global population (Paillard et al., 2014).

This productivity improvement was due to higher use of chemical inputs, the development of irrigation schemes, large-scale adoption of mechanization, progress in genetics, extensive use of fossil energy and the recent introduction of technological devices (OECD, 2021). Those inputs made possible the dissociation between agricultural production and land use for agricultural purposes: production is no longer correlated with cultivated surfaces (see Fig. 3.2).

Neither the global population grew as fast as the production. As a consequence, the average caloric availability per capita reached unprecedent levels, around 2950 Kcal/cap/day, in 2017 (FAOStat).

At the same time, upstream (credit, inputs, mechanization, irrigation, etc.) and downstream corporates (supply chains, agri-food processing, retailing, etc.) involved with agriculture became bigger and more powerful. Food market chains (including infrastructure such as roads, storage facilities, slaughterhouses, etc.) got longer and more complex, and concentrated on a large part of the food processing that used to take place at the farm or consumer levels. Many processes were industrialized and normalized, food safety was regulated, and huge multinational firms in logistics and distribution emerged (McMichael, 2009). Food trade has

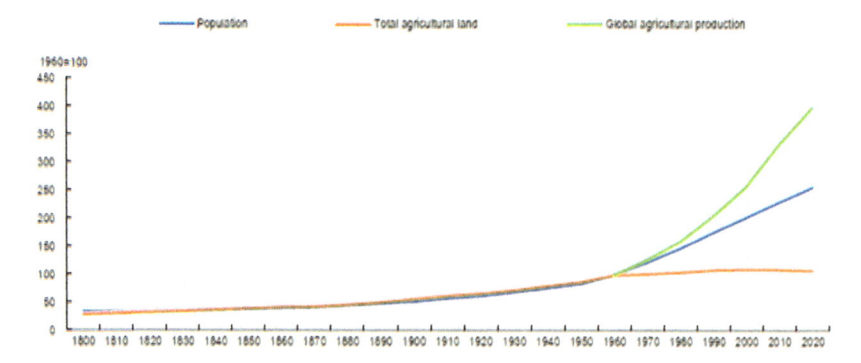

Fig. 3.2 Global population, food production and agricultural land use in the long run (OECD, 2021) (*Source* Population data from Maddison's historical statistics for 1820–1940; UN Population Division for 1950–2010; 1800 and 1810 extrapolated from Maddison. Agricultural [crops and pasture] land data for 1800–2010 from the History Database of the Global Environment [HYDE 3.2], Klein Glodewijk et al. [2017]. Global agricultural production data for 1960–2010 from FAOSTAT (Net Agricultural Production Index); data for 2020 from OECD/FAO [2020], "OECD-FAO Agricultural Outlook", OECD Agriculture statistics [database], http://dx.doi.org/10.1787/agr-outl-data-en

never reached such high levels (Krausmann & Langthaler, 2019). One has to recognize that they performed well to deliver food in time everywhere (or almost), even in time of crisis (Béné et al., 2021). This modern farming sector and the powerful international agri-food companies have proven to be successful and resilient by many criteria, as previously mentioned (caloric production and trade).

As a result, hunger (or food insecurity) as it used to be defined for years by the FAO—a lack of calories compared to individual requirements, estimated on the basis of national food balances—declined from the 1960s to mid-2010s at the global scale. The massive famines that occurred until the 1980s and affected many countries in Africa and Asia were effectively eradicated. While the first edition of the FAO State of Food and Agriculture report was published in 1947 and estimated the prevalence of undernourishment in 1945 to be 50% of the world population,[1] this figure dropped to 23% in 1990 with 980 million people suffering from hunger. In 2019, 688 million people (8.9% of the world population) were undernourished. Even if the calculation method has evolved[2] and even if the trend has reversed since 2018, the improvement was massive.

In addition, the economic burden of food provisioning was substantially reduced for households in recent decades in many countries. Today, the share of the household budget allocated to food consumption represents 8 to 15% in rich countries (Fig. 3.3). While in low-income countries this burden is often still above 50%, in middle-income countries it is now between 20 and 35%.

Considering the population increase as a huge and unique stressor for the humanity and the planet, one can thus acknowledge that the food system has been resilient at the worldwide scale.

The global food system has also created some tools and institutions dedicated to take care of the most destitute people, in the worst or most vulnerable contexts. Based on specific international institutions whose role is to deal with emergency assistance in low-income countries, the global

---

[1] http://www.fao.org/3/ap635e/ap635e.pdf.

[2] The number of undernourished was initially calculated by the FAO on the basis of the availability of food in each country and compared to its population and its nutritional needs owning to age, sex, status and physical activity. Today, as caloric availability is no longer the number one problem, the FAO has launched a new indicator, the Food Insecurity Experience Scale (FIES; http://www.fao.org/in-action/voices-of-the-hungry/en/), to better monitor food security in terms of access, in alignment with the 1996 definition of food security.

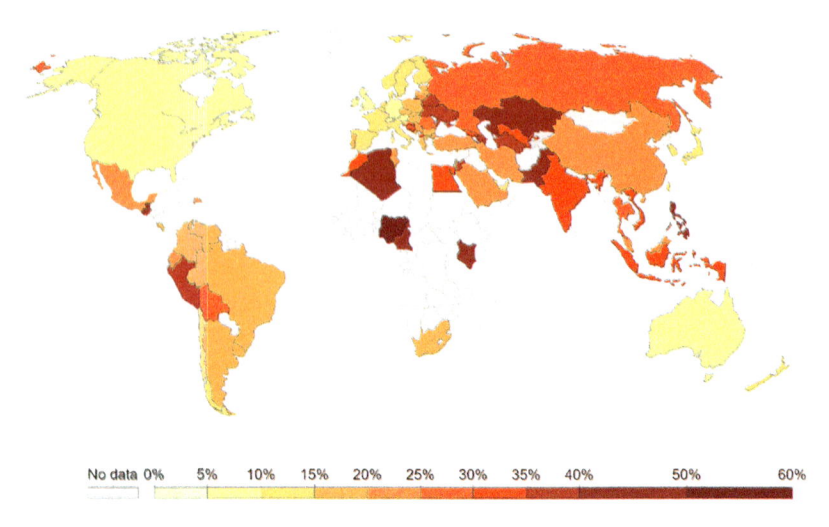

**Fig. 3.3** Share of consumer expenditure spent on food worldwide in 2015 (*Data source* United States Department for Agriculture, Economic Research Service, USDA, ERS, 2017. Chart produced by Our World in Data consulted online https://ourworldindata.org/grapher/share-of-consumer-expenditure-spent-on-food)

food system has been able to cope with localized or temporary crises. For example, the World Food Programme (WFP) assisted 97 million people in 2019 in 88 countries. Moreover, in all countries (low-, middle- but also high-income), systems of food assistance were developed to address risk of hunger when other mechanisms fail to deliver food to vulnerable populations. In 2019, for example, 10.5% of U.S. households were food insecure for at least some time during the year (Coleman-Jensen et al., 2020). Most of them are entitled to specific food aid in different forms (e.g. coupons). In Europe, food banks distributed 768,000 tons of food and assisted 9.5 million people in 24 European countries in 2019 (European Food Banks Federation, 2020). Institutions dealing with food deprived people are parts of the food system and not just post-crisis coping mechanisms.

Therefore, the question is no longer limited to a problem of improving and smoothing food availability but of healthy food access all over the world. Resilience of food supply and global markets is not sufficient to eradicate undernutrition as the global number of affected persons has remained approximatively the same in the past 40 years. Rather than being driven by supply shortage, the issue is most often demand related and is caused by poverty. It has been exacerbated by the COVID-19 crisis and by conflicts that hinder the elimination of food insecurity and hunger.

## MANY REASONS WHY THE GLOBAL FOOD SYSTEM NEEDS A PROFOUND TRANSFORMATION

Poor quality diets are among the top risk factors contributing to the global burden of disease (Afshin et al., 2019). Not only have current food systems failed to eradicate hunger (despite preventing global food shortages), they have also incentivized the spread of diet-related diseases. New nutrition problems have emerged with the increase of supply and the new nature of the food products. The expansion of ultra-processed foods rich in salt, sugar and fat threatens public health in many countries (Popkin, 2017). One billion people will soon be obese, most of them eating too many calories compared to their needs, and at risk of non-communicable diseases such as diabetes, cardio-vascular disorders and cancer. Nowadays, malnutrition in all its forms (overweight/obesity, micronutrient deficiencies and undernutrition including stunting and wasting) affects all countries in the world and most are affected by multiple forms of malnutrition (Fig. 3.4). This triple burden of malnutrition coexists at all levels: global, national, local and even at family level (HLPE, 2017a). It affects urban areas as well as rural.

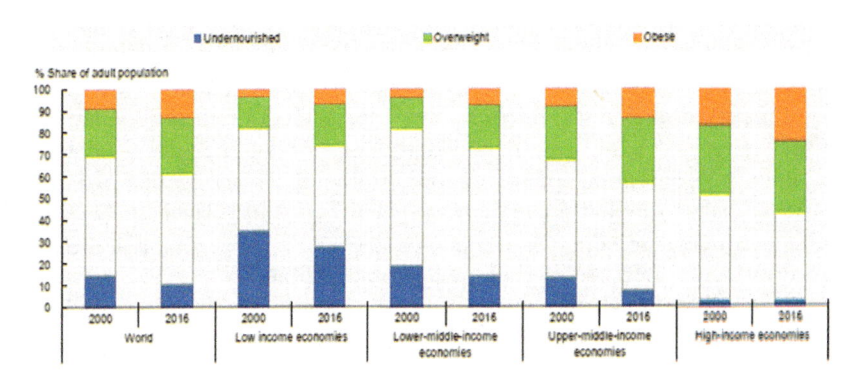

**Fig. 3.4** Undernourishment, overweight and obesity, 2000–2016 (OECD, 2021)

Several socio-economic and environmental factors, like global trade, demographic and economic transitions, rapid urbanization and increasing availability and affordability of poor quality ultra-processed food, have led to changes in dietary patterns and consumers' preferences (Béné et al., 2020b; HLPE, 2017a). More women involved in economic life has led to an increase in the demand for ready-to-eat, convenient, ultra-processed food which is often of poor quality. Nutritious foods are more expensive than energy-dense foods with poor nutritional qualities, and their total cost (defined as the sum of the cost of food items and preparation time) is also much higher than less-healthy ready-to-eat alternatives (FAO et al., 2020). "It has been estimated that the labour costs of a healthy diet for a single-headed household recipient of the Supplemental Nutrition Assistance Program (SNAP, formerly the Food Stamp Program) in the United States of America would represent 60% of the total cost of food" (FAO et al., 2020, p. 130). On the other hand, globalization and industrialization have allowed big companies to make economies of scale thus distributing worldwide cheap and convenient processed food, yet with low nutrient density (Haddad et al., 2016; Willett et al., 2019). Unhealthy foods become more attractive while nutritious food is less affordable (FAO et al., 2020).

Poverty and inequalities are other underlying causes of all forms of malnutrition all around the world. High-income countries have succeeded in producing cheap calories. However, a recent study estimates that, in those countries, the cost of a healthy diet is on average 6 times more than that of an energy-sufficient diet. As for the Global South, the cost of a healthy diet is higher than the national average food expenditure for most countries. "A healthy diet is not affordable in lower-middle-income countries, and it is far from being affordable – almost 3 times the average food expenditure– in low-income countries" (FAO et al., 2020). In the same report, it has been estimated that, based on an analysis of incomes, 3 billion people around the world could not afford a healthy diet in 2017.

If globalization has increased the availability and diversity of food while reducing seasonal shortages, not everyone benefits from these improvements. Because of their geographical situation (remote areas) or social status (gender, ethnic, economic situation), vulnerable groups have limited access to diverse and quality food. This is the case, for example, in food deserts[3] (HLPE, 2017a). Local production (including family farming for own consumption) remains then an important part of the source of food and is not always sufficient to cover the nutrient needs. People living from traditional food systems (rural or indigenous communities, for example) might still experience "hunger seasons" (HLPE, 2017a).

Transforming food systems so that everyone in the world has access to sufficient quality food is key for current and future generations. Indeed, there is an intergenerational cycle of malnutrition. Because of inadequate nutritional status of women and inadequate infant and young children child-caring and feeding practices, malnutrition has consequences across generations (CFS, 2021).

In addition to generating poor health conditions, malnutrition also has economic consequences by reducing labour productivity and incomes, thus affecting people's livelihood through their lifetime (HLPE 2017a). Moreover, poor nutrition increases health expenditure at country level, accounting for a significant burden on national healthcare systems (Global Panel, 2016).

Furthermore, the environmental and social drawbacks of the existing food system are threatening the sustainability and resilience of the system (Caron et al., 2018; OECD, 2021; Dury et al., 2019). This was also very much discussed during the UN Food Systems Summit in 2021.

As already mentioned in several scientific papers (Willett et al., 2019) and reports (Dury et al., 2019, among others), existing food systems are under pressure and face many threats. They contribute in return to exacerbated risks through unreasonable use of natural resources and abuse and disrespect of human fundamental rights and dignity (Caron et al., 2018).

Food systems are responsible for an irreversible loss in biodiversity. The dramatic evolution of agriculture in the past century in industrialized and some low- and middle-income countries, based on improved varieties and synthetic inputs, greatly increased production but also led to the artificialization of agroecosystems and great losses of specific and genetic biodiversity. In turn, these losses have hampered food systems in different ways: degraded ecosystem services affecting crop yields and resilience, reduced crop biodiversity and highly specialized industrialized

---

[3] Geographic areas where residents' access to food is restricted or non-existent due to the absence or low density of "food entry points" within a practical travelling distance.

food processing, which has decreased the diversity of the food supply and its nutritional value (Hainzelin, 2019; HLPE, 2017b).

Plateauing yields have been reported in several crops and 20% of the world's cultivated land has lost productive capacity (FAO, 2019). Insect species loss or sharp decline of species have been documented and linked to agricultural intensification (Wagner, 2020), including pesticide use (Van der Sluijs et al., 2015), destruction of habitat, changes in land use and so on. In low- and middle-income countries, commercial agriculture is the most important driver of deforestation, followed by subsistence agriculture (FAO, 2016; Feintrenie et al., 2019).

Food systems are responsible for up to one-third of anthropogenic greenhouse gas (GHG) emissions and are therefore a major driver of climate change (Xu et al., 2021). These emissions include carbon dioxide ($CO_2$), methane ($CH_4$) and nitrous oxide ($N_2O$). The environmental pressures of food systems are likely to intensify, as humanity is arguably already operating beyond planetary boundaries (Steffen et al., 2015; Vermeulen et al., 2012). Food production and thus the livelihoods of billions of people, especially the most vulnerable, including small farmers, are impacted and will be even more in the coming decades by the effects of climate change (Demenois et al., 2019; FAO, 2018; IPCC, 2018).

Moreover, existing food systems are threatening the social and territorial balances, deepening economic inequalities and fuelling social unrest (Caron et al., 2018; Giordano et al., 2019; HLPE, 2013, 2017b). In low- and middle-income countries, large-scale land and water acquisition for food production and large investment projects are, for example, considered as drivers of conflicts since they deprive local communities (Anseeuw et al., 2019).

Food systems are the backbone of economies in many countries, but their relative importance in the GDP is shrinking when countries become richer. In those countries, the number of farmers has fallen dramatically since World War II. They were replaced by machines and, as a consequence, the labour productivity has increased sharply; much of the workforce switched to other economic sectors (including the agrifood sector). In low- and middle-income countries, many questions are raised regarding the capacity of food systems to accompany the demographic and economic growth. For example, in sub-Saharan Africa, the food economy represents two-third of total employment for both men and women (Allen et al., 2018) and there are uncertainties regarding the inclusion of smallholders or micro- and small food enterprises from the informal economic sector (HLPE, 2013), in modern upgrading food chains (Soullier et al., 2019).

Considering that problems and concerns also offer opportunities, food systems should also be looked at as strong levers towards implementing the Agenda 2030 for sustainable development and its 17 sustainable development goals (HLPE, 2020). More and more reports and authors call for engaging in their transformation, which is looked upon as a priority avenue to prevent major disruptions and to address sustainability concerns.

## How to Move Towards such a Transformation?

### Reticence and Obstacles Despite Alerts

Scientific evidence is considered sufficient by a great majority of scientists and policymakers to demonstrate that "business as usual is not an option" (IAASTD, 2009), to pay due attention to whistle blower alerts and to call for deep changes in order to prevent catastrophes and the worst from coming. Evidence reported in previous sections has contributed to shaping the global political agenda, as illustrated by the 2030 Agenda for Sustainable Development and the Paris Agreement on climate. This has led the UN Global Sustainable Development Report to identify food systems and nutrition patterns as one of the six entry points to achieve the 2030 Agenda (United Nations, 2019). This has also led the UN Secretary General to convene a UN Food systems summit in September 2021 to deliver progress on all 17 SDGs, beyond food security issues.

Despite such shared observations, alerts and engagements, transformation is not taking place with the necessary pace to address sustainability concerns (HLPE, 2017a; Webb et al., 2020; IPES-Food and ETC Group 2021). This would indeed question the paradigm and social order that have prevailed and made evolution possible in the twentieth century (Brundtland, 1987; Meadows et al., 1972). The necessary paradigm shift (Caron et al., 2018) is about intellectual and political framing, in particular to agree on the functions of the agricultural sector, the role it may play in development and the way we accordingly measure its performances. Increasing productivity to contribute to production and to global supply cannot be the only way to account for expressed future expectations. Such a paradigm shift generates resistance. It would imply a revolution that perturbs previous political equilibria and social agreements. Paradoxically, the word revolution has been used in the twentieth century to describe a transformation process, i.e. the Green Revolution, that essentially relied on technological advances and not on social and political changes. The breadth and depth of the transformation required conversely suggest disruptions that are at the moment not agreed, nor even discussed. The resistance to the acknowledgement of the notion of multifunctionality of agriculture (Caron et al., 2008) and the incapacity to organize global discussion about the best way to shape trade and to transform the World Trade Organization to contribute to sustainability and to embed technological advances into a political project rather than advocating for a supposed neutrality offer perfect examples of such resistance. The identification of the cost of inaction (Stern, 2007) is useful to advocate about the need to engage in a transformation process and to call for courage, but it is not sufficient.

How is it possible to explain the paradox of such a gap between the awareness of the need to act and the incapacity to do it? Is this just path dependency and the difficulty of coping with the cost of change, blindness or lack of political will, as regretted by most experts and scientists in their reports? We would rather advance that procrastination relies on obstacles and barriers to be understood and removed, as highlighted by the HLPE report on food systems and nutrition (HLPE, 2017a). Many countries fail to recognize the right to food and to implement rights-based approaches that target the most vulnerable persons. In addition, and as highlighted by the HLPE report, "power struggles present challenges as transnational food corporations use their economic power to hinder political action to improve food systems and diets" (HLPE, 2017a, p. 16). As evidenced in the industry (Hawkes, 2002), conflicts of interest are also reported as such obstacles that may affect health and nutrition goals: "salient examples include food and beverage marketing in unhealthy food environments and advertising foods high in fat, sugar and salt" (HLPE, 2017a, p. 16).

Such barriers and obstacles are more and more documented. The recent OECD report "Making Better Policies for Food Systems" (2021), for example, clearly identifies frictions and tensions around facts, interests and values that make food systems transformation difficult. It also points out obstacles to be removed in order to progress towards the needed agreements to undertake changes beyond diverging interests. The report provides examples in the seed, the ruminant livestock and the processed food sectors.

Such a transformation is even more difficult because of potential disparity and divergence between local transitions and global expectations, and the resulting trade-offs. Many changes are taking place at the local level through place-specific arrangements and modalities, and the coexistence of diverse pathways and their convergence (or lack thereof) raise important questions in terms of coordination, arbitration and regulation (Béné et al., 2020a). We are at a crossroads, where the long-lasting international consensus on a public support for a trade liberalization agenda is weakening, while more and more countries are adopting food sovereignty policies (HLPE, 2017b).

The COVID-19 pandemic and its subsequent impacts may play a triggering role to generate such a transformation. Not so much because it offers an opportunity to celebrate the reterritorialization of food systems (see Losch and May, Chapter 10 in this volume) and claims for sovereignty, but mainly because of the economic crisis that will call for addressing controversial issues. Will the global food system be able to adapt once again by orchestrating a great transformation with the breadth and depth of the one suggested by Karl Polanyi (1944)?

### *The Engagement of Science to Help Moving Beyond Obstacles*

Evidence and alerts from scientists are not sufficient to generate the expected transformation, since science and policy interactions are much more complex and dynamic than just a linear monodirectional relationship, where science would produce and provide knowledge that is used by policymakers (Louafi, 2021). When looking more closely at these interactions, one also realizes that such alerts are not a recent process, as highlighted by Mathis (2021) when looking at the premises of the industrial revolution. Malthus's name for instance is also resurfacing to remind us that alerts already emerged long ago.

As a consequence, more and more scientists are calling for a strong investment in identifying and understanding obstacles to transformation of our development model and for valuing emerging schools of thought. This field had already been implicitly explored by the emergence of system thinking and approaches in the 1970s (Le Moigne, 1990; Morin, 1986) which looked at the complexity of processes and the rationale of human decision and behaviour through interdisciplinary lenses. With the acceleration of technological advances in genetics and digital sciences and the occurrence of related crises, growing attention by scholars has been paid from the late 1980s onward, to socio-technical controversies (IHEST, 2015; Latour, 1987; Lemieux, 2007). Better understanding of Genetically Modified Organisms-related debates and disputes, for example, has been an issue for many authors. Similarly, and as the motto "win–win" was gaining traction in development spheres, scientists became involved with the identification of trade-offs. This was a way to balance and critically look at the capacity to generate synergetic options for action, considered by some as naïve assumptions (Cheyns et al., 2017).

Specifically looking at obstacles in order to transcend them is thus emerging as an intellectual and operational field that builds upon the legacy of the above-mentioned schools of thought and intends to provide actionable knowledge. Looking at controversies as a fertile field is, for example, gaining more traction (IHEST, 2015). This is also the implicit assumption that resulted in the creation in 2010 of the High-Level Panel of Experts (CFS/HLPE[4]) by the UN Committee on World Food Security. The reports of this Panel aim at providing policymakers with analyses that explain divergences in viewpoints in a balanced way and point to the weight of scientific evidence on all sides of contentious issues. The HLPE reports provide recommendations that are considered as entry points for political negotiation around these issues. They are supposed to help moving from polemical disagreements towards agreement on disagreements, in order to further contribute to agreements to be designed and implemented.

Gathered during the 4th International Conference on Global Food Security[5] in December 2020, 900 scientists thus called for intensifying investments "in research to analyse transformation, its political economy and the power relationship that shape or prevent transformation, its patterns and consequences, and what makes it difficult". Those scientists urged to identify "obstacles and resistance to change, with a specific focus on conflicts of interests among different actors and contexts, the enforcement of rights, lock-ins, and path dependencies" (Caron et al., 2021, p. 2).

Reflections from this conference have been instrumental in discussions about the event "Bonding science and policy to accelerate food

---

[4] http://www.fao.org/cfs/cfs-hlpe/.

[5] http://www.globalfoodsecurityconference.com.

systems transformation" that was held on 4 February 2021 as a contribution to the 2021 United Nations Food Systems Summit. Hainzelin et al. (2021) concluded that understanding such obstacles, their consequences for transformation and their impact on management of shock responses, risk and uncertainty, including the reasons why scientific evidence is not being used, might facilitate the elaboration of a shared vision of desired changes and the formulation of explicit pathways to achieve them.

### *Two Avenues to Illustrate the Journey From Obstacles to Transformation*

Positions are increasingly polarized when it comes to food-related issues. Be they about technological advances, market mechanisms, environmental concerns, quality of food, consumption of animal source products, these issues are often not just about food, but rather about values and distinct perceptions of development models and expectations (Béné et al., 2019; Eakin et al., 2017; OECD, 2021). This polarization is amplified by exclusive rhetoric that often looks at disqualifying opposed views and their tenants, rather than constructively contributing to a common project. As stated by Caron (2020, p. 558), an irreducible dualism "has taken place between those who deny sustainability concerns and oppose any change and those who advocate for a revolution to prevent an announced collapse. All are convinced they defend the general interest. Strategies to delegitimize opponents of all sides, as well as doubt or certainty selling behaviours will not make transformation easy. All knowledge resources will be required as well as reshaping the role of technology to move beyond binary opposites between positivism versus reject stances".

The growing mediatization of such issues makes any attempt to reach agreement and embark on transformation through collective action even more difficult. We believe that science can play a role in contributing to dialogues through a mediation process that builds upon the characterization of opposed views. We are thus suggesting here two avenues to be explored based on the identification of obstacles.

### *Impacting at Scale Through "Cross Scales Contamination"*

While opposition is being instrumentalized between local and global processes, respectively considered as virtuous or devilish by some (Smith et al., 2016), a relevant and consistent articulation between locally driven and place-specific processes, national policies and international frameworks is required to put the great transformation in motion. Change cannot actually occur through the mere scaling up of local success stories; those stories are in most cases not reproducible because of the context specificity in which they take place and because of the efforts and investments that they rely on.

The pandemic may play a detonating role towards an important transformation, because of the economic crisis and trade disruption it may generate in the near future. Local processes and the search for both local and national sovereignty have undoubtedly gained traction (see, however, Hoddinott, Chapter 6 in this volume). Such a rediscovery relies on two different processes. On the one hand, it emerges as a consequence of the rejection by many civil society organizations of undesirable negative economic, social, environmental and political effects of globalization, corporate concentration and long-distance value chains, and on the sentiment of loss of social and political control they generate. On the other hand, it builds upon the assumption that the proximity dimension of local processes would be synonymous of sustainability,[6] because of the generation of decent employment and livelihoods, control of the quality of products through adoption of environmentally friendly practices, valorization of territorial assets, etc. There are four main reasons to celebrate such a rediscovery, although this requires a critical analysis. It first helps in addressing both market and state failures through the strengthening of the capacity to govern the commons (Ostrom, 1995). Secondly, such a capacity is pivotal for the design of collective projects that might be promoted through and contribute to the formulation of public policies. Thirdly, it is essential to design place-based specific solutions to wicked problems (see Losch and May, Chapter 10 in this volume). And finally, it values important conceptual and operational experiences that have been conducted during the last 30 years regarding territorial agri-food systems (Muchnik & de Sainte Marie, 2010).

While celebrating such a movement, omitting the importance of national and international contexts, processes and regulations would be a tremendous mistake. They are not only essential to influence, boost or hamper the potential of local initiatives, but also represent important levels to act. In a context of weakened multilateralism, their virtue

---

[6] https://tii.unido.org/news/strengthening-resilience-food-systems-role-short-food-supply-chains.

should not be forgotten, as they offer the space to shape public policies, norms and values, to organize stocks and exchanges and regulate prices, including for preventing food loss and waste, and to implement international co-operation, including the prevention of identity closure and for addressing global concerns such as pandemics or climate change.

Rather than opposing them, articulating local, national and international processes, arrangements and frameworks, and organising interdependencies and regulations among scales are thus key to promote the diversity and the coexistence of context-specific pathways while addressing the global challenges (Carlsson et al., 2017; Gasselin et al., 2021). In addition, as they depend on a paradigm shift and often meet resistance, such transformations cannot take place spontaneously and then require an incentivizing or arbitration framework that is needed at a supra-level. Such frameworks are pivotal to ensure consistency between local and global levels and the coexistence of differentiated pathways.

Whatever the level, transformation most often relies on an agreement to be crafted by the many stakeholders, often characterized by vested and sometimes divergent interests. As already mentioned, addressing disagreement and misunderstanding explicitly may help moving beyond disqualifying and non-constructive rhetoric (Meijer & Jong, 2020). It may help in elaborating a collective project for the future that reconnects production and consumption to address sustainability concerns. Yet, such pacts are not possible simultaneously at all scales since they require specific and favourable political configurations, arrangements and conditions to do so. As a consequence, as a political strategy towards transformation, we suggest identifying, designing and implementing such pacts where and when possible, be they locally, regionally, nationally or internationally. This might be the case of a city or metropolitan area, which often stand as very innovative and dynamic spaces, as illustrated by the Milan Urban Food Policy Pact.[7] This might be the case at the global level, as illustrated by the recent adoption by the UN Committee on World Food Security of Voluntary Guidelines on Food Security and Nutrition (CFS, 2021). This might be the case at the national level as well, through the adoption of food acts and policies, as, for example, the 2016 "Loi Garrot" enacted in France to reduce food losses and waste. These pacts can then serve as supports to "contaminate" other scales. A local experience can pave the

---

[7] https://www.milanurbanfoodpolicypact.org/.

road to the design of a public policy, while international guidelines can be mobilized by local stakeholders to advocate for a change or to co-design a collective project. "Think globally and act locally" has been instrumental in the 1990s to open an avenue for the acknowledgement of global environmental issues, but time has come to think and act locally and globally, one for each other and consistently, in order to generate the expected great transformation. Identifying and removing obstacles through knowledge production and specific mediation and foresight methods should be key.

### *Generating a New Action Regime to Ensure the Convergence Between the Production of Private and Public Goods*

The substantial increase in food production that was achieved during the twentieth century mainly relied on the delivery of private goods. Any technology that would contribute to increasing production would be economically profitable (Sebillote, 1996) and consequently generate wealth and act in return as a lever for positive social and economic transformation including in other sectors. Until the 1980s, employment was generally not an issue and the environment was not yet a central one. In sum, the production of private and public goods was somehow aligned and synergetic and pacts between producers and consumers were made easy in such a context.

This is no longer the case, because of hidden costs and externalities on the one hand (Hendriks et al., 2021), and unfair distribution of wealth and increasing inequity on the other hand. Being private activity domains in most countries, agriculture and the agri-food sector thus lie at the heart of very complex and increasing tensions as concerns are growing regarding the health, social and environmental costs and impacts of production. The macro-political and economic environment that was in place in the twentieth century is no longer adapted to stimulate the production of public goods by agriculture, and a structural transformation is thus required to address sustainability concerns. This is necessary to facilitate a realignment between the production of private and public goods, at all levels including the global (climate change, biodiversity, global health, etc.).

The need for a new political framing to acknowledge and reward the production of public goods invites trade regulation to be reshaped accordingly. As we were able to take into consideration zoo- and phytosanitary concerns to organize or prohibit exchanges in the twentieth century, e.g. the importance of foot and mouth disease for livestock exchanges, we should be able to do so to address sustainability challenges in the twenty-first century. This means regulations and standards, including patterns for framing economic competition and redirecting taxes and subsidies to promote sustainable practices and prevent non-sustainable ones (HLPE, 2017b). This should also consider reviewing intellectual property rights and adapting innovation processes to reconcile the stimulation of innovation and sustainability criteria (UNCTAD, 2017).

## Conclusion

We have shown that food systems have been resilient in the past. They have not completely eliminated food security problems in some regions or for some vulnerable groups, but they have proven able to adapt and transform themselves to address the unprecedented demographic pressure without major disruption of their capacity to provide food at the global level. The targets set in the 1974 food security definition have been met while the population has grown from 3 to 7 billion people on the planet.

However, the twenty-first-century challenges, as set in the 2030 Agenda for sustainable development, question such a capacity to adapt to current and future shocks, as they call for huge paradigm shifts and obstacles to be removed. Most of the current obstacles to address future challenges are intrinsically linked with and generated by what made the twentieth-century series of transformations a success. This results in an increasing polarization of debates and the need to better identify and make obstacles and disagreements explicit in order to move beyond them and address sustainability concerns. This might be considered as a necessary step to design scale-specific and global agreements that are required, including the structural transformation to align the delivery of private and public goods and the cross-scale contamination process to ensure consistency among scales and sectors.

To conclude, one could then question if the past resilience would not be an obstacle to resilience in the future, because of path dependency and asymmetry of powers that have been generated through recent transformation. As a consequence, the interaction between food systems and resilience might be considered in two ways: the need for resilience to contribute to food systems transformation and resilience as an emerging property of past food system transformation to enable addressing further challenges and undertaking transformation in the future.

## REFERENCES

Afshin, A., Sur, P. J., Fay, K. A., Cornaby, L., Ferrara, G., Salama, J. S., ... & Murray, C. J. L. (2019). Health effects of dietary risks in 195 countries, 1990–2017: A systematic analysis for the Global Burden of Disease Study 2017. *The Lancet, 393* (10184), 1958–1972. https://doi.org/10.1016/s0140-6736(19)30041-8

Allen, T., Heinrigs, P., & Heo, I. (2018). *Agriculture, food and jobs in West Africa* (Vol. 14). OECD Publishing.

Anseeuw, W., Hertzog-Adamczewski, A., Jamin, J. -Y., & Farolfi, S. (2019). Large-scale land and water acquisitions: What implications for food security? In S. Dury, P. Bendjebbar, E. Hainzelin, T. Giordano & N. Bricas (Eds.), *Food systems at risk. New trends and challenges* (pp. 67–69). CIRAD, European Commission, FAO. https://doi.org/10.19182/agritrop/00095

Baldos, U. L. C., & Hertel, T. W. (2016). Debunking the 'new normal': Why world food prices are expected to resume their long run downward trend. *Global Food Security, 8,* 27–38. https://doi.org/10.1016/j.gfs.2016.03.002

Béné, C., Bakker, D., Chavarro, M. J., Even, B., Melo, J., & Sonneveld, A. (2021). Global assessment of the impacts of COVID-19 on food security. *Global Food Security, 31,* 100575. https://doi.org/10.1016/j.gfs.2021.100575

Béné, C., Fanzo, J., Haddad, L., Hawkes, C., Caron, P., Vermeulen, S., Herrero, M., Oosterveer, P. (2020a). Five priorities to operationalize the EAT–Lancet Commission report. *Nature Food, 1*(8), 457–459. https://doi.org/10.1038/s43016-020-0136-4

Béné, C., Fanzo, J., Prager, S. D., Achicanoy, H. A., Mapes, B. R., Alvarez Toro, P., & Bonilla Cedrez, C. (2020b). Global drivers of food system (un)sustainability: A multi-country correlation analysis. [Research Support, Non-U.S. Gov't]. *PLoS One, 15*(4), e0231071. https://doi.org/10.1371/journal.pone.0231071

Béné, C., Oosterveer, P., Lamotte, L., Brouwer, I. D., de Haan, S., Prager, S. D., Talsma, E. F., & Khoury, C. K. (2019). When food systems meet sustainability—Current narratives and implications for actions. *World Development, 113,* 116–130. https://doi.org/10.1016/j.worlddev.2018.08.011

Brundtland, G. H. (1987). *Our common future: The World Commission on environment and development*. United Nations. Retrieved December 21, 2021, from http://www.un-documents.net/wced-ocf.htm

Burchi, F., & De Muro, P. (2016). From food availability to nutritional capabilities: Advancing food security analysis. *Food Policy, 60*, 10–19. https://doi.org/10.1016/j.foodpol.2015.03.008

Carlsson, L., Callaghan, E., Morley, A., & Broman, G. (2017). Food system sustainability across scales: A proposed local-to-global approach to community planning and assessment. *Sustainability, 9*(6). https://doi.org/10.3390/su9061061

Caron, P. (2020). From crisis to utopia: Crafting new public-private articulation at territorial level to design sustainable food systems. *Agriculture and Human Values, 37*, 557–558. https://doi.org/10.1007/s10460-020-10065-1

Caron, P., Reig, E., Roep, D., Hediger, W., Cotty, T. L., Barthelemy, D., Hadynska, A., Hadynski, J., Oostindie, H., & Sabourin, E. (2008). Multifunctionality: Refocusing a spreading, loose and fashionable concept for looking at sustainability? *International Journal of Agricultural Resources, Governance and Ecology, 7*(4/5), 301. https://doi.org/10.1504/ijarge.2008.020078

Caron, P., Ferrero y de Loma-Osorio, G., Nabarro, D., Hainzelin, E., Guillou, M., Andersen, I., ... & Verburg, G. (2018). Food systems for sustainable development: Proposals for a profound four-part transformation. *Agronomy for Sustainable Development, 38*(4), 41. https://doi.org/10.1007/s13593-018-0519-1

Caron, P., van Ittersum, M., Avermaete, T., Brunori, G., Fanzo, J., Giller, K., Hainzelin, E., Ingram, J., Korsten, L., Martin-Prével, Y., Osiru, M., Palm, C., Rivera Ferre, M., Rufino, M., Schneider, S., Thomas, A., & Walker, D. (2021). Statement based on the 4th international conference on global food security—December 2020: Challenges for a disruptive research Agenda. *Global Food Security, 30*, 100554. https://doi.org/10.1016/j.gfs.2021.100554

Cheyns, E., Daviron, B., Djama, M., Fouilleux, E., & Guéneau, S. (2017). The standardization of sustainable development through the insertion of agricultural global value chains into international markets. In R. A. L. D. Biénabe Estelle (Ed.), *Sustainable development and tropical agri-chains* (pp. 283–303). Springer. https://doi.org/10.1007/978-94-024-1016-7_23

Coleman-Jensen, A., Rabbitt, M., Gregory, C., & Singh, A. (2020). *Household Food Security in the United States in 2019* (Vol. ERR-275): USDA Economic Research Service. Retrieved December 23, 2021, from https://www.ers.usda.gov/webdocs/publications/99282/err-275.pdf?v=785

CFS, Committee on World Food Security. (2021). Voluntary guidelines on food systems and nutrition. Retrieved December 23, 2021, from http://www.fao.org/fileadmin/templates/cfs/Docs2021/Documents/CFS_VGs_Food_Systems_and_Nutrition_Strategy_EN.pdf

Demenois, J., Chaboud, G., & Blanfort, V. (2019). Food systems emission and climate change consequences. In S. Dury, P. Bendjebbar, E. Hainzelin, T.

Giordano & N. Bricas (Eds.), *Food systems at risk. New trends and challenges* (pp. 35–37). Montpellier, France: CIRAD, European Commission, FAO. https://doi.org/10.19182/agritrop/00084

Dury, S., Bendjebbar, P., Hainzelin, E., Giordano, T., & Bricas, N. (Eds.). (2019). Food systems at risk. New trends and challenges. Cirad, UE, FAO.

Eakin, H., Connors, J. P., Wharton, C., Bertmann, F., Xiong, A., & Stoltzfus, J. (2017). Identifying attributes of food system sustainability: Emerging themes and consensus. *Agriculture and Human Values, 34*(3), 757–773.

European Food Banks Federation. (2020). Annual report 2019. Retrieved from https://www.eurofoodbank.org/annual-reports/

FAO. (2016). State of the World's Forests. *Forests and agriculture: Land use challenges and opportunities* (p. 125). Food and Agriculture Organization.

FAO. (2018). The state of agricultural commodity markets. *Agricultural trade, climate change and food security* (p. 92). Food and Agriculture Organization.

FAO. (2019). The state of world's biodiversity for food and agriculture. In J. Bélanger & D. Pilling (Eds.), *FAO commission on genetic resources for food and agriculture assessments* (p. 572). Food and Agriculture Organization.

FAO, IFAD, UNICEF, WFP, & WHO. (2020). *The state of food security and nutrition in the world 2020. Transforming food systems for affordable healthy diets*.

Feintrenie, L., Betbeder, J., Piketty, M.-G., & Gazull, L. (2019). Deforestation for food production. In S. Dury, P. Bendjebbar, E. Hainzelin, T. Giordano & N. Bricas (Eds.), *Food systems at risk. New trends and challenges* (pp. 43–46). FAO, CIRAD & European Commission.

Gasselin, P., Lardon, S., Cerdan, C., Loudiyi, S., & Sautier, D. (2021). *Coexistence et confrontation des modèles agricoles et alimentaires : un nouveau paradigme du développement territorial ?* Éditions Quae.

Giordano, T., Losch, B., Sourisseau, J.-M., & Girard, P. (2019). Risks of mass unemployment and worsening of working conditions. In S. Dury, P. Bendjebbar, E. Hainzelin, T. Giordano & N. Bricas (Eds.), *Food systems at risk. New trends and challenges* (pp. 75–77). CIRAD, European Commission, FAO. https://doi.org/10.19182/agritrop/00097

The Global Network Against Food Crisis. (2021). Global report on food crises: Joint analysis for better decisions, September 2021 Update: EU, FAO, WFP.

Global Panel. (2016). The cost of malnutrition. Why policy action is urgent. *London (UK): Global panel on agriculture and food Systems for nutrition*. Technical Brief N°3. Retrieved December 23, 2021, from https://glopan.org/sites/default/files/pictures/CostOfMalnutrition.pdf

Haddad, L., Hawkes, C., Waage, J., Webb, P., Godfray, C., & Toulmin, C. (2016). *Food systems and diets: Facing the challenges of the 21st century*. Global Panel on Agriculture and Food Systems for Nutrition. Retrieved December 23, 2021, from https://openaccess.city.ac.uk/id/eprint/19323/

Hainzelin, E. (2019). Risks of irreversible biodiversity loss. In S. Dury, P. Bendjebbar, E. Hainzelin, T. Giordano, & N. Bricas (Eds.), *Food systems at risk. New trends and challenges* (pp. 59–62). CIRAD, European Commission, FAO.

Hainzelin, E., Caron, P., Place, F., Alpha, A., Dury, S., Echeverria, R., & Harding, A. (2021). How could science-policy interfaces boost food system transformation? *Food System Summit Brief* (pp. 16). IFPRI. https://doi.org/10.48565/scfss2021-4y32

Hawkes, C. (2002). Marketing activities of global soft drink and fast food companies in emerging markets: A review. In World Health Organization (Ed.), *Globalization, diets and noncommunicable diseases* (pp. 98–185): WHO. Retrieved December 23, 2021, from https://apps.who.int/iris/bitstream/handle/10665/42609/?sequence=1

Hendriks, S. L., de Groot Ruiz, A., Herrero Acosta, M., Baumers, H., Galgani, P., Mason-D'Croz, D., Godde, C., Waha, K., Kanidou, D., von Braun, J., Benitez, M., Blanke, J., Caron, P., Fanzo, J., Greb, F., Haddad, L., Herforth, A., Jordaan, D., Masters, W., Sadoff, C., Soussana, J.-F., ... & Watkins, M. (2021). The true cost and true price of food. *A paper from the Scientific Group of the UN Food Systems Summit 2021*, 42. Retrieved January 4, 2022, from https://sc-fss2021.org/wp-content/uploads/2021/06/UNFSS_true_cost_of_food.pdf

HLPE. (2013). Investing in smallholder agriculture for food security Vol. 6. A report by the High Level Panel of Experts on Food Security and Nutrition of the CFS-Committee on World Food Security. (pp. 112). Retrieved from https://www.fao.org/3/i2953e/i2953e.pdf

HLPE. (2017a). *Nutrition and Food Systems Vol. 12. A report by the high level panel of experts on food and nutrition of the CFS-Committee on World Food Security* (pp. 151). Retrieved from http://www.fao.org/3/ai7846e.pdf

HLPE. (2017b). *2nd Note on critical and emerging issues for food security and nutrition* (p. 23). Retrieved from https://www.fao.org/fileadmin/user_upload/hlpe/hlpe_documents/Critical-Emerging-Issues-2016/HLPE_Note-to-CFS_Critical-and-Emerging-Issues-2nd-Edition__27-April-2017b_.pdf

HLPE. (2020). *Food Security and nutrition building a global narrative towards 2030. Vol. 15. High Level Panel of Experts on Food and Nutrition of the CFS-Committee on World Food Security* (pp. 112). Retrieved from https://www.fao.org/3/ca9731en/ca9731en.pdf

IAASTD. (2009). Agriculture at a crossroads: Global report. In B. MacIntyre, H. Herren, J. Wakhungu, & R. Watson (Eds.), *International assessment of agricultural knowledge, science and technology for development* (p. 606). Retrieved from http://apps.unep.org/publications/pmtdocuments/Agriculture_at_a_Crossroads_Global_Report.pdf

IHEST. (2015). *Au coeur des controverses. Des sciences à l'action*. ACTES SUD/IHEST.
IPCC. (2018). Global Warming of 1.5°C. An IPCC Special Report on the impacts of global warming of 1.5°C above pre-industrial levels and related global greenhouse gas emission pathways, in the context of strengthening the global response to the threat of climate change, sustainable development, and efforts to eradicate poverty. In V. Masson-Delmotte, P. Zhai, H.-O. Pörtner, D. Roberts, J. Skea, P.R. Shukla, A. Pirani, W. Moufouma-Okia, C. Péan, R. Pidcock, S. Connors, J.B.R. Matthews, Y. Chen, X. Zhou, M.I. Gomis, E. Lonnoy, T. Maycock, M. Tignor, and T. Waterfield (Eds.) (p. 630). Retrieved from https://www.ipcc.ch/site/assets/uploads/sites/2/2019/06/SR15_Full_Report_High_Res.pdf
IPES-Food & ETC Group. (2021). *A long food movement: Transforming food systems by 2045* (p. 176). Retrieved from http://www.ipes-food.org/pages/LongFoodMovement
Klein Goldewijk, K., Beusen, A., Doelman, J., & Stehfest, E. (2017). Anthropogenic land use estimates for the Holocene—HYDE 3.2. *Earth System Science Data, 9*(2), 927–953. https://doi.org/10.5194/essd-9-927-2017
Krausmann, F., & Langthaler, E. (2019). Food regimes and their trade links: A socio-ecological perspective. *Ecological Economics, 160*, 87–95. https://doi.org/10.1016/j.ecolecon.2019.02.011
Latour, B. (1987). *Science in action: How to follow scientists and engineers through society*. Harvard University Press.
Le Moigne, J. (1990). *La modélisation des systèmes complexes*. Dunod.
Lemieux, C. (2007). À quoi sert l'analyse des controverses? *Mil neuf cent. Revue d'histoire intellectuelle, 1*(25), 191–212.
Louafi, S. (2021, April 25). Les scientifiques alertent mais les politiques ne font rien… Est-ce vraiment si simple. *The Conversation*. Retrieved January, 2022, from https://theconversation.com/les-scientifiques-alertent-mais-les-politiques-ne-font-rien-est-ce-vraiment-si-simple-158205
McMichael, P. (2009). A food regime genealogy. *The Journal of Peasant Studies, 36*(1), 139–169. https://doi.org/10.1080/03066150902820354
Mathis, C, F. (2021). *La civilisation du charbon. En Angleterre, du règne de Victoria à la Seconde Guerre mondiale*. Vendémiaire.
Meadows, D., Meadows, D., Randers, J., & Behrens, W. (1972). *The limits to growth*. Universe Books.
Meijer, A., & De Jong, J. (2020). Managing value conflicts in public innovation: Ostrich, chameleon, and dolphin strategies. *International Journal of Public Administration, 43*(11), 977–988.
Morin, E. (1986). *La Connaissance de la connaissance (t. 3), La Méthode*. Seuil. Nouvelle édition, coll. « Points » 19.

Muchnik, J., & De Sainte Marie, C. (2010). *Le temps des Syal. Techniques, vivres et territoires*. UPdate Sciences & Technologies.

OECD. (2021). *Making better policies for food systems*. OECD Publishing. https://doi.org/10.1787/ddfba4de-en.

Ostrom, E. (1995). *Governing the commons. The evolution of institutions for collective action*. Cambridge University Press.

Paillard, S., Treyer, S., & Dorin, B. (2014). *Agrimonde–scenarios and challenges for feeding the world in 2050*: Springer Science & Business Media.

Popkin, B. M. (2017). Relationship between shifts in food system dynamics and acceleration of the global nutrition transition. *Nutrition Reviews, 75*(2), 73–82.

Porkka, M., Kummu, M., Siebert, S., & Varis, O. (2013). From food insufficiency towards trade dependency: A historical analysis of global food availability. *PLoS ONE, 8*(12), e82714.

Polanyi, K. (1944). *The great transformation: The political and economic origins of our time*. Beacon Press.

Roser, M., & Ritchie, H. (2021). "Food Prices". Published online at OurWorldInData.org. Retrieved from: https://ourworldindata.org/food-prices

Schmidt-Traub, G., Obersteiner, M., & Mosnier, A. (2019). *Fix the broken food system in three steps*. Nature Publishing Group.

Sebillote, M. (1996). L'agriculture face à l'évolution de la société. In M. Sebillote (Ed.), *Les mondes de l'agriculture. Une recherche pour demain* (pp. 43–67). Éditions Quæ, « Sciences en questions ».

Smith, K., Lawrence, G., MacMahon, A., Muller, J., & Brady, M. (2016). The resilience of long and short food chains: A case study of flooding in Queensland, Australia. *Agriculture and Human Values, 33*(1), 45–60. https://doi.org/10.1007/s10460-015-9603-1

Soullier, G., Moustier, P., & Lançon, F. (2019). Risks of smallholder exclusion from upgrading food chains. In S. Dury, P. Bendjebbar, E. Hainzelin, T. Giordano, & N. Bricas (Eds.), *Food systems at risk. New trends and challenges* (pp. 79–82). FAO, CIRAD & European Commission.

Steffen, W., Richardson, K., Rockström, J., Cornell, S. E., Fetzer, I., Bennett, E. M., Biggs, R., Carpenter, S. R., de Vries, W., De Wit, C. A., Folke, C., Gerten, D., Heinke, J., Mace, G. M., Persson, L. M., Ramanathan, V, Reyers, B., & Sörlin, S. (2015). Planetary boundaries: Guiding human development on a changing planet. *Science, 347*(6223). https://doi.org/10.1126/science.1259855

Stern, N. (2007). *The economics of climate change: The stern review*. Cambridge University Press. https://doi.org/10.1017/CBO9780511817434

UNCTAD. (2017). New innovation approaches to support the implementation of the sustainable development goals. United Nations Conference on Trade and Development. Retrieved from https://unctad.org/system/files/official-document/dtlstict2017d4_en.pdf

United Nations. (2019). GSDR, global sustainable development report 2019: The future is now—Science for Achieving Sustainable Development. Retrieved from https://sustainabledevelopment.un.org/gsdr2019

Van der Sluijs, J. P., Amaral-Rogers, V., Belzunces, L. P., Van Lexmond, M. B., Bonmatin, J.-M., Chagnon, M., Downs, C. A., Furlan, L., Gibbons, D. W., & Giorio, C. (2015). Conclusions of the Worldwide Integrated Assessment on the risks of neonicotinoids and fipronil to biodiversity and ecosystem functioning. *Environmental Science and Pollution Research, 22*(1), 148–154.

Vermeulen, S. J., Campbell, B. M., & Ingram, J. S. (2012). Climate change and food systems. *Annual Review of Environment and Resources, 37*, 195–222.

Wagner, D. L. (2020). Insect declines in the Anthropocene. *Annual Review of Entomology, 65*, 457–480.

Xu, X., Sharma, P., Shu, S., Lin, T.-S., Ciais, P., Tubiello, F. N., Smith, P., Campbell, N., & Jain, A. K. (2021). Global greenhouse gas emissions from animal-based foods are twice those of plant-based foods. *Nature Food, 2*(9), 724–732. https://doi.org/10.1038/s43016-021-00358-x

Webb, P., Benton, T. G., Beddington, J., Flynn, D., Kelly, N. M., & Thomas, S. M. (2020). The urgency of food system transformation is now irrefutable. *Nature Food, 1*(10), 584–585. https://doi.org/10.1038/s43016-020-00161-0a

Willett, W., Rockström, J., Loken, B., Springmann, M., Lang, T., Vermeulen, S., ... & Murray, C. J. L. (2019). Food in the Anthropocene: The EAT–Lancet Commission on healthy diets from sustainable food systems. *The Lancet*. https://doi.org/10.1016/s0140-6736(18)31788-4

**Open Access** This chapter is licensed under the terms of the Creative Commons Attribution 4.0 International License (http://creativecommons.org/licenses/by/4.0/), which permits use, sharing, adaptation, distribution and reproduction in any medium or format, as long as you give appropriate credit to the original author(s) and the source, provide a link to the Creative Commons license and indicate if changes were made.

The images or other third party material in this chapter are included in the chapter's Creative Commons license, unless indicated otherwise in a credit line to the material. If material is not included in the chapter's Creative Commons license and your intended use is not permitted by statutory regulation or exceeds the permitted use, you will need to obtain permission directly from the copyright holder.

CHAPTER 4

# Food Security and the Fractured Consensus on Food Resilience: An Analysis of Development Agency Narratives

*Karl-Axel Lindgren and Tim Lang*

## INTRODUCTION: CLARIFYING THE POLICY LANGUAGE

The concept of food security is decades-old, its initial definition established in the 1974 FAO World Food Conference and further refined in the 1996 World Food Summit (Shaw, 2007). Its plasticity has been much remarked (see chapters "Food security from a food system perspective" by Jess Fanzo, and "Towards an integrated model of food security and resilience" by Mark Constas), with questions about whether a term which means all things to all people can retain policy and research value (Candel et al., 2013; Carolan, 2013). Yet the term has not disappeared, and instead, Maye and Kirwan argue, a 'fractured consensus' has arisen

K.-A. Lindgren (✉) · T. Lang
Centre for Food Policy, City, University of London, London, UK
e-mail: karlaxel.lindgren@gmail.com

T. Lang
e-mail: T.Lang@city.ac.uk

around food security—highlighting that, while each interested party may agree that the outcome of food security, the absence of hunger, is a goal worth pursuing, how to actually achieve that state is far more contentious (Maye & Kirwan, 2013). There may be consensus on the broad objective of achieving food security without agreement on how to do so.

The notion of food resilience is a more recent arrival into this policy milieu (Tendall et al., 2015) and is explored in more depth in Mark Constas' chapter in this volume. In the present chapter, food resilience is utilised as a catch-all term to capture the broad range of definitions and narratives exhibited by development agencies (see also the De Pinto, Islam and Katic chapter). Some adopt 'climate change resilience' as part of the wider food security narrative; others see 'resilience' as implicit within a food systems approach; and others are interested in 'agricultural resilience' as part of the climate change discourse.

The notion of resilience derives from material science—before it was applied to natural ecology—to convey the capacity of materials to be elastic and to return to form after external pressure. The notion was borrowed and incorporated into modern (post-mid-twentieth century) environmental sciences (Béné et al., 2018; Moser et al., 2010). Its usefulness for environmental analysis was its intellectual flexibility. It could capture processes as varied as slow ecosystems adaptation or recovery from sudden shocks and external stress (Bahadur et al., 2010). While its initial appearance was limited to natural sciences, its use spread to the social sciences (Berkes & Folke, 1998) and today has become common across disciplines and policy. In this spread, resilience is perhaps following the path taken by ecosystems and sustainability, being picked up by business analysts and economists who now speak of resilience in financialised terms completely divorced from the complexity and nuances of its original material or even ecological meaning. In this sense, the tight approach of Tendall and colleagues (2015) is clear; they see food resilience as a two-dimensional concept: the capacity to bounce back from shock, plus durability over time. While those two core features remain central to our present conceptualisation, it has been argued elsewhere that a social dimension is also needed for the term to be of comprehensive value to policy-makers (Lang, 2020). Unless people, habits and cultures are included, the term has a hole at its heart. How can a food system be resilient unless it is also part of how people live and eat?

The risk of an unreflective use of 'resilience' is real; meaning and value lessen if used loosely by institutions. Already academics fear its potential is being dissipated (Bahadur et al., 2010; Béné et al., 2016; Leach, 2008). Optimistically, however, the research reported here still sees its value as a 'code' for key challenges widely agreed to be facing humanity, backed by hard data, such as climate change, biodiversity loss, natural disasters and human conflict. Furthermore, the notion of resilience has a critical edge in discussions that encompass access to justice, women's empowerment, economic issues such as livelihoods, as well as to political processes and governance. The concept of resilience has value for development policy if institutions and agencies, such as those appraised here, retain this 'edge' (Béné et al., 2018; Chelleri, 2012; Twigg, 2007). The term's applicability across diverse discourses means we need to ask if its utility is being blunted as it becomes an interdisciplinary 'pluralistic' term with many definitions and meanings (Pearson, 2013; Béné et al., 2018). Its core meaning, the capacity to bounce back after shock, remains central to the narratives of agencies despite, as Béné and colleagues have said, the lack of a rigorous framework that marries the theoretical and the practical (Béné et al., 2014, 2018).

In this chapter, we report on our exploration of how food security and food resilience, these two rich yet stretched concepts, are being used by international development agencies. The motivation for the study was simple and pragmatic. How, we ask, are both terms presented—not in speeches by political leaders or in protestations of civil society or in erudite academic studies, but by a sample of key policy actors in the relevant policy zone? How, in other words, do development agencies, who have considerable financial reach and influence, utilise or present these two policy concepts? Do they mean the same thing? What are their variations, if any, in the policy narrative to which the agencies contribute?

These are deceptively simple questions. In practice, the meaning of policy terms can and do change over time, and resilience is no exception (Béné et al., 2018). While we initially decided to focus on a narrow slice of time, setting the cut-off at the emergence of resilience in the 2017 State of Food Security report titled "Building Resilience for Peace and Food Security" (FAO, IFAD, UNICEF, WFP and WHO, 2017), many of the formative reports that still influence the various agencies predated this cut-off, so we instead focused on identifying the significant reports of each agency around the keywords of food security and resilience. The goal was not to collate what we think they mean, but what the agencies themselves say they mean. The aim of this study was not to provide a historical perspective, such as the one provided by Mark Constas in this book, nor to track the evolution of the usage of 'resilience' within food security, such as done by de Pinto, Islam and Katic, also in this book, but instead to provide a snapshot view of a wide range of international development agencies' own interpretations of, and narratives on, resilience. The appraisal was concluded before the UN Food Systems Summit, the COP15 conference on biodiversity and COP26 on climate change events held in late 2021, all of which add to international debate on both food security and food resilience.

## Aims, Methods and Approach

The study reported here is an attempt to understand the meaning of two key concepts situated within particular narratives and a contribution to how policy language can be used to justify or legitimise particular decisions or policy orientations (Béné et al., 2016, 2018; Goldstein et al., 2012). Our approach drew upon experience of earlier work by one of us (Tim Lang) for the World Health Organisation, which reviewed major food companies for how seriously they took diet-related health matters and whether they followed the WHO's agreed approach to diet and health (Lang et al., 2006; Nestle, 2006). This, in turn, drew on established methods by bodies such as the US Government Accountability Office and the UK National Audit Office that specifically ask agency programmes what their goals and processes are in relation to outcomes.

In the present study, we report only the first step. We explore what development agencies say they mean when referring to food security and resilience, particularly in relation to economic, environmental, societal, political and public health issues. We then contrast three types of international development agency—bilateral, regional and multilateral—to derive trends, commonalities and differences from their narratives. Our study thus only considers agencies' meanings and intentions, not their delivery.

16 international development agencies and regional banks were chosen to represent three different types of agencies: (i) bilateral development agencies; (ii) multilateral or regional development banks; and (iii) UN-based international development agencies. Of our sample, six national donor agencies were selected as illustrations of bilateral development agencies. Of these six, it should be noted that five are from countries whose main policy and operational language is English (Australia, Canada,

Ireland, UK, US), and one whose English-language website was their secondary one (Sweden). A further five regional agencies were selected to highlight the narratives of regional, multilateral development banks: Asian Development Bank (ADB), African Development Bank (AfDB), European Bank for Reconstruction and Development (EBRD), the Inter-American Development Bank (IDB) and the Islamic Development Bank (IsDB), to cover five regions of the world engaged in the development community. A further three UN agencies (Food and Agricultural Organisation [FAO], United Nations Children's Fund [UNICEF] and the World Food Programme [WFP]) and two Bretton-Woods finance institutions—the World Bank Group and the International Monetary Fund (IMF)—were also appraised.

It should be noted that the UK's DfID—long criticised by elements of UK Conservatism for being too liberal, globalist and independent, and insufficiently linking aid to perceived national economic self-interest (Mitchell, 2020)—was dissolved in September 2020 and its responsibilities were merged with the Foreign Office, creating the Foreign, Commonwealth and Development Office (FCDO). The present study does not report on this new hybrid body. This restructuring of the UK development agency echoes changes made elsewhere. The Canadian International Development Agency (CIDA) had been an independent-standing government body until 2013, when it was merged with the Department of Foreign Affairs, rebranded in 2015 to Global Affairs Canada (GAC). AusAID was similarly an autonomous government-funded development agency until 2014, when merged with the Department of Foreign Affairs and Trade (DFAT). The FCDO, GAC and DFAT thus share similarities, not least the reframing of humanitarian development support under national trade policy. These agencies are thus different to Irish Aid, SIDA and USAID, for example, who retain singular mandates on development work, rather than a merging of development with trade.

The websites of the 16 agencies, their yearly reports, policy documents, white papers, and prominent research programmes were searched for information on resilience, food security, and food security and resilience (as a paired concept). With large organisations, one can often find deep and thoughtful documents which go against the main grain of the organisation, so by searching within the agencies' main documents and their formal self-presentations, we tried to capture where their focus lies. We searched for relevant keywords, including 'food security', 'resilience', 'food resilience', 'food security resilience', 'food systems resilience', 'agriculture resilience' and 'resilient'. This form of narrative analysis assumes that language doesn't mirror the world, but actively reflects and shapes it (Fischer & Forester, 1993; Roe, 1994). By establishing a scoping template set in a broad framework, each agency was put through the same 'snapshot' case study method. This process could be endless, as extensive and deep as each organisation's administrative structure and connections; we therefore limited the task by focusing on how 'accessible' and 'prominent' the concept appears in the agencies' own literature and on their website, being able to be accessed by searching on their websites or through search engines. This qualitative process allows for a relatively rapid appraisal of how agencies present their narratives, even if it may well be that other more nuanced or divergent internal positions also exist within those agencies. What we report here is a mirror of what they say they do at a certain point in time. We concur with Béné et al. (2018) that a 'snapshot' analysis has limitations due to the malleability and evolution of narratives but, given the policy influence wielded by the agencies we report on here, there is value in capturing, comparing and contrasting what they say they do and want to do.

## FINDINGS: 'FRACTURED CONSENSUS' IN WHAT THE AGENCIES SAY THEY MEAN

Our findings reinforce the thinking that development agencies frequently invoke resilience as an umbrella term when addressing many different sectors and ways of thinking, using different frameworks of development (Bahadur et al., 2010; Béné et al., 2012, 2018; Chelleri, 2012; Leach, 2008; Twigg, 2007). Our study suggests that there is a 'fractured consensus' not only within food security, as noted earlier, but also within the concept of resilience. We detect a similar disagreement around food security and resilience as a paired concept. A fractured consensus arises when there is the agreement that food security, resilience or food resilience might be important objectives, but disagreement on how these objectives should be achieved, or even what those objectives entail. The narrative-creation and framing around these concepts thus produce differentiated interpretations and meanings to related issues, with previous underlying ideological narratives and policy commitments by agencies, governments and expert bodies leading to specific 'policy solutions' that may well exacerbate the issues identified through data and contribute to the obfuscation of the underlying issues (Lang & Barling, 2012).

Several agencies, particularly the multilateral agencies such as World Bank and UN agencies, have placed 'resilience' as a central direction for many development sectors. 'Resilience' is a scaffolding in their frameworks (Béné et al., 2016), becoming 'Resilience Development' (FAO, 2019a; World Bank, 2020). Resilience has thus become a core concept that permeates every other aspect of their agendas, utilised in many different contexts and with a broad range of implications. Table 4.1 gives examples. Although far from exhaustive, the list illustrates broad usage.

Notions of food resilience vary greatly: from in-depth, complex frameworks, such as FAO's landmark document *Food and Nutrition Security Resilience Framework* (FAO, 2015), as well as related institutions, such as in USAID's rebranding in 2019 of their core food security institution to the *Bureau for Resilience and Food Security* (USAID, 2020), to the bare minimum of linking agriculture with climate resilience, such as IDB's Climate Resilience Sustainability Report (IDB, 2020a) or even complete omission of the term, such as DFAT's lack of discourse around food resilience, focusing instead on 'food and trade' as an interlinked core concept (DFAT, 2020). The scaffolding approach many agencies take to resilience means that even when it is not explicitly mentioned, the intent

Table 4.1 Types of resilience and examples of agencies who address it

| Type of resilience | Agencies using it in their approach to development |
| --- | --- |
| Climate change resilience | ADB, AfDB, DFAT, DfID, GAC, IDB, World Bank |
| Resilience in disaster management | ADB, DfID, USAID |
| Livelihood resilience | AfDB, FAO, Irish Aid, USAID |
| Infrastructure resilience | ADB, IDB, World Bank |
| Nation-state fragility | IsDB, World Bank |
| Sustainable agriculture resilience | ADB, AfDB, SIDA, USAID |
| Food systems resilience | World Bank |
| Resilience of children | UNICEF |

*Source* Authors ADB (2016), DFAT (2021b), DfID (2016), FAO (2019a, 2019b) GAC (2018), IDB (2019), IsDB (2019b), UNICEF (2020a), and World Bank (2016)

implies that the concept is still there in some fashion. In regional development bank reports, for example, resilience in relation to food security was mainly used in terms of a broader framework (ADB, 2015; AfDB, 2016; IDB, 2020a; IsDB, 2020b). Previous analysts have highlighted how this systematic discourse on resilience, in lieu of a rigorous, unified framework, may be misleading (Béné et al., 2016; Leach, 2008). A gradient of use ranges from broad application of the concepts ('Resilience as a Central Paradigm'; Béné et al., 2016) to specific focus, such as institutions that exist to solely address food security and resilience. This variability can be shown in, for example, how resilience is used in relation to climate change, where resilience is attempted to be quantified and addressed, to the term being thrown in to a longer text on food systems, where there are no quantifiable or concrete goals in relation to resilience specifically. In this respect, resilience seems to be exhibiting some of the plasticity used for 'sustainability' in relation to agriculture (ADB, 2009; DfID, 2016; IDB, 2019, 2021; SIDA, 2017; World Bank, 2020). A more detailed exploration of the utilisation of resilience within a climate change context can be found in de Pinto, Islam and Katic's chapter in this book.

Our initial analysis of the narratives around food security and resilience suggested some themes. These were grouped into five: Environment, Economy, Society, Politics and Public Health (see Table 4.2). In each of them, there was substantial complexity. Under Environment, for example, concerns such as agriculture, climate change, and water are frequently invoked, while others such as ecosystems, soil or biodiversity were less frequently cited. Some agencies such as the FAO attempt to capture as many themes as possible, while others focus narrowly on specific aspects. The EBRD focused almost exclusively on economic aspects of resilience in food security (EBRD, 2018). In complex terrain such as this, lending itself to interdisciplinary analysis, it is not surprising that the food resilience discourse should cross-reference to matters such as nutrition security, water security and so on. In this sense, resilience is a new lens to address old issues.

The following sections take a closer look at the three categories of development agencies appraised and compare them, going from bilateral development agencies, regional development banks, to the multilateral development agencies.

Table 4.2 Emerging themes and sub-themes around food resilience

| Environment | Economy | Society | Politics | Public Health |
|---|---|---|---|---|
| Agriculture | Livelihoods | Gender | Access to services | Nutrition |
| Climate change | Corporations | Age | Access to infrastructure | Diet |
| Biodiversity | Cost of food | Disability | Institutions | Infectious disease |
| Ecosystems | Scientific research | Vulnerable communities | Social safety nets | Non-communicable Disease |
| Soil | Technological innovation | | Conflict | Chronic illness |
| Water | Trade | | Governance | |

*Source* Authors

### Bilateral Development Agencies

Six bilateral development agencies were selected for appraisal. Table 4.3 lays out the broad narratives on food security and resilience, either as specific definitions or how these agencies broadly engage with the terms.

### Food Security and Food Resilience Narratives

Four out of the six agencies had explicit definitions on food security, though none overlapping, leaving the GAC and Irish Aid without any official definitions of those specific concepts. Irish Aid and GAC instead framed their discourse around hunger. Each bilateral agency also utilised its own interpretation of resilience, albeit often heavily borrowed from or influenced by existing frameworks. Despite DfID, GAC, SIDA and USAID all being members of the Global Resilience Partnership (GRC), a network of public and private organisations established in 2014 (Global Resilience Partnership, 2018) ostensibly to collate and share knowledge and policy on resilience, there is little overlap in their definitions. Appendices 1 and 2 give additional detail on the key aspects of the food security and resilience narratives among bilateral development agencies—including elements around focal points, policy objectives, delivery mechanisms, target groups, budget and policy scale.

As mentioned, Irish Aid utilised the concept of hunger as the underpinning of their aid work, which is perhaps unsurprising considering Ireland's history with the 1845–1849 Great Famine, having had a lasting impact on the Irish political psyche (Jordon, 1998; Kinealy, 2002; Mokyr & Ó Gráda, 2002).

**Table 4.3** Bilateral agencies—narratives on food security & resilience

| Bilateral agency | Narrative on food security | Narrative on resilience |
| --- | --- | --- |
| GAC (Canada) | "Hunger and malnutrition degrade health and reduce resiliency to disease and shock. This means the world's poorest people struggle the most to live long and healthy lives"—GAC website (2020a) | "Hunger and malnutrition degrade health and reduce resiliency to disease and shock. This means the world's poorest people struggle the most to live long and healthy lives"—GAC website (2020a) |
| DFAT (Australia) | "Improving food security by investing in agricultural productivity, infrastructure, social protection and the opening of markets is one of the ten development objectives... Food security underpins all other development, as without it, food insecure populations prioritise food and sustaining their own lives and those of their families over everything else. Australia's approach to food security is centred on increasing the availability of food through increased production and improving trade, while also increasing the poor's ability to access food"—DFAT website (2020) | "External shocks, including natural disasters, conflict, and economic shocks (such as food and fuel price spikes) severely undermine growth, reverse hard-won development gains and increase poverty and insecurity. Disaster preparedness, risk reduction and social protection help build the resilience of countries and communities"—DFAT website (2020) |
| DfiD (UK) | "Access to sufficient food for dietary needs and food preferences defines food security"—DfiD website (2019) | "The ability of countries, communities and households to manage change by maintaining or transforming living standards in the face of shocks or stresses without compromising their long-term prospects"—DfiD What is Resilience? paper (2016) |

(continued)

**Table 4.3** (continued)

| Bilateral agency | Narrative on food security | Narrative on resilience |
|---|---|---|
| Irish Aid | Narrative and discourse centred on 'hunger' | "Building resilience empowers people, communities, institutions and countries to anticipate, absorb, adapt to, or transform, shocks and stresses"—Irish Aid website (2021) |
| SIDA (Sweden) | "Guaranteed access to nutritious and safe food is a right and fundamental prerequisite for a decent life and the opportunity for people to contribute to the economy"—SIDA Aid Policy Framework (2017) | "Resilience should, wherever possible, be mainstreamed in activities to ensure that humanitarian aid helps to strengthen the resilience, recovery and adaptation capacity of populations affected by natural disasters, conflicts or health threats, such as epidemics, without compromising on the humanitarian principles of neutrality and impartiality"—Strategy for SIDA (2017–2020 [2017]) |
| USAID (USA) | "Food security means having, at all times, both physical and economic access to sufficient food to meet dietary needs for a productive and healthy life. A family is food secure when its members do not live in hunger or fear of hunger. Food insecurity is often rooted in poverty and has long-term impacts on the ability of families, communities and countries to develop and prosper"—USAID website (2020) | "For USAID, resilience is the ability of people, households, communities, countries, and systems to mitigate, adapt to, and recover from shocks and stresses in a manner that reduces chronic vulnerability and facilitates inclusive growth"—USAID website (2020) |

*Source* Authors

> Ireland's history and experience of famine echoes through the generations and influences our approach to helping those with whom we share our humanity in the fight against poverty and hunger. — Brian Cowen, Irish Aid Hunger Task Force Report (2008, p. 4).

This quote from the former Prime Minister of Ireland illustrates how context can define what is meant by food security. The deep impact of the Irish Great Famine permeates Irish Aid's humanitarian policy language and goals. The urgency of hunger prevention is a seam running through its approach to maternal and child malnutrition and smallholder agriculture improvement, with twenty per cent of Irish Aid's budget in 2019 allocated to 'hunger-related activities' (Irish Aid, 2020). Similarly, Ireland's resilience discourse is around 'empowering' communities.

In direct contrast, Australia's DFAT had an extensive definition of food security centred on trade and economic opportunities, with the goal of further opening of markets a central component, along with increasing agricultural production and access to food (DFAT, 2020). Australia has positioned itself as a regional economic powerhouse with long-standing neoliberal policies focused on 'improving trade' and 'opening markets'. Sixty per cent of the Australian agricultural market is export-oriented (Lawrence et al., 2013). Its resilience discourse, similarly, highlights how shocks 'undermine growth'.

SIDA's definition on food security leans heavily on food being a right that is central to a 'decent' life (SIDA, 2017). SIDA cites the Sustainable Development Goals (SDGs) as shaping its approach to development. Furthermore, its definition of resilience holds a particular emphasis on neutrality and impartiality, which have been Sweden's guiding principles since at least World War II.

> Resilience should, wherever possible, be mainstreamed in activities to ensure that humanitarian aid helps to strengthen the resilience, recovery and adaptation capacity of populations affected by natural disasters, conflicts or health threats, such as epidemics, without compromising on the humanitarian principles of neutrality and impartiality— Strategy for SIDA (2017–2020 [2017])

USAID, in contrast, utilises the 1996 World Food Summit (WFS) definition of food security, expanding it to include language of economic prosperity. USAID also utilises the UN definition of resilience at its core, expanding upon it to facilitate economic growth.

> For USAID, resilience is the ability of people, households, communities, countries, and systems to mitigate, adapt to, and recover from shocks and stresses in a manner that reduces chronic vulnerability *and facilitates inclusive growth*—USAID website (2020 [emphasis ours])

These examples help illustrate how both food security and resilience take on their parent countries' political and ideological traditions, both historically and contemporaneously.

Utilising a gradient scale, from a 'simple' to a 'complex' engagement with food resilience, the findings suggest that USAID leads the field, followed by the now-defunct DfID, with GAC, SIDA and Irish Aid relying more on partnership programmes with multilateral agencies, and DFAT, due to its isolated location, focusing on its own partnerships with regional NGOs and interest groups.

> Through its Bureau for Resilience and Food Security, USAID works with a host of partners to advance inclusive agriculture-led growth, resilience, nutrition, and water security, sanitation and hygiene in priority countries to help them accelerate and protect development progress—USAID website (2020)

USAID has promulgated the *Bureau for Resilience and Food Security* (RFS) in 2019, driven by its agenda of 'self-reliance' and strengthening resilience in the nations that it targets. Other agencies vary greatly to the degree that they consider food resilience, if at all. Our finding is that food resilience was mainly invoked through narratives around agriculture and climate change, with DFAT, DfID and SIDA lacking substantial narratives around resilience as a paired concept with food security beyond agriculture and climate change (DFAT, 2017; DfID, 2016; SIDA, 2017). Irish Aid was more occupied with livelihood resilience in its 'hunger' discourse (Irish Aid, 2016). These disparities highlight how a humanitarian framework, such as Irish Aid's, changes the foci of how resilience is applied, in contrast to development frameworks that utilise more technocratic approaches. SIDA's discourse, for example, is still centred around sustainable agriculture as a core policy goal in food security (SIDA, 2017). 'Resilience in food systems' is 'Principle 4' of its five principles for attaining 'sustainable agriculture'; resilience is utilised within a sustainable agriculture discourse rather than as an overarching, standalone, component.

> Enhanced resilience of people, communities and ecosystems is key to sustainable agriculture. In the context of sustainable food and agriculture, resilience is the capacity of agro-ecosystems, farming communities, households or individuals to maintain or enhance system productivity by preventing, mitigating or coping with risks, adapting to change, and recovering from shocks—SIDA (2017)

The GAC, who are core funders of the Global Resilience Partnership alongside USAID and SIDA, utilise a partnership approach when addressing their programmes—as such, their website lacks a clear definition or discourse on resilience, as well as a lack of transparency of the breakdown of their budget towards these goals (Brown, 2021). The Canadian government contributed US$38 million to a joint FAO, WFP and IFAD project, the 'Rome-based Agencies Resilience Initiative'. This was focused on DRC, Niger and Somalia and applied the UN's resilience framework for development work (FAO, 2019b). It reflects a novel approach among bilateral agencies, where Canada's own bilateral agency has taken a backseat to the Canadian government's direct contributions to the UN on resilience and food security (FAO, 2019b). Another example is the Canadian government, rather than the GAC, being a core funder of the Global Resilience Partnership (GRP, 2018). This may lie behind GAC's light website attention to food resilience; there is more attention to climate resilience and economic resilience, particularly in the Caribbean (GAC, 2018, 2020b). The FAO report on the Rome-based Agencies Resilience Initiative clearly outlines that this is a pilot for the Canadian government, who may be interested in pursuing this approach to development in the future. However, despite the disparate way food resilience was addressed on the GAC website, it was the only agency appraised that utilised 'resilience' on its website when discussing 'hunger and malnutrition'.

Broad target groups could be indicated without much specificity as to whose resilience was being addressed. Bilateral agencies, by and large, kept it vague—with USAID mentioning 'vulnerable communities', DfID considering 'rural families', and DFAT and SIDA highlighting farmers and fishermen. Irish Aid was one agency that actively considered 'whose' resilience, without reaching a conclusion, raising the question in a 2016 document without ultimately specifying a target group (Irish Aid, 2016). Consequently, the current engagement on target groups by bilateral agencies in general seems superficial and rhetorical.

USAID has by far the largest budget, as well as the most developed policy goals and discourse on food resilience, of all the bilateral agencies. USAID is the only bilateral agency that has developed ways of measuring resilience (excepting the now-defunct DfiD), with a fully-fledged discourse around food security and resilience (USAID, 2017). USAID's Global Food Security Strategy (GFSS) is focused on the 'Feed the Future Initiative'. The goal of GFSS, like RFS's remit, is to promote *"global food security, resilience and nutrition"* (USAID, 2020). Due to the vast resources that USAID has at its disposal, we interpret its relative influence as being more comparable to multilateral agencies, being able to operate globally, formulate clear definitions and measurement indices, as well as operationalise programmes independently. Due to more limited budgets, other bilateral agencies rely on partnerships or otherwise take a targeted approach to where their money goes, tending to choose a 'topic' to focus on, such as GAC focusing on nutrition security, SIDA on sustainable agriculture and DFAT aiming its funding towards facilitating trade, or a region, such as Irish Aid predominantly active in Africa, while DFAT's sphere of influence is South-East Asia and the Pacific.

### *Regional Development Banks*

*Food Security and Food Resilience Narratives*
The regional development banks, in contrast to the bilateral and multilateral development agencies, are broad funders, not having specific mandates in the same way UN agencies might have. The ability to command greater economic influence lends itself to be influential in shaping policy in their respective regions, making their narratives around food resilience important to highlight. The regional banks appraised in our study showed more consistent narratives and definitions (among themselves) than the wide-ranging definitions seen with bilateral agencies. Table 4.4 lays out the food security and resilience narratives fostered by the regional development banks. The ADB, AfDB, IDB and EBRD have fairly strict geographical remits—ADB covering Asia (with particular focus on South and South-East Asia), AfDB covering Africa, IDB covering Central and South America, and EBRD covering Eastern Europe and Central Asia. In contrast, the IsDB cover Islamic nations regardless of geographic location, and while most of their work is centred on MENA (Middle East and North Africa), IsDB also find themselves in South-East Asia and South Asia.

**Table 4.4** Regional development banks—narratives of food security & resilience

| Development bank | Narrative of food security | Narrative of resilience |
| --- | --- | --- |
| ADB (Asia) | "ADB's efforts and strategy to achieve food security in the region emphasizes the integration of agricultural productivity, market connectivity, and resilience against shocks and climate change impacts as the three pillars to achieve sustainable food security"—ADB website (2020) | "The ability of a system, community or society to pursue its social, ecological and economic development and growth objectives, while managing its disaster risk over time in a mutually reinforcing way (Keating et al. 2016)"—ADB's "Generating Multiple Resilience Dividends from Managing Unnatural Disasters in Asia" (2018, p. 5) |
| AfDB (Africa) | No concrete definition, but frequent reference to 'food security' and the SDGs (AfDB Website, 2020) | No concrete definition, frequent references to 'climate resilience' and 'resilient livelihoods' (AfDB, 2013) |
| EBRD (Europe) | "Food security is the state achieved when sufficient food of adequate quality is consistently available to meet consumer demand at affordable prices"—EBRD website (2020) | "A resilient market economy supports growth while avoiding excessive volatility and lasting economic reversals […] the ability of markets and market-supporting institutions to resist shocks"—EBRD's Transition Concept Review (2019) |
| IDB (Central and South America) | "The situation that exists when all people, at all times, have physical and economic access to sufficient, safe and nutritious food to meet their dietary needs for an active and healthy life"—IDB Food Security and COVID-19 Report (2020) | "Resilience is the process of adjustment to actual or expected climate and its effects…adaptation seeks to moderate harm or exploit beneficial opportunities. In natural systems, human intervention may facilitate adjustment to expected climate and its effects"—Climate Change at the IDB Report (2014) |

(continued)

Table 4.4 (continued)

| Development bank | Narrative of food security | Narrative of resilience |
| --- | --- | --- |
| IsDB (Islamic) | "Food security at the individual, household, national, regional, and global levels is achieved when all people, at all times, have physical and economic access to sufficient safe and nutritious food that meets their dietary needs and food preferences for an active and healthy lifestyle"—OIC Agriculture and Food Security Paper (2020) | "Resilience is the ability to adapt positively and to transform household, community and state structures and means to respond to risks, stresses and shocks (also drawing on OECD definitions). Resilient states and communities are characterized by stable social and political contracts, functional, inclusive and accountable institutions, and the provision of basic services. They are able to maintain political stability and prevent conflict"—IsDB Fragility and Resilience Policy (2019) |

*Source* Authors

The general focal points of the regional development banks can be simplified as three main themes—economic development, agricultural production and climate resilience. These 'big ideas' have dominated the narrative on food security for decades and are in line with the larger multilateral agencies' development agendas, who are seen as 'influential' (Shaw, 2007). Development banks place a central emphasis on economic development, particularly in terms of 'developing markets', with strong focus on economic and technical solutions (ADB, 2015; AfDB, 2016; EBRD, 2018; IDB, 2015; IsDB, 2020b).

Yet even among the development banks, this discourse varies. Akin to the bilateral agencies, the interpretation of resilience through the specific historical, political/ideological or other lens of each regional development bank provides space for disparities in narratives to emerge. The ADB, for instance, looks at resilience from the perspective of natural disasters and climate change (ADB, 2016)—with many areas of Asia having a recent history of being severely impacted by natural disasters, this focus on disaster management resilience is perfectly in line with the priorities of the region. With the increasing impact of climate change becoming apparent across Africa, it stands to reason that the AfDB places climate change resilience at the heart of its 'Feed Africa Strategy' (AfDB, 2016), with some acknowledgments of 'resilience in livelihoods' and 'socio-economic shocks' (AfDB, 2016). In contrast, the EBRD's resilience narrative is exclusively focused on market resilience (EBRD, 2019a, b), which is in line with the European region mainly consisting of high-income countries with highly developed agribusinesses, while the IsDB's broader resilience narrative is centred around conflict resolution and 'fragility' of nation-states, reflecting the history of political instability and conflict in MENA, IsDB's priority region (IsDB, 2019a, 2019b). These narratives around food security and food resilience are reflected in the overall strategic goals of the regional development banks. Four of the five regional development banks have extensive discourses on food security and resilience, showing a more elaborated approach than the bilateral agencies.

Despite the greater consistency in definitions, this shows there is still variation, even in its food security definitions, from the IDB utilising the 1996 WFS definition and the IsDB utilising the World Bank's definition (also derived from the 1996 WFS definition, albeit with some changes), with the ADB and the EBRD utilising their own definitions. AfDB, surprisingly, lack any concrete definition of food security on their website, instead frequently referencing the Sustainable Development Goals SDGs (AfDB, 2020).

The divergences in discourse, budgeting and policy priorities, as laid out in Appendices 3 and 4, reflect the diverging aims and goals of the respective development banks, such as the EBRD's entire focus on the economic aspects of food security, while the ADB focuses heavily on climate change resilience and agricultural productivity. It should be noted that ADB was the only agency appraised, from any category, where 'resilience' was utilised in their working definition of food security.

> Food security is the state achieved when sufficient food of adequate quality is consistently available to meet consumer demand at affordable prices—EBRD website (2021b)

> ADB's efforts and strategy to achieve food security in the region emphasizes the integration of agricultural productivity, market connectivity, and resilience against shocks and climate change impacts as the three pillars to achieve sustainable food security—ADB website (2020)

As one example, the ADB places resilience against shocks and climate change impacts as one of its core emphases, alongside agricultural productivity and market connectivity, in its resilience development framework

(ADB, 2015, 2019). Further, the ADB has an extensive discourse on sustainable food security, called the 'Operational Plan for Sustainable Food Security in Asia and the Pacific' (OPSFS) (ADB, 2009, 2015). This highlights the centrality of resilience to the ADB's overall strategy.

In contrast, for the AfDB, 'Climate Resilience Funding' is one of seven different initiatives laid out, among increasing productivity, infrastructure investment, agricultural financing and agribusiness (AfDB, 2016). While the AfDB sets 'Food Security and Agriculture' as a core part (along with 'Gender' and 'Fragile States') of their 'areas of special emphasis', resilience is not a specific focal point and lacks a concrete definition. The term, however, is still used throughout AfDB reports, such as in AfDB's 'Strategy for 2013–2022' report (AfDB, 2013).

For the IDB, 'resilience' and 'food security' are conceptualised separately, with the Food Security and COVID-19 Report in 2020 having only two mentions of resilience (IDB, 2020b) and the IDB's own internal 2015 Review of the IDB's food security strategy critical of the lack of clear conceptualisation of food security; we can also note that food security is grouped together with environment and climate change, but with no clear explanation of the relationships between these concepts (IDB, 2015, p. 6), and with no clear reference to resilience. As such, the IDB kept the concepts separate. The 2019 Sustainability Report, for example, does not have a single mention of food security, despite plenty of attention towards climate resilience (IDB, 2020a). While the IDB's Climate Change Action Plan makes passing references to food security and agriculture, it is secondary to its discourse on climate resilience (IDB, 2021).

The IsDB, in turn, leans heavily on existing narratives around food security (IsDB, 2018), while the EBRD exclusively emphasises market forces in a purely economic discourse (EBRD, 2021a, 2021b). For them, food resilience means resilience of agribusiness and markets, with attention given to access to finance and 'improving' trade:

> Resilience in food security is improved by making agriculture more productive and more resilient to external shocks including climate change. Making value chains more efficient, reducing losses and improving trade both increase food availability and reduce price volatility. Resilience is also strengthened through improving access to finance and private sector investment—EBRD website (2021a, 2021b)

The actual delivery mechanism of these broad policy goals can also differ substantially, with the IsDB as an example of innovative approaches to micro-financing and focusing on water infrastructure (IsDB, 2018), to the EBRD, whose focus is more on funding and developing agribusinesses (EBRD, 2018). These different delivery mechanisms are thus tailored to needs of the regions that these development banks target.

In terms of target groups, the regional development banks by and large focus their food resilience discourse on providing for 'smallholder' farmers. The usage of the term 'smallholder' was not apparent when looking at bilateral agencies, yet incredibly common among regional development banks, being used by ADB, AfDB and IDB. IsDB and EBRD focus their resilience narrative on agribusiness, although IsDB also acknowledges the need for resilience among 'rural communities'.

In its totality, resilience has become a 'core belief' for regional development banks, on the same level as market connectivity (economic growth) and agricultural productivity (productivism). However, 'climate resilience' remains the key concept in their resilience narratives, with much of the discourse around climate change and natural disasters, and in the IDB's case, exclusively so (IDB, 2021). In 2019, many of the large international development banks, led by the IDB and including the World Bank Group, EBRD, AfDB, ADB, IsDB, among others, published a joint report titled 'A Framework and Principles for Climate Resilience Metrics in Financing Operations', signalling a broad cooperative approach to measuring resilience, at least in the context of climate and finance (IDB, 2019). However, their quantitative approaches to measuring resilience as of writing (2021) remain separate. The ADB has the largest food security budget of the regional development banks assessed, followed by the IsDB (ADB, 2019; IsDB, 2020b). The AfDB has the smallest food security budget, despite being such a core component of their narrative. The EBRD has no specific budget for food security, instead rolling their food security initiatives into their finance investment budget.

## *Multilateral Development Agencies*

### *Food Security and Food Resilience Narratives*
Multilateral agencies are arguably the most influential agencies appraised here, as they can solidify definitions, expand terminologies and offer narratives that set global agendas (Shaw, 2007). While many nation-states have their own bilateral agencies, with varying degrees of economic

and political influence, the parent countries of the agencies appraised also provide additional funding to multilateral agencies. Furthermore, while regional development banks may have international stakeholders (particularly the United States and Japan), their scope is limited to the regions that they represent. Multilateral agencies such as those in the United Nations (UN) and the World Bank do not have these limitations, being central institutions influencing global policy formation since the late 1940s (Shaw, 2007). The UN agencies all have specific mandates, while the World Bank, not operating on any specific mandate, has been highly involved with food security discourse since the 1980s (Shaw, 2007).

Perhaps unsurprisingly, these multilateral agencies have complex, exhaustive definitions of food security and resilience, as well as extensive discourses around these topics (see Table 4.5, and Appendices 5 and 6). Apart from the IMF, the multilateral agencies appraised shared the 1996 WFS definition of food security, with a slight modification in the World Bank definition, dropping 'social' from its definition and focusing on 'economic' and 'physical' access. Of note, each agency has a slightly different definition of resilience, showing the malleability and lack of strong consensus on how to concretely define resilience in the same way that the 1996 WFS set a core definition for food security (one that bilateral agencies have by and large not utilised, barring USAID). It should be noted here that, while the ADB utilises 'resilience' in their working definition of food security, none of the multilateral agencies do. While this may be a vestigial result of utilising a now-antiquated definition of food security, the lack of a more contemporary definition of food security problematises the current paradigm shift to a resilience discourse as central and systematic to every framework and agenda (Béné et al., 2016).

As is to be expected, the UN agencies' narratives (as well as targeting) focus on their own respective mandates—UNICEF with ensuring child and maternal nutrition, and thus highly focused on targeting women and children; FAO with agriculture and food security research, with its resilience discourse focused on rural communities and rural livelihoods; and the WFP with emergency food assistance and 'Zero Hunger', the mission to 'end global hunger' through a broad range of programmes as well as providing technical expertise and policy advice, invoking food resilience when focusing on the role of governments to develop 'resilient institutions and systems' (WFP, 2021b). The World Bank set forth wide-ranging, ambitious, policy goals with an international scope that invokes

Table 4.5 Multilateral agencies—narratives of food security & resilience

| Multilateral agency | Definition of food security | Definition of resilience |
|---|---|---|
| FAO | "Food security exists when all people, at all times, have physical, social and economic access to sufficient, safe and nutritious food which meets their dietary needs and food preferences for an active and healthy life"—FAO website (2020) | "The ability to prevent disasters and crises as well as to anticipate, absorb, accommodate or recover from them in a timely, efficient and sustainable manner. This includes protecting, restoring and improving livelihood systems in the face of threats that impact agriculture, nutrition, food security and food safety"—FAO website (2020) |
| UNICEF | "Food insecurity [...] refers to limited access to food, at the level of individuals or households, due to lack of money or other resources"—State of Food Security (SOFI) (2020) | "Resilience is the ability to withstand threats or shocks, or the ability to adapt to new livelihood options, in ways that preserve integrity and that do not deepen vulnerability. The resilience of a household is related to the available resources (e.g. financial, assets, human capital, social resources etc.) and household's ability to use these resources (e.g. access to markets, access to public services and social protection)"—UNICEF website (2020) |
| WFP | "Food security exists when all people, at all times, have physical, social and economic access to sufficient, safe and nutritious food which meets their dietary needs and food preferences for an active and healthy life"—WFP website (2020) | "The ability of a system, community or society exposed to hazards to resist, absorb, accommodate to and recover from the effects of a hazard in a timely and efficient manner, including through the preservation and restoration of its essential basic structures and functions"—Committee on World Food Security (2015) |
| World Bank | "Food security exists when all people, at all times, have physical and economic access to sufficient, safe and nutritious food to meet their dietary needs and food preferences for an active and healthy life"—World Bank website (2020) | "The ability of a system and its component parts to anticipate, absorb, accommodate or recover from the effects of a hazardous event in a timely and efficient manner, including through ensuring the preservation, restoration or improvement of its essential basic structures and functions"—World Bank website (2020) |

| Multilateral agency | Definition of food security | Definition of resilience |
| --- | --- | --- |
| IMF | None given on website | "Three-pillar strategy for building structural, financial, and post-disaster resilience. Enhancing structural resilience requires infrastructure and other investments to limit the impact of disasters (Pillar I); building financial resilience involves creating fiscal buffers and using pre-arranged financial instruments to protect fiscal sustainability and manage recovery costs (Pillar II); and post-disaster (including social) resilience requires contingency planning and related investments ensuring a speedy response to a disaster (Pillar III)"—IMF Policy Paper on Building Resilience (2019) |

*Source* Authors FAO (2015), FAO (2019b), UNICEF (2020b), WFP (2015, 2017, 2020), and World Bank (2020)

each of the environmental, social, economic and public health themes, with their targeting most similar to the regional development banks, highlighting the necessity of resilience for 'vulnerable households'. Interestingly, a central aspect of the World Bank's food resilience discourse is specifically on resilient food systems, one of the few agencies that explicitly focuses on a food systems discourse, reflecting a different set of conceptualisations and terminology (World Bank, 2021a, 2021b, 2021c, 2021d).

The IMF has no set definition of food security on its website, despite having multiple papers on food affordability and pricing, as well infographics, blog posts and podcasts for their Finance & Development Magazine on food security—indicating that IMF research is both familiar with and conversant in the discourse around food security (IMF, 2020a, 2020b). One such example is a podcast available on the IMF website where two IMF economists discuss food security in sub-Saharan Africa, mentioning 'raising climate change resilience' as a major factor in addressing food security (IMF, 2020c). Otherwise, discourse around food security available on the IMF website is sporadic and linked with shocks, such as several articles in the wake of the 2007–2008 World Food Price Crisis (IMF, 2008).

The UN agencies all utilise the Resilience Index Measurement and Analysis (RIMA-II) for resilience measurement and analysis within food security, an index developed for practical application (FAO, 2016). RIMA has been around in some form since 2008, pioneering the utilisation of resilience in a food security context (FAO, 2016). With resilience seen as fundamental to the FAO's and WFP's work and given particular emphasis through RIMA-II (FAO, 2016), the lack of a clear engagement with resilience within the existing food security definition might indicate either that the term 'food security' has become unhelpful and outdated, or that 'food resilience' may replace it, or it might signal that resilience is not seen as a critical element of food security. Whichever emerges, there seems to be delay in integrating the concept of resilience wholly into food security.

The World Bank's resilience definition, in contrast, is derived wholly from the 2012 (IPCC) definition, heavily focused on climate change and agriculture (World Bank, 2013, 2016). The World Bank, in turn, has its own measure of resilience called the Resilience Rating System. This system is focused on rating World Bank projects, with its criteria based on tracking the effects of climate change (FAO, 2016; World Bank, 2021a, 2021b, 2021c, 2021d). The Resilience Rating System Report, however, does mention food security repeatedly (eight times). The World Bank's Resilience Rating System was first introduced in February 2021.

The World Bank and the WFP have much larger budgets than the FAO, while half of UNICEF's proposed programme budget will go towards child and maternal nutrition (UNICEF, 2021). Of note, almost half of FAO's budget in 2020–21, that is US$930 million of its US$2.8 billion total, went towards livelihood resilience, being a cornerstone in the FAO's strategic framework (FAO, 2019a, 2019b). This reflects how the FAO has prioritised a food resilience approach as a core component of its food security strategies. While the World Bank has a significantly larger budget, its role in broad funding means that funds go to many different considerations. Their narrative on resilience within food systems highlights their funding efforts towards agricultural production and mitigating climate change, which remains the key consideration in the World Bank's resilience discourse.

### COVID-19, Food Resilience and Agency Responses

Due to the scale and speed of the COVID-19 pandemic as the present study was being conducted, we decided to consider early narratives on food resilience that emerged related to the pandemic. We, let alone the development agencies, were aware of the need to distinguish between initial short-term policy reactions to the pandemic and any long-term legacy and responses. With the initial responses being reactive rather than planned, the main observation has been that the notion of resilience became again widely used, with both growing concern and data on the impact of climate change on agri-food systems and the impact of the pandemic likely encouraging its increased usage. The chapter by Joanna Upton et al. in this book looks more closely at the impact of COVID-19 on household resilience.

The UN agency leading the response to COVID-19, the World Health Organisation (WHO), was not appraised for the purposes of this chapter. The most prominent frontline agencies of those appraised were the multilateral agencies. The influence of the UN agencies was visible in several ways. First, they set the narrative of the impact of the pandemic on food security for the bilateral agencies, regional development banks and the global community to respond to. They particularly highlighted the need for financial support and for humanitarian relief to mitigate effects of the pandemic (FAO, IFAD, UNICEF, WFP and WHO, 2021;

WFP, 2021b). The WFP, for example, headlined the 'Hunger Pandemic' when introducing the 2021 SOFI (WFP, 2021b). David Beasley, WFP Executive-Director, firmly stated that:

> The path to zero hunger is being stopped dead in its tracks by conflict, climate and COVID-19—David Beasley, WFP (2021b)

Concerns were expressed at the UN level as the pandemic took grip at the prospect of food insecurity being exacerbated by the pandemic. The FAO was concerned about existing conflicts being worsened (FAO, 2020; UN, 2020). A year later, still lacking solid statistical analysis as to whether millions were plunged into hunger because of the pandemic, the agencies readily admitted that the early fears might not fully have been met (FAO, IFAD, UNICEF, WFP and WHO, 2021; WFP, 2021b). Two years on, however, a more sober assessment was being voiced—that inequalities of access to resources (not least vaccines) might exacerbate food and health-related economic inequalities within and between countries. In this sense, the initial, sometimes emotional, concerns might turn out to have been justified (Béné et al., 2021).

Multilateral agencies were characteristically practical in launching new programmes and operations in response to the pandemic, particularly the WFP, while UNICEF and FAO have published reports and analyses of pandemic impacts on food security and food resilience (UNICEF, 2020c; UNICEF, 2020d; FAO, 2021; WFP, 2021a, 2021b). UNICEF, for example, published an extensive report on child food security and nutrition in June 2020 (UNICEF, 2020c). WFP recast training and insurance schemes for smallholder farms, with the long-term goal of "build[ing] resilience" and "adapt[ation] to better link to food systems" (WFP, 2021b). The World Bank has pushed for a 'Resilient and Inclusive Recovery', and echoing the calls from world leaders articulated the recovery as a 'rebuilding better' (World Bank, 2021b), adapting 'Building Back Better' (Harley & Acheampong, 2021). The World Bank has emphasised the importance of 'resilient recovery', and although world food prices remained fairly stable in the first two years, 2022 saw concerns regarding global food inflation and potential shortages come to fruition with the added shock to the global system with the war in Ukraine (World Bank, 2021b; World Bank, 2022).

The regional development banks shared the same overall narrative as the multilateral agencies, though they took different approaches. For example, the ADB had a 16-page brief on the impact of COVID-19 on food security by June 2020, while the AfDB issued a report on the changes they would undertake in their 'Feed Africa' initiative to help tackle COVID-19 and food insecurity in July 2020, with a heavy emphasis on resilience. The IDB issued a slew of reports and research on the impact of COVID-19 and food security in Central and South America (IDB, 2020b, 2020c, 2020d), while the IsDB formed a working group to explore the impact of COVID-19 on 'Islamic Finance' (IsDB, 2020c), as well as a press release in April 2021 outlining a US$500 million commitment to addressing climate change and food security, putting a heavy emphasis on resilience, to respond to the challenges they saw arising in the 'post-COVID-19 world'.

> The partnership we are signing today will allow us to co-create financing and investment programs that will address these challenges, but also help our Member Countries tap into emerging global value chain opportunities to build resilience and create wealth in a post COVID-19 world—IsDB Press Release (2021)

The bilateral agencies were slower to respond than the regional development banks and multilateral agencies. While they all recognised the impact of the pandemic on food security, they followed the same narrative path laid out by the UN agencies. Some minor examples include Australia's DFAT providing US$5.5 million to WFP towards its food security work in the Pacific region and USAID launching multiple projects that seek to address food insecurity arising from the pandemic, with the US government reorienting its food security strategy to centre around COVID-19 recovery and climate change, highlighting how the pandemic has "had an unprecedented impact on food security" (USAID, 2021a, 2021c), with Samantha Power, the USAID Administrator, calling it a "pivotal moment for global food security" (USAID, 2021c).

## Discussion

In this final section, we offer four reflections on how the concepts studied here are being articulated and addressed by development agencies.

The first reflection is on our own methodology and approach. Reviewing existing literature and data for this chapter, it was decided a rapid appraisal of what agencies mean by the key terms of this book would be useful, building on the definition of food systems resilience by Tendall et al. (2015). Rapid research snapshots are often conducted in policy analysis and other scientific endeavours, not least in emergencies. While the methodology can be varied, many interventions in public health, rural planning and social, environmental and humanitarian emergencies have to be based on rapid appraisals of evolving situations (Crawford, 1997; Kumar, 1993; McNall & Foster-Fishman, 2007). Despite limitations, having a reasoned overview can help policy formation of, and for, a situation. The findings from such rapid appraisals may be altered or questioned later, but their value lies in capturing the moment. It is better to have some pinpoints of light than to justify inaction as waiting for floodlights from lengthier research, and a rapid appraisal can also contribute to critical review of policy positions held by powerful, influential agencies.

The second point is that while agencies could, and probably will, head in divergent policy directions, the terms food security and resilience will lead a life of their own in international food policy. Given the importance of agri-food systems for employment, health, environmental and social well-being, these two terms have considerable symbolic significance locally, nationally and globally. National and global agri-food systems are already under considerable pressure in relation to public health, environmental and socio-economic divisions (Searchinger et al., 2018; UNEP, 2020; Willett et al., 2019). Analysts within as well as outside development agencies know paradigm shifts are highly likely. In that context, our finding that both food security and food resilience are subject to significant difference of interpretation is important. We have used the notion of fractured consensus to capture those tensions. Resilience and security might be invoked to convey a shared vision of the future, when the reality is that divergent paths are being pursued by, and shaped by, lobbies and sectoral interests. The terms are thus 'codes' for what can subsequently be taken in different directions. A world that took feeding everyone seriously, or that had already ensured that humanity would genuinely tame the drivers of ecosystem destruction and reduce the severity of future shocks, would already be building food system capacities for change. Judging by the brevity of the UN Food Systems Summit in September 2021—given just one day—and with food consumption barely featuring at the UN Climate Change COP26 in October 2021, that centrality of purpose cannot yet be assumed, even though it is sorely needed.

To press the point further, the practicalities of building food resilience are having to compete for support in already contested and crowded policy space. If food resilience becomes more fractured, the auguries for achieving effective food system change might not be good. The critics of development—particularly of Western foreign policy from the 1970s to the 1990s—created a narrative that development was excessively framed by conventional politics and dominated by US-based agencies such as the World Bank, IMF and USAID (Shaw, 2007). These agencies, characterised as ideologically right-of-centre and tacitly supporting neoliberal, market-oriented solutions, were contrasted with European agencies such as Irish Aid or SIDA, pitched as being left-of-centre and reflecting the expressed values of European social democracies, oriented around a discourse of human rights and dignity. The UK's DfID, in this narrative, fell in between, initially aligned with the Europeans, but more recently shifting its alignment, particularly in its subsuming into the FCDO, becoming more mercantilist, overtly yoked to UK post-Brexit economic aspirations. In fact, the present appraisal has suggested that this may be an overtly simplistic over-politicised reading of how agencies approach food security. True, there are 'hard' neoliberal approaches, such as Australia's DFAT and the EBRD's relentlessly economic language. Yet the ingraining of concepts such as sustainability and resilience, finding itself on par with economic and productivist goals such as market access and improved agricultural productivity, signals that these concepts, even when co-opted, provide a useful contrast to more simplistic narratives. The politics of food resilience might in fact be becoming more subtle, particularly with the arrival on the global development terrain of China's state capitalist investment in infrastructure across low and middle-income nations through the Belt and Road Initiative (Chatzky & McBride, 2020; Jie & Wallace, 2021). With China's assertion of 'softer' humanitarian and social support in the wake of the COVID-19 pandemic, particularly through vaccine diplomacy (Lee, 2021; Su et al., 2021), this may further challenge these agencies on how they act and present themselves. For example, the World Bank, a financial body that in recent years has worked hard to counter decades of criticisms that it imposed a narrow, neoliberal model of economic progress (George, 1976, 1988; Holman, 1984; McMichael, 2012), has built tackling climate change, climate change resilience, and environmental degradation into its core narrative on global policy and governance, and thus into its criteria for investments in food resilience, particularly through their food systems approach, which in turn has influenced the regional development banks.

The coupling of 'food security' and 'resilience' belies different foci and different starting points. As such, bilateral agencies, regional development banks and multilateral agencies may share similarities in the overall narrative on food resilience, yet their policy objectives can differ greatly. Resilience can be pitched as a return to normalcy, or as a new direction altogether, with the findings presenting a mixed picture of, on the one hand, a broad consensus on the importance of food resilience and, on the other hand, differing overt policy objectives. Agencies differ even on whose resilience is being addressed, from as broad as 'food insecure' to as specific as 'smallholder farmers'. Barely a decade from resilience entering common policy terminology, divergent meanings have emerged. Governments may have collectively signed up to the 2015 SDGs, but what the agencies mean by food resilience cuts across this. The SDGs are often invoked as the new global framework (from SIDA, to IDB, to AfDB), but our findings suggest these invocations can be light on details. Resilience as a term is being re-interpreted to fit wider historical contexts and disparate priorities, depending on the region, country, culture, origin and issues the agencies have historically focused on. The risk is that the terms are being pulled into ubiquitous vagueness, a fate long noted for sustainability (Lang & Barling, 2012). There is an expanding literature concerned about how adequately, concretely and quantitatively to measure food security and resilience (Ansah et al., 2019; Barrett & Constas, 2014; Cissé & Barrett, 2018; d'Errico & Di Giuseppe, 2018; Knippenberg et al., 2017; Maxwell et al., 2003; Serfilippi & Ramnath, 2018; Upton et al., 2016). Perhaps this is inevitable at the global or UN level, where compromise is the whole purpose of institutional decision-making. The post-war reconstruction did not envisage an era of climate change or ecosystem destruction reshaping humanity's planetary role. The focus was on human rights and to build agreement where self-interests crossed borders. This certainly happened for public health institutional structures from the 1930's International Health Organisation to the World Health Organisation of today (Rayner & Lang, 2012). We should not be surprised if there are difficulties in the face of contemporary threats, but this is why common purpose and clarity of meaning matter.

A third consideration thus arises—the legitimacy of public understanding. The funds disbursed by these agencies are mostly taxpayer-derived; democratic accountability is thus implicit. To many citizens and taxpayers, the existence of development agencies is probably as far as their knowledge goes, with only some among them knowing, or even being interested, in more. The appraisal here in many ways explores the outward-facing nature of these agencies, and what they present to their citizens and other interested parties. If, as happened in some agencies discussed here, that humanitarian internationalism is reined back, this suggests a failure of pro-development movements to win sufficient public support to stop 'aid' becoming financialised. While the present study did not explore agencies' website layout, search functions, usability or access policies, we did note how some agencies such as Irish Aid utilised accessible language to allow for a broader audience to be engaged, while other agencies were tailored towards their own niche and specific interest groups. As Watkins noted decades ago, neither aid nor development are truly neutral (Watkins, 2001). The long-standing historical tensions of debate in the 1970s and 1980s—aid as an extension of colonialisation, etc. (Hayter, 1971; Raikes, 1988; Tudge, 1977)—might be returning considering China's well-funded push into development in the name of South-South cooperation (Cohen, 2020).

A fourth point concerns the purpose of the types of agencies we have reported on. These agencies have substantially different budgets, reach and remit, which influences their narratives, policy approaches and internal capacity to quantify food resilience. The UN agencies and the World Bank have a global focus and broader, holistic policy goals—the WFP, for example, aims for 'Zero Hunger', while the World Bank has goals covering political, economic, environmental, social and public health themes. The bilateral agencies, apart from USAID, tend to focus on one policy objective, while the regional development banks have broader policy goals but a limited geographical reach. The field of development agency practice covers, on the one hand, the World Bank's complex, detailed and multifaceted policy objectives, while, on the other hand, Sweden's SIDA, one of the smallest agencies appraised, has chosen to champion sustainable agriculture as its focal point within its food security narrative. USAID's budget is incomparably large when placed next to other bilateral agencies, having an influence and presence more akin to the multilateral agencies. As such, the way USAID defines food resilience, and their approach to food resilience, may have a more noticeable impact on how NGOs and countries who work with USAID, and who depend on USAID funding, shape their own food resilience narratives. Perhaps part of the fluidity of food resilience is that they are framed by the scale of whoever is applying them.

A recurring theme in the resilience literature is the problem of data. On the one hand, there are well-known methodological difficulties in quantifying and analysing resilience. On the other hand, everyone agrees on the importance of accurate data in measuring resilience (Ansah et al., 2019; Barrett & Constas, 2014; Béné et al., 2016; Cissé & Barrett, 2018; d'Errico et al., 2018; Knippenberg et al., 2017; Serfilippi & Ramnath, 2018; Upton et al., 2016). Compounding these methodological difficulties, USAID, DfID, the regional development banks, the UN agencies and the World Bank all have (or had) different ways of measuring resilience, utilising their own frameworks and approaches. This is a reflection, a microcosm, of the broader fractured consensus we found in the resilience discourse. While the UN Agencies offer a concrete, quantitative methodology to specifically measure food resilience (RIMA-II), agencies with enough economic and political reach choose to go their own way. USAID established the Resilience Evaluation, Analysis and Learning (REAL) Award, formulating the USAID/REAL Resilience Measurement and Analysis Framework (REAL, 2018), while the IDB, in collaboration with other regional development banks, have formulated a framework to specifically measure and quantify climate resilience, with less attention given to food resilience (IDB, 2019). Lastly, the World Bank utilises its own framework and methodology to measure climate and disaster resilience (World Bank, 2017). The battle over which measurement methodology is to dominate reflects the tensions embedded in the terminology. The operationalisation and conceptualisation are fractured by related fissures, with as many approaches and tools as there are sectors and focal points (Béné et al., 2016). The diverse interpretations of food resilience, and the diversity in measurement, reflect how what each agency means by resilience frames, how they measure it, undermining any eventual formulation of a unified framework and common understanding of what is needed in measuring resilience (Barrett & Constas, 2014; Béné et al., 2016).

## Conclusion

Our conclusion reflects on the value of food resilience, as explored, in its current usage among agencies. Food security has become a term that, when used at its face value, is more about its apparent meaning than its content and depth (Carolan, 2013). The value of the newer term of food resilience is that it offers an opportunity to move away from the fluidity of the 'food security' term in public policy (Lang & Barling, 2012), transcending it in its role as a 'mobilizing metaphor', capturing the necessity for a more systematic, comprehensive and complex direction and approach (Béné et al., 2016). One that can address both short-term and long-term interventions, and one more suited for the complexities and challenges posed by inevitable consequences of climate change (Upton et al., 2016) and the aftermath of the COVID-19 pandemic. There is, however, the risk that it could be rendered meaningless, yet another buzzword that falls in and out of vogue, particularly if subsumed by pre-existing motivations, rather than reflecting a fundamental rethink in each agency remit. This appraisal has shown that development agencies can be placed on a type of gradient, reflecting the depth and complexity that food resilience is addressed with—from a facile, surface-level engagement, light on details with only nascent arguments, heavily leaning on terms and definitions from other agencies, to a more complex, multifaceted approach, one that contains or implies qualitative arguments, quantitative analysis and institutional restructure. For the Australians and the EBRD, aid is to lever business growth, utilising the language of food resilience to prise open trade and forge economic partnerships, while for Irish Aid, the motivations seem on the surface more altruistic, built around the concept of 'combating hunger', reflecting Ireland's lasting and dreadful legacy of famine (Jordon, 1998; Kinealy, 2002; Mokyr & Ó Gráda, 2002).

There may be many factors that determine where agencies lie on such a gradient: budget limitations, existing ideological approaches, lack of policy focus or sector engagement, or simply because institutions take time to react to and integrate new approaches (the 'trickle-down' effect of policy discourse). While discourse around climate change resilience and sustainable agriculture can be found on every one of the agencies' websites reviewed, food resilience had a high degree of variability—with the USAID's *Bureau of Resilience and Food Security* and the UN agencies' highly detailed food resilience development framework leading the scene, fully engaged with food resilience as a paired concept. Most of the regional development banks reflected a simpler 'institutional discourse' (early adopters of 'trickle-down' food resilience?). The ADB is noteworthy, as it utilises the concept of resilience in its definition of food security, being the only agency appraised to do so. Some agencies had continued focus on earlier terms (the SDGs, sustainable agriculture, climate change resilience) such as the IDB or SIDA. One can speculate that the crisis event of the pandemic will lead to a rapid adoption in the coming years of a broader food resilience discourse than currently found, a further entrenchment of resilience as a core term and a widely adopted definition akin to the 1996 WFS settling the definition of food security (which, in turn, now seems lacking). The pandemic has highlighted the importance of a food resilience narrative, almost certainly helping shape the perception of the necessity of these agencies for future funding (Béné et al., 2021).

Finally, we stress again that the snapshot we have conducted was at a time of disruption not only because of the COVID-19 pandemic. With the changing nature of development agencies—illustrated by Canada, Australia and the UK integrating development more closely into foreign and trade policy—it is yet unclear if the prime narrative of international humanitarianism will remain. Even USAID, which has extensive discourse and funding favouring food resilience, bluntly emphasises that its central goal is to help countries help themselves, viewing resilience as a form of self-sufficiency, leaving it unclear whether 'donor' governments will rise to the enormity of the looming challenges on the horizon. The disruption and retreat to self-interest illustrated in the early responses to the pandemic cannot be ignored. Self-interest triumphed on vaccine and equipment procurement (Lancet, 2021; Torres et al., 2021). In that respect, the initial years of the pandemic echoed the inequalities and distortions of the global food system, with the pressures of looming economic instability, rising geopolitical tensions, the unrelenting consequences of climate change and an unresolved public health crisis, all highlight how the narratives around food security and food resilience, in this sense, are both a test and reflection of what sort of societies influential, high-income countries envisage ahead.

## Appendix 1: Bilateral Development Agencies—Key Aspects of Food Security Narrative

| Donor agency | GAC (Canada) | DFAT (Australia) | DfID (UK) | Irish aid | SIDA (Sweden) | USAID (USA) |
|---|---|---|---|---|---|---|
| Key focus | Welfare and Health | Investment and Trade | Diet and Nutrition | Alleviating Hunger | Nutrition as a fundamental right | Physical and Economic Access to Food |
| Primary policy objective | Nutrition Security | Developing Economic Partnerships in South-East Asia and Pacific Regions | Food Security Research and Analysis, Addressing SDGs holistically | Improving smallholder agriculture, preventing maternal and child undernutrition | Sustainable Agriculture | Feed the Future Initiative to "strengthen agriculture-led growth, nutrition and resilience" |
| Primary target groups | Women, Children | Small-scale farmers and fishermen in the Pacific region | Rural families reliant on agriculture | Recognises the need for targeting but doesn't offer any specific target group (2016) | Farmers, fishermen, foresters and rural communities in low- and middle-income countries | Vulnerable communities |

(continued)

(continued)

| Donor agency | GAC (Canada) | DFAT (Australia) | DfID (UK) | Irish aid | SIDA (Sweden) | USAID (USA) |
|---|---|---|---|---|---|---|
| Delivery mechanism | Emergency & Development Food Assistance, International Partnerships, Food Security Research | Funding NGOs in Pacific Region, Partnerships with other bodies | Partnerships with other bodies, investments in international climate finance, commercial agricultural programmes | Development programmes to reduce global poverty and contribute to Sustainable Development Goals | Supporting "more productive and climate and environmentally sustainable agriculture" | Collaboration with U.S. government agencies and departments, private sector, civil society, researchers, universities and partner governments; Providing emergency food assistance in crisis |
| Budget | Have not committed to any specific budget target | 2020–2021 Budget Allocation: AUS$242.2 million (US$188.2 million) | 2018–2019 budget allocation for resilience building: £807 million | 20% of Irish Aid's budget goes to 'hunger-related activities' – €130 million (US$157.4 million) | 955 Million SEK (US$144 million), 5% of SIDA's total development assistance, goes to sustainable agriculture | Agricultural budget of US$807.6 million, Nutrition budget of US$24.6 million in 2019, and $30 million in 'Resilience Challenge Funds' in 2020 |

(continued)

(continued)

| Donor agency | GAC (Canada) | DFAT (Australia) | DfID (UK) | Irish aid | SIDA (Sweden) | USAID (USA) |
|---|---|---|---|---|---|---|
| Policy scale | International | Regional—South-East Asia & Pacific Focus | International, National | Regional —Focus on African Nations | International, National | International |

*Source* Authors

## Appendix 2: Bilateral Development Agencies—Key Aspects of Resilience/Food Resilience Narratives

| Bilateral agency | GAC (Canada) | DFAT (Australia) | DfID (UK) | Irish aid | SIDA (Sweden) | USAID |
|---|---|---|---|---|---|---|
| Key focus | Climate Resilience, Livelihood Resilience | Natural disasters, conflict and economic shocks—disaster preparedness, risk reduction and social protection | Countries, communities and households managing change | Empowering people, communities, institutions and countries to protect against shocks and stresses | Recovery and adaptation capacity against natural disasters, health and conflict | Mitigation and adaption to shocks and stresses to reduce chronic vulnerability |
| Part of agency's approach to food security? | Yes | No | Yes, to an extent | Yes | Yes | Yes, through the Bureau of Resilience and Food Security |
| Focus within Food Resilience | Climate Change Adaptation | None | 'Building *climate change resilience*' is mentioned as a core component in addressing insecurity within food and agriculture, water, sustainable energy and cities—**What is resilience?, 2016 Environment:** Climate Change | Resilience is mentioned alongside food security, agriculture, climate change, livelihoods—**Irish Aid website, 2020** | Yes—one of four main goals within sustainable agriculture—SIDA, (2017) | **Environment:** Agriculture, Climate Change **Economic:** Food Affordability, Science & Technology; **Political:** Access to Services, Social Safety Nets, Infrastructure, Institutions **Health:** Diet & Nutrition, Food Governance |

*Source* Authors

# Appendix 3: Regional Development Banks—Key Aspects in Food Security Narratives

| Regional agency | ADB (Asia) | AfDB (Africa) | EBRD (Europe) | IDB (Central and South America) | ISDB (Islamic) |
|---|---|---|---|---|---|
| Key focus | Sustainable Food Security—Agriculture, Economy, Resilience | Agricultural Transformation, Sustainable Agriculture | Economic and Financial Development to address Food Security | Food Availability, Food Access, Food Use and Food Stability | Sustainable, Inclusive, Equitable Agriculture and Rural Development |
| Primary policy objective(s) | Agricultural productivity, market connectivity, resilience against shocks and climate change impacts | Agriculture and rural development, poverty reduction | 'Overcoming constraints to supply and efficiency', and to 'develop markets, logistics infrastructure, and technology'—EBRD website, 2020 | Three Lines of Action: Trade and Market Access, Productivity through Production Intensification, and Environmental Planning and Sustainable Agriculture —IDB Food Security Review, 2015 | Agriculture and rural development, population resilience through 'natural resource management, increased productivity and connectivity' |
| Primary target groups | Smallholder farmers, women farmers | 'Most vulnerable households'— smallholders, stock breeders, agro-pastoralists, fishermen, women and children | Agribusiness | Smallholder farmers, women, indigenous people | Agribusiness, rural communities |

(continued)

(continued)

| Regional agency | ADB (Asia) | AfDB (Africa) | EBRD (Europe) | IDB (Central and South America) | ISDB (Islamic) |
|---|---|---|---|---|---|
| Delivery mechanism | Multisector approach, 'increasing positive linkages', mix of public and private sector, regional cooperation | Loans, grants, technical assistance and expert advice under the "Feed Africa" initiative | Investments, public–private policy dialogues and coordinating actions with partners. Private sector seen as playing a decisive role in "helping bridge the growing divide between demand and supply" | Multi-donor financing for technical cooperation projects through AgroLac 2025 | Developing rural micro-finance industry, off-farm enterprises and funding rural water infrastructure |
| Budget | Budget around US$6.8 billion between 2015 and 2017 | US$1.1 billion in 2017 | None specifically for food security | None specifically for food security—16% of budget towards 'climate change & environmental sustainability—est. $US91 million, 2020 | The current active portfolio is more than US$6 billion, but no single source or figure for each project |
| Policy scale | Regional, National—Asia-Pacific Region | Regional, National— Africa Region | Regional, National —Europe and Central Asia | Regional, National—Nations across Central and South America (Latin American Countries) | Regional, National—Islamic Nations across Asia, Africa |

*Source* Authors

## Appendix 4: Regional Development Banks—Key Aspects in Resilience/Food Resilience Narratives

| Regional Bank | ADB (Asia) | AfDB (Africa) | EBRD (Europe) | IDB (Central and South America) | IsDB (Islamic) |
|---|---|---|---|---|---|
| Key focus | Social, Ecological, Economic Development and growth through disaster risk management | Climate resilience and livelihood resilience | Economic Resilience—'Resilient Market Economy' | Climate resilience: building adaptive capacity and strengthening mitigation | Household, community, state adaption to risks, stresses and shocks—stable social and political stability and conflict prevention |
| Part of agency's approach to food security? | Yes | No | Yes | No | Yes |
| Approach within food resilience discourse | **Environmental:** Agriculture, Climate Change; **Societal:** Livelihoods, Gender Focus; **Economic:** Food Costs, Food Affordability, Science & Technology; **Political:** Infrastructure, Institutions | **Environmental:** Agriculture, Climate Change; **Societal:** Livelihoods; **Economic:** Food Affordability; **Political:** Access to Services, Social Safety Nets, Infrastructure, Institutions | **Economic:** Food Governance, Science & Technology, Private Enterprise | **Environmental:** Agriculture, Climate Change; | **Environmental:** Agriculture, Climate Change; **Economic:** Science & Technology, Finance/Private Industry |
| Food resilience & COVID-19 | Yes—June 2020 Brief | Yes—FAREC (Feed Africa Response to Covid 19) July 2020 Report | Yes—EBRD and FAO US$3 Million Technical Support Package for Agrifood Sector—October 2020 | Yes—Multiple Reports on Impact of COVID-19 on Food Security (IDB, 2020c, 2020d) | Yes—US$84 Million Emergency Food Security Assistance to Senegal under COVID-19 Response Plan. Additional US$500 Million agreement to tackle Climate Change and Improve Food Security April 2021 (Mentions Covid 19) |

*Source* Authors

## Appendix 5: Multilateral Agencies—Key Aspects in Food Security Narratives

| International agency | FAO | IMF | UNICEF | WFP | World Bank |
|---|---|---|---|---|---|
| Key focus | Physical, Social and Economic Access to Food, Resilient Agriculture and Rural Livelihoods | Agricultural Productivity and Household Resilience | Individual and household access to food of women and children | 'Zero Hunger' policy goal, aiming for Physical, Social and Economic Access to Food | Physical and Economic Access to Food, as well as agricultural, economic, social, public health and political focus |
| Primary policy objective(s) | Eradication of hunger, food insecurity, malnutrition, as well as promoting the sustainable management and utilisation of natural resources | Policy Research and Policy Recommendations | Ensuring childhood nutrition as an integral part of broader programmes targeted at women and children | Sustainable Development Goal 2: End Hunger, Achieve Food Security, Improve Nutrition and Promote Sustainable Agriculture | Five Main Focal Points: Agriculture, Economy, Society, Public Health and Politics |
| Primary Target Groups | Rural Communities | No specific target groups mentioned | Children, women | Governments | Vulnerable households |
| Delivery mechanism | Strategic framework from central institution, then regional plans to translate strategy into practice—results-based approach (FAO, 2012) | IMF has no specific programmes or plans regarding food security specifically | Programmes, raising awareness, funding NGOs | Broad range of policies and programmes—primarily emergency food distribution, but also cash/non-food transactions, technical expertise, policy advice, etc | Funding projects and programmes within agriculture and related services |

(continued)

(continued)

| International agency | FAO | IMF | UNICEF | WFP | World Bank |
|---|---|---|---|---|---|
| Budget | US$2.8 billion in 2020–2021 | No specific budget | Maternal and child nutrition budget in 2019 was US$687 million | WFP raised US$8 billion in 2020 | US$10 billion—US$5.8 billion through IBRD/IDA, and US$4.2 billion through IFC financing |
| Policy scale | International, Regional, Household | International, Regional, Household | International, National | International, Regional, National | International, National, Household |

*Source* Authors

## Appendix 6: International Agencies—Key Aspects of Resilience/Food Resilience Narratives

4 FOOD SECURITY AND THE FRACTURED CONSENSUS ... 129

| International agency | FAO | IMF | UNICEF | WFP | World Bank |
|---|---|---|---|---|---|
| Key focus | Anticipating, preventing and recovering from disasters and crises in a timely, efficient and sustainable manner—particularly protecting, restoring and improving livelihoods in the face of threats that impact agriculture, nutrition, food security and food safety | Three-Pillar Strategy: Structural, Financial, Post-Disaster Resilience | Withstanding threats, shocks and adapting to new livelihood options for a household— availability and access to resources (financial, human capital, social capital) | Adapting to shocks and stressors from conflict, political instability and natural hazards such as climate change, environmental degradation, water scarcity, disease, and population growth | Anticipate, absorb, accommodate, and recover from hazardous events by preserving, restoring, improving a system's essential basic structures and functions |
| Measures of resilience | Yes (RIMA) | No | Yes (RIMA) | Yes (RIMA) | Yes |
| Part of agency's approach to food security? | Yes | No | Yes—in the sense that all their policies are now through 'building resilience' through all 'Strategic Plan' areas | Yes; Core part of WFP 'country strategic plans', putting resilience-building at the core of 'long-term planning frameworks' | Yes, as in, food security and resilience are often mentioned together, although a specific concept of food resilience— and a definition of such—remains elusive |
| Approach/focus within food resilience | **Environmental**: Agriculture, Ecosystems, Climate Change; **Societal**: Livelihoods; **Political**: Access to Services, Social Safety Nets, Infrastructure, Institutions | **Environmental**: Climate Change; **Economic**: Finance; **Political**: Access to Services, Social Safety Nets, Institutions | **Environmental**: Climate Change; **Societal**: Gender Focus (Women), Age Focus (Children) **Political**: Access to Services, Social Safety Nets, Infrastructure, Institutions **Public Health**: Nutrition | **Environmental**: Agriculture, Climate Change; **Societal**: Livelihoods, Gender; **Political**: Access to Services, Social Safety Nets, Infrastructure, Institutions **Public Health**: Diet, Nutrition | **Environmental**: Agriculture, Climate Change; **Societal**: Livelihoods, Gender, Age; **Economic**: Food Costs, Science & Technology; **Political**: Institutions; |

(continued)

(continued)

| International FAO agency | IMF | UNICEF | WFP | World Bank |
|---|---|---|---|---|
| Food resilience & COVID-19 | Yes—Publication: COVID-19 and Pacific Food system Resilience: Opportunities to build a robust response | Yes—April 2020 Policy Analysis for Sub-Saharan Africa | Yes—UNICEF Lives Upended Report: How the Pandemic is affecting South Asian Children | Yes—New programmes and other operations to address food resilience during COVID-19 Pandemic | Yes |

*Source* Authors

## REFERENCES

ADB. (2016). *Organizational resilience*. Asian Development Bank. Retrieved 22 January, 2022, from https://www.adb.org/sites/default/files/institutional-document/189052/organizational-resilience.pdf

ADB. (2020). *Timor-Leste: COVID-19 food security emergency response*. Asian Development Bank. Retrieved 22 January, 2022, from https://www.adb.org/projects/54310-001/main

ADB. (2009). *Operational plan for sustainable food security in Asia and the Pacific*. Asian Development Bank. Retrieved 22 January, 2022, from https://www.adb.org/documents/operational-plan-sustainable-food-security-asia-and-pacific

ADB. (2015). *Operational plan for agriculture and natural resources: Promoting sustainable food security In Asia and The Pacific 2015–2020*. Asian Development Bank. Retrieved 22 January, 2022, from https://www.adb.org/sites/default/files/institutional-document/175238/op-agriculture-natural-resources.pdf

ADB. (2019). *Rural development and food security forum 2019 highlights and takeaways*. Asian Development Bank. Retrieved 22 January, 2022, from https://www.adb.org/sites/default/files/publication/672751/rural-development-food-security-forum-2019.pdf

AfDB. (2013). *At the center of Africa's transformation strategy for 2013–2022*. African Development Bank. Retrieved 22 January, 2022, from https://www.afdb.org/fileadmin/uploads/afdb/Documents/Policy-Documents/AfDB_Strategy_for_2013-2022_-_At_the_Center_of_Africa's_Transformation.pdf

AfDB. (2016). *Feed Africa: Strategy for agricultural transformation in Africa 2016–2025*. Abidjan: African Development Bank Group. Retrieved 22 January, 2022, from https://www.afdb.org/fileadmin/uploads/afdb/Documents/Policy-Documents/Feed_Africa-Strategy-En.pdf

AfDB. (2015). *For a better life: Four rural projects which transformed the lives of people*. African Development Bank Group. Retrieved 22 January, 2022, from https://www.afdb.org/fileadmin/uploads/afdb/Documents/Generic-Documents/Agriculture_for_a_better_life.pdf

AfDB. (2020). *Feed Africa response to Covid-19—Brief*. African Development Bank Group. Retrieved 22 January, 2022, from https://www.afdb.org/en/documents/feed-africa-response-covid-19-brief

AfDB. (2021a). *Central African Republic—Resilience, food and nutrition security support project in Kémo and Ouaka Prefectures (PARSANKO)—Project Appraisal Report*. African Development Bank Group. Retrieved 22 January, 2022, from https://www.afdb.org/en/documents/central-african-republic-resilience-food-and-nutrition-security-support-project-kemo-and-ouaka-prefectures-parsanko-project-appraisal-report

AfDB. (2021b). *African development Bbnk Unveils strategy roadmap to safeguard food security against impacts of COVID-19—Press Release*. African Development Bank Group. Retrieved 22 January, 2022, from https://www.afdb.org/en/news-and-events/press-releases/african-development-bank-unveils-strategy-roadmap-safeguard-food-security-against-impacts-covid-19-36012

Ansah, I., Gardebroek, C., & Ihle, R. (2019). Resilience and household food security: A review of concepts, methodological approaches and empirical evidence. *Food Security, 11*(6), 1–17. https://doi.org/10.1007/s12571-019-00968-1

Barrett, C., & Constas, M. (2014). Toward a theory of development resilience for international development applications. *Proceedings of the National Academy of Sciences of the United States of America, 111*(40), 14625–14630. https://doi.org/10.1073/pnas.1320880111

Bahadur, A. V., Ibrahim, M. & Tanner, T. (2010). *The resilience renaissance? Unpacking of resilience for tackling climate change and disasters* (Strengthening Climate Resilience Discussion Paper 1). Institute of Development Studies. Retrieved 22 January, 2022, from https://opendocs.ids.ac.uk/opendocs/handle/20.500.12413/2368.

Béné, C. (2013). *Towards a quantifiable measure of resilience* (IDS Working Paper 434). Retrieved 22 January, 2022, from www.ids.ac.uk/publication/towards-a-quantifiable-measure-of-resilience.

Béné, C., Godfrey-Wood, R., Newsham, A. & Davies, M. (2012). *Resilience: New utopia or new tyranny? -Reflection about the potentials and limits of the concept of resilience in relation to vulnerability reduction programmes* (IDS working Paper 405). Institute of Development Studies.

Béné, C., Newsham, A., Davies, M., Ulrichs, M., & Godfrey-Wood, R. (2014). Review article: Resilience, poverty and development. *Journal of International Development, 26*(5), 598–623. https://doi.org/10.1002/jid.2992

Béné, C., Mehta, L., McGranahan, G., Cannon, T., Gupte, J., & Tanner, T. (2018). Resilience as a policy narrative: Potentials and limits in the context of urban planning. *Climate and Development, 10*(2), 116–133. https://doi.org/10.1080/17565529.2017.1301868

Béné, C., Headey, D., Haddad, L., & von Grebmer, K. (2016). Is resilience a useful concept in the context of food security and nutrition programmes? Some conceptual and practical considerations. *Food Security, 8*(1), 123–138. https://doi.org/10.1007/s12571-015-0526-x

Béné, C., Bakker, D., Chavarro, M. J., Even, B., Melo, J., & Sonneveld, A. (2021). Global assessment of the impacts of COVID-19 on food security. *Global Food Security, 31.* https://doi.org/10.1016/j.gfs.2021.100575

Berkes, F., & Folke, C. (1998). *Linking social and ecological systems: Management practices and social mechanisms for building resilience*. Cambridge University Press.

Bridgman, T., & Barry, D. (2002). Regulation is evil: An application of narrative policy analysis to regulatory debate in New Zealand. *Policy Sciences, 35*, 141–161. https://doi.org/10.1023/A:1016139804995

Brown, S. (2021). *Federal budget 2021: Foreign aid*. Open Canada. Retrieved 22 January, 2022, from https://opencanada.org/federal-budget-2021-foreign-aid/

Candel, J., Breeman, G., & Stiller, & S., Termeer, C. (2013). Disentangling the consensus frame of food security: The case of the EU Common Agricultural Policy reform debate. *Food Policy, 44*, 47–58. https://doi.org/10.1016/j.foodpol.2013.10.005

Carolan, M. (2013). *Reclaiming food security*. Taylor and Francis.

Chelleri, L. (2012). From the resilient city to urban resilience. A review essay on understanding and integrating the resilience perspective for urban systems. *Documents d'Anàlisi Geogràfica, 58*(2), 287–306. doi:https://doi.org/10.5565/rev/dag.175

Chatzky A., & McBride, J. (2020). *China's massive belt and road initiative*. Council on Foreign Relations. Retrieved 22 January, 2022, from https://www.cfr.org/backgrounder/chinas-massive-belt-and-road-initiative

Cissé, J., & Barrett, C. (2018). Estimating development resilience: A conditional moments-based approach. *Journal of Development Economics*, 135272–135284. S0304387818303511. https://doi.org/10.1016/j.jdeveco.2018.04.002

Cohen, J. (2020). China's vaccine gambit. *Science, 370*(6522), 1263–1267. https://doi.org/10.1126/science.370.6522.1263

Constas, M., Frankenberger, T. R., Hoddinott, J., Mock, N., Romano, D., Béné, C., & Maxwell, D. (2014). *A common analytical model for resilience measurement—Causal framework and methodological options* (Resilience Measurement Technical Working Group, FsiN Technical Series Paper No. 2). United Nations World Food Program and United Nations Food and Agriculture Organization.

Crawford, I. M. (1997). *Marketing research and information systems—Chapter 8 rapid rural appraisal*. Food & Agriculture Organisation Regional Office for Africa.

D'Errico, M., & Di Giuseppe, S. (2018). Resilience mobility in Uganda: A dynamic analysis. *World Development, 104*, 78–96. https://doi.org/10.1016/j.worlddev.2017.11.020

D'Errico, M., Romano, D., & Pietrelli, R. (2018). Household resilience to food insecurity: Evidence from Tanzania and Uganda. *Food Security, 10*(4). 1033–1054. https://doi.org/10.1007/s12571-018-0820-5

DFAT. (2017). *Market systems development: Operational guidance note*. Department of Foreign Affairs and Trade. Retrieved 22 January, 2022, from https://www.dfat.gov.au/sites/default/files/operational-guidance-note-market-systems-development.pdf

DFAT. (2020). *Australian NGO cooperation program (ANCP) thematic review: Agricultural development and food security*. Department of Foreign Affairs and Trade. Retrieved 22 January, 2022, from https://www.dfat.gov.au/sites/default/files/ancp-thematic-review-agricultural-development-and-food-security.pdf

DFAT. (2021a). *Agricultural development and food security initiatives*. Department of Foreign Affairs and Trade. Retrieved 22 January, 2022, from https://www.dfat.gov.au/aid/topics/investment-priorities/agricultural-development-and-food-security/initiatives

DFAT. (2021b). *Building resilience: humanitarian assistance, disaster risk reduction and social protection*. Canberra: Department of Foreign Affairs and Trade. Retrieved 22 January, 2022, from https://www.dfat.gov.au/aid/topics/investment-priorities/building-resilience

DfID. (2016). *What is resilience?* Department for International Development. Retrieved 22 January, 2022, from https://www.gov.uk/research-for-development-outputs/what-is-resilience

DfID. (2020). *Principles of health system resilience in the context of the COVID-19 response*. Department for International Development. Retrieved 22 January, 2022, from https://www.gov.uk/research-for-development-outputs/principles-of-health-systems-resilience-in-the-context-of-covid-19-response

DfID. (2015). *Global agriculture and food security program (GAFSP)*. Department for International Development. Retrieved 22 January, 2022, from https://www.gov.uk/government/publications/global-agriculture-and-food-security-program-gafsp

DfID. (2018). *Food systems in protracted crises: Strengthening resilience against shocks and conflicts. Overview of evidence on food systems in protracted crises and interventions to build resilient food systems against shocks*. Department for International Development. Retrieved 22 January, 2022, from https://www.gov.uk/research-for-development-outputs/food-systems-in-protracted-crises-strengthening-resilience-against-shocks-and-conflicts

EBRD. (2021a). *The EBRD's private sector for food security initiative*. European Bank for Reconstruction and Development. Retrieved 22 January, 2022, from https://www.ebrd.com/what-we-do/sectors-and-topics/private-sector-food-security-initiative.html

EBRD. (2021b). *What is Food Security?* London: European Bank for Reconstruction and Development. Retrieved 22 January, 2022, from https://www.ebrd.com/what-we-do/sectors-and-topics/food-security.html

EBRD. (2018). *Agribusiness sector strategy 2019–2023: Report on the invitation to the public to comment*. European Bank for Reconstruction and Development.

Retrieved 22 January, 2022, from https://www.ebrd.com/documents/agribusiness/agribusiness-strategy-comments.pdf?blobnocache=true

EBRD. (2019a). *EBRD financial report*. European Bank for Reconstruction and Development. Retrieved 22 January, 2022, from https://www.ebrd.com/news/publications/financial-report/ebrd-financial-report-2019a.html

EBRD. (2019b). *EBRD annual review*. European Bank for Reconstruction and Development. Retrieved 22 January, 2022, from https://www.ebrd.com/news/publications/annual-report/ebrd-annual-review-2019b.html

FAO. (2015). *Strengthening resilience for food security and nutrition: A conceptual framework for collaboration and partnership among the Rome-based agencies*. United Nations Food and Agricultural Organisation. Retrieved 22 January, 2022, from http://www.fao.org/fileadmin/user_upload/emergencies/docs/RBA%20Conceptual%20Framework%20for%20Resilience_Final%2023April2015.pdf

FAO. (2016). *RIMA-II: Resilience index measurement and analysis—II*. United Nations Food and Agricultural Organisation. Retrieved 22 January, 2022, from http://www.fao.org/3/a-i5665e.pdf

FAO. (2019a). *Increase the resilience of livelihoods*. United Nations Food and Agricultural Organisation. Retrieved 22 January, 2022, from http://www.fao.org/3/ca3922en/ca3922en.pdf

FAO. (2019b). *The director-general's medium term plan for 2018–21 (reviewed) and programme of work and budget 2020–21*. United Nations Food and Agricultural Organisation. Retrieved 22 January, 2022, from http://www.fao.org/3/my734en/my734en.pdf

FAO. (2020). *COVID-19 ad the impact on food security in the Near East and North Africa: How to respond?* United Nation's Food and Agricultural Organisation. Retrieved 22 January, 2022, from https://www.fao.org/policy-support/tools-and-publications/resources-details/en/c/1278069/

FAO. (2021a). *FAO COVID-19 response and recovery programme*. United Nation's Food and Agricultural Organisation. Retrieved 22 January, 2022, from http://www.fao.org/partnerships/resource-partners/covid-19/en/

FAO, IFAD, UNICEF, WFP & WHO. (2017). *The state of food security and nutrition In the world: Building resilience for peace and food security*. United Nations and World Bank Group. Retrieved 22 January, 2022, from https://docs.wfp.org/api/documents/WFP-0000022419/download/

FAO, IFAD, UNICEF, WFP & WHO. (2020). *The state of food security and nutrition in the world: Transforming food systems for affordable healthy diets*. United Nations and World Bank Group. Retrieved 22 January, 2022, from https://www.unicef.org/media/72676/file/SOFI-2020-full-report.pdf

FAO, IFAD, UNICEF, WFP & WHO. (2021). *The state of food security and nutrition in the world: Transforming food systems for food security, improved nutrition and affordable healthy diets for all*. United Nations and World

Bank Group. Retrieved 22 January, 2022, from https://www.fao.org/publications/sofi/2021/en/

Fischer, F., & Forester, J. (1993). Editors' introduction. In F. Fischer & J. Forester (Eds.), *The argumentative turn in policy analysis and planning* (pp. 1–17). Duke University Press.

George, S. (1976). *How the other half dies: The real reasons for world hunger*. Penguin.

George, S. (1988). *A fate worse than debt*. Grove Press.

Global Affairs Canada. (2018). *Canadian action on building resilience in the Caribbean*. Global Affairs Canada. Retrieved 22 January, 2022, from https://www.international.gc.ca/gac-amc/news-nouvelles/climate_change_action-action_changements_climatiques.aspx?lang=eng

Global Affairs Canada. (2020a). *Global affairs Canada page on global resilience partnership*. Global Affairs Canada. Retrieved 22 January, 2022, from https://www.globalresiliencepartnership.org/partner/global-affairs-canada/

Global Affairs Canada. (2020b). *Project browser*. Global Affairs Canada. Retrieved 22 January, 2022, from https://w05.international.gc.ca/projectbrowser-banqueprojets/?lang=eng

Global Affairs Canada. (2020c). *Canada working with CARICOM to build climate and economic resilience*. Global Affairs Canada. Retrieved 22 January, 2022, from https://www.canada.ca/en/global-affairs/news/2020c/02/backgrounder---canada-working-with-caricom-to-build-climate-and-economic-resilience.html

Global Resilience Partnership. (2018). *Centre new host for global resilience partnership secretariat*. Global Resilience Partnership. Retrieved 22 January, 2022, from https://www.stockholmresilience.org/news—events/general-news/2018-03-23-centre-new-host-for-global-resilience-partnership-secretariat.html#:~:text=In%202014%2C%20The%20Global%20Resilience,and%20prosperous%20future%20for%20all

Goldstein, B. E., Wessells, A. T., Lejano, R., & Butler, W. (2012). Narrating resilience: Transforming urban systems through collaborative storytelling. *Urban Studies, 52*(7), 1285–1393. https://doi.org/10.1177/0042098013505653

Hayter, T. (1971). *Aid as imperialism*. Penguin.

Harley, G., & Acheampong, Y. (2021). *Building back better from the crisis: What will it take for the poorest countries?* World Bank Blog—Voices. Retrieved 22 January, 2022, from https://blogs.worldbank.org/voices/building-back-better-crisis-what-will-it-take-poorest-countries

Holman J. (1984). Underdevelopment aid: A critique of the international monetary fund and the World Bank [electronic version]. *Berkeley Journal of Sociology* 29: 119–152. Retrieved 22 January, 2022, from http://www.jstor.org/stable/41035336.

IDB. (2015). *Review of the bank's support to agriculture, 2002–2014: Evidence from key thematic areas; food security in the LAC: Evaluation of the focus of the IDB's agriculture portfolio.* Inter-American Development Bank Group. Retrieved 22 January, 2022, from https://publications.iadb.org/publicati ons/english/document/Food-Security-in-LAC-Evaluation-of-the-Focus-of-the-IDB-Agriculture-Portfolio-Annex-1.pdf

IDB. (2019). *A framework and principles for climate resilience metrics in financing operations.* Inter-American Development Bank Group. Retrieved 22 January, 2022, from https://www.isdb.org/sites/default/files/media/documents/2021-01/Climate%20Resilience%20Metrics%20in%20Financing%20Operations%20-%20English%20Version.pdf

IDB. (2020a). *Inter-American development bank sustainability report 2019.* Inter-American Development Bank Group. Retrieved 22 January, 2022, from https://publications.iadb.org/en/inter-american-development-bank-sus tainability-report-2019

IDB. (2020b). *Ensuring food security in LAC in the context covid-19: Challenges and interventions.* Inter-American Development Bank Group. Retrieved 22 January, 2022, from https://publications.iadb.org/publications/english/doc ument/Ensuring-Food-Security-in-LAC-in-the-Context-of-Covid-19-Challe nges-and-Interventions.pdf

IDB. (2020c). *The unequal impact of the coronavirus pandemic: Evidence from seventeen developing countries.* Inter-American Development Bank Group. Retrieved 22 January, 2022, from https://publications.iadb.org/en/the-unequal-impact-of-the-coronavirus-pandemic-evidence-from-seventeen-develo ping-countries

IDB. (2020d). *The unequal burden of the pandemic: Why the fallout of covid-19 hits the poor hardest.* Inter-American Development Bank Group. Retrieved 22 January, 2022, from https://publications.iadb.org/en/unequal-burden-pandemic-why-fallout-covid-19-hits-poor-hardest

IDB. (2020e). *Agriculture and rural development.* Inter-American Development Bank Group. Retrieved 22 January, 2022, from https://www.iadb.org/en/sector/agriculture/overview

IDB. (2021). *Climate change action plan: 2021–2025.* Inter-American Development Bank Group. Retrieved 22 January, 2022, from https://publicati ons.iadb.org/publications/english/document/Inter-American-Development-Bank-Group-Climate-Change-Action-Plan-2021-2025.pdf

IMF. (2008). *Ensuring Food Security.* Finance and Development Magazine, December 2008. Washington, D.C.: International Monetary Fund. Retrieved 22 January, 2022, from https://www.imf.org/external/pubs/ft/fandd/2008/12/ivanic.htm

IMF. (2019). *Building resilience in developing countries vulnerable to large natural disasters.* International Monetary Fund. Retrieved 22 January, 2022, from https://www.imf.org/en/Publications/Policy-Papers/Issues/2019/06/24/Building-Resilience-in-Developing-Countries-Vulnerable-to-Large-Natural-Disasters-47020

IMF. (2020a, September). *Resilience: Healing the fractures*. Finance And Development Magazine. International Monetary Fund. Retrieved 22 January, 2022, from https://www.imf.org/external/pubs/ft/fandd/2020a/09/pdf/fd0920.pdf

IMF. (2020b, June). *Safeguarding Africa's food security in the age of COVID-19*. IMFBlog. International Monetary Fund. Retrieved 22 January, 2022, from https://blogs.imf.org/2020b/06/04/safeguarding-africas-food-security-in-the-age-of-covid-19/

IMF. (2020c, June). *Safeguarding food security in sub-Saharan Africa*. IMF Podcasts. International Monetary Fund. Retrieved 22 January, 2022, from https://www.imf.org/en/News/Podcasts/All-Podcasts/2020c/06/03/afr-reo-food

IMF. (2021). *Policy responses to COVID-19*. https://www.imf.org/en/Topics/imf-and-covid19/Policy-Responses-to-COVID-19. Washington, D.C.: International Monetary Fund. Retrieved 22 January, 2022, from https://www.imf.org/en/Topics/imf-and-covid19/Policy-Responses-to-COVID-19

Irish Aid. (2008). *Hunger task force: Report to the government of Ireland*. Irish Aid. Retrieved 22 January, 2022, from https://www.irishaid.ie/media/irishaid/allwebsitemedia/20newsandpublications/publicationpdfsenglish/hunger-task-force.pdf

Irish Aid. (2009). *Hunger envoy report*. Irish Aid. Retrieved 22 January, 2022, from https://www.irishaid.ie/media/irishaid/allwebsitemedia/20newsandpublications/publicationpdfsenglish/hunger-envoy-report.pdf

Irish Aid. (2014). *Building resilience for food and nutrition security*. Irish Aid. Retrieved 22 January, 2022, from https://www.irishaid.ie/news-publications/news/newsarchive/2014/may/building-resilience-for-food-nutrition-security/

Irish Aid. (2016). *Irish Aid policy brief: Building resilience*. Irish Aid. Retrieved 22 January, 2022, from https://www.climatelearningplatform.org/sites/default/files/resources/2016_irish_aid_policy_brief_-_building_resilience.pdf

Irish Aid. (2021). *Our priority areas: Hunger*. Irish Aid. Retrieved 22 January, 2022, from https://www.irishaid.ie/what-we-do/our-priority-areas/hunger/

Irish Aid. (Undated). *Launch of a better world: Ireland's policy for international development*. Irish Aid. Retrieved 22 January, 2022, from https://www.irishaid.ie/media/irishaid/aboutus/abetterworldirelandspolicyforinternationaldevelopment/A-Better-World-Irelands-Policy-for-International-Development.pdf

IsDB. (2018). *Agriculture & rural development sectors climate change adaptation guidance note*. Islamic Development Bank. Retrieved 22 January, 2022, from https://www.isdb.org/sites/default/files/media/documents/2020-09/Agriculture%20Sector%20Climate%20Change%20IsDB%20Guidance%20Note%20-Print%20version.pdf

IsDB. (Undated). *Agriculture*. Islamic Development Bank. Retrieved 22 January, 2022, from https://www.isdb.org/sector/agriculture

IsDB. (2019a). *Disaster risk management and resilience policy*. Islamic Development Bank. Retrieved 22 January, 2022, from https://www.isdb.org/sites/default/files/media/documents/2020-02/Disaster%20Risk%20Management%20Resilience%20Policy.pdf

IsDB. (2019b). *Fragility and resilience policy*. Islamic Development Bank. Retrieved 22 January, 2022, from https://www.isdb.org/sites/default/files/media/documents/2020-02/Fragility%20and%20Resilience%20Policy.pdf

IsDB. (2020a). *Agriculture and food security in OIC member countries 2020a*. Organisation of Islamic Cooperation. Retrieved 22 January, 2022, from https://www.sesric.org/files/article/748.pdf

IsDB. (2020b). *Rebuilding resilient agri-food chains for the future*. Organisation of Islamic Cooperation. Retrieved 22 January, 2022, from https://www.isdb.org/news/rebuilding-resilient-agri-food-value-chains-for-the-future

IsDB. (2020c). *The COVID-19 crisis and Islamic finance*. Organisation of Islamic Cooperation. Retrieved 22 January, 2022, from https://www.isdb.org/sites/default/files/media/documents/2020c-10/1.%20IsDB%20Group%20Report%20on%20Covid-19%20and%20Islamic%20Finance__FINAL.pdf

IsDB. (2021). *IsDB and IFAD sign $500 million agreement to tackle climate change and improve food security*. Islamic Development Bank Group. Retrieved 22 January, 2022, from https://www.isdb.org/news/isdb-and-ifad-sign-500-million-agreement-to-tackle-climate-change-and-improve-food-security

Jie, Y., Wallace, J. (2021). *What is China's belt and road initiative?* Chatham House. Retrieved 22 January, 2022, from https://www.chathamhouse.org/2021/09/what-chinas-belt-and-road-initiative-bri

Jordon, W. C. (1998). *The great famine*. Princeton University Press.

Kinealy, C. (2002). *The Great Irish famine: Impact, ideology and rebellion*. Palgrave.

Knippenberg, E., Jensen, N., & Constas, M. (2017). *Resilience, shocks and the dynamics of well-being evidence from Malawi*. Retrieved 22 January, 2022, from https://www.semanticscholar.org/paper/Resilience%2C-Shocks-and-the-Dynamics-of-Well-being-Knippenberg-Jensen/6facf0b44239fd283e98b9645c9c2127e2d46933

Kumar, K. (1993). Rapid appraisal methods. *World Bank regional and sectoral studies*. World Bank Group. Retrieved 22 January, 2022, from https://documents1.worldbank.org/curated/en/888741468740959563/pdf/multi0page.pdf

Lancet. (2021). Operation warp speed: Implications for global vaccine security. *The Lancet, 9*(8), 1017–1021. https://doi.org/10.1016/S2214-109 X(21)00140-6

Lang T., Rayner G., & Kaelin E. (2006). *The food industry, diet, physical activity and health: A review of reported commitments and practice of 25 of the world's largest food companies*. Report to the World Health Organisation. London: City University Centre for Food Policy.

Lang, T., & Barling, D. (2012). Food security and food sustainability: Reformulating the debate. *The Geographical Journal, 178*, 313–326.

Lang, T. (2020). *Feeding Britain: Our food problems and how to fix them*. Pelican.

Lawrence, G., Richards, C., & Lyons, K. (2013). Food security in Australia in an era of neoliberalism, productivism and climate change. *Journal of Rural Studies, 29*, 30–39.

Leach, M. (2008). *Re-framing resilience: A symposium report* (STEPS Working Paper 13). Brighton: Institute of Development studies.

Lee, S. T. (2021). Vaccine diplomacy: Nation branding and China's COVID-19 soft power play. *Place Brand Public Diplomacy*. https://doi.org/10.1057/s41254-021-00224-4

Maye, D., & Kirwan, J. (2013). Food security: A fractured consensus. *Journal of Rural Studies, 29*, 1–6.

Maxwell, D., Watkins, B., Wheeler, R., & Collins, G. (2003). *The coping strategies index: A tool for rapidly measuring food security and the impact of food programmes in emergencies*. CARE/WFP. Retrieved 22 January, 2022, from https://www.researchgate.net/profile/Ben-Watkins-4/publication/265059159_The_Coping_Strategies_Index_A_tool_for_rapidly_measuring_food_security_and_the_impact_of_food_aid_programmes_in_emergencies_1/links/54d9d2c70cf24647581f8437/The-Coping-Strategies-Index-A-tool-for-rapidly-measuring-food-security-and-the-impact-of-food-aid-programmes-in-emergencies-1.pdf

McMichael, P. (2012). *Development and social change: A global perspective*. Pine Forge Press.

McNall, M., & Foster-Fishman, P. G. (2007). Methods of rapid evaluation, assessment, and appraisal. *American Journal of Evaluation, 28*, 151–168.

Mitchell A. (2020). Merging DfID with foreign office is an act of vandalism. *The Observer*. London: Guardian News and Media. 21 June. Retrieved 22 January, 2022, from https://www.theguardian.com/politics/2020/jun/20/merging-dfid-with-foreign-office-is-an-act-of-vandalism-says-andrew-mitchell

Mokyr J & Gráda, C. Ó. (2002). Famine disease and famine mortality: lessons from the Irish experience, 1845–1850. In T. Dyson & C. Ó. Gráda (Eds.), *Famine demography: Perspectives from past and present* (pp. 19–43). Oxford University Press.

Moser, C., Norton, A., Stein, A. & Georgieva, S. (2010). *Pro-poor adaptation to climate change in urban centers: Case studies of vulnerability and resilience in Kenya and Nicaragua* (Report No. 54947-GLB). World Bank, Social Development Department.

Nestle, M. (2006). Food industry and health: Mostly promises, little action. *The Lancet, 368*, 564–565.

Pearson, L. (2013). In search of resilient and sustainable cities: Prefatory remarks. *Ecological Economics*, 86222–86223. S0921800912004612. https://doi.org/10.1016/j.ecolecon.2012.11.020

Raikes, P. L. (1988). *Modernising hunger: Famine, food surplus & farm policy in the EEC & Africa*. Catholic Institute for International Relations in collaboration with James Currey.

Rayner G., & Lang, T. (2012). *Ecological public health*. Abingdon: Routledge Earthscan.

REAL. (2018). *Resilience measurement practical guidance series: Guidance Note 5—Design and planning for resilience monitoring and evaluation at the activity level*. USAid. Retrieved 22 January, 2022, from https://www.fsnnetwork.org/resource/resilience-measurement-practical-guidance-series-guidance-note-5-design-and-planning

Roe, E. M. (1994). *Narrative policy analysis: Theory and practice*. Duke University Press.

Searchinger, T., Waite, R., Hanson, C., Ranganathan, J., Matthews, E. (2018). *Creating a sustainable food future: A menu of solutions to feed nearly 10 billion people by 2050*. World Resources Institute.

Serfilippi, E., & Ramnath, G. (2018). Resilience measurement and conceptual frameworks: A review of the literature. *Annals of Public and Cooperative Economics, 89*, 645–664.

SIDA. (2013a). *Risk reduction and resilience*. Swedish International Development Agency. Retrieved 22 January, 2022, from https://www.sida.se/English/publications/109872/risk-reduction-and-resilience/

SIDA. (2013b). *Enhancing resilience of coastal and marine biosphere reserves in Vietnam*. Swedish International Development Agency. Retrieved 22 January, 2022, from https://www.sida.se/English/publications/109836/enhancing-resilience-of-coastal-and-marine-biosphere-reserves-in-vietnam/

SIDA. (2015). *Portfolio overview: Food security and agriculture*. Swedish International Development Agency. Retrieved 22 January, 2022, from https://publikationer.sida.se/contentassets/0c87f63535074aafa53008aac5611af8/21661.pdf

SIDA. (2016). *Food security and agriculture secure food for a growing world and end hunger*. Swedish International Development Agency. Retrieved 22 January, 2022, from https://www.sida.se/contentassets/7083a0e371dc416eb35497c081d0aaf0/22443.pdf

SIDA. (2017). *Climate smart agriculture*. Swedish International Development Agency. Retrieved 22 January, 2022, from https://www.sida.se/contentas sets/faa0179914204ac48931adb1f93d7d2b/22115.pdf

SIDA. (2021). *SIDA's Response to COVID-19*. Stockholm: Swedish International Development Agency. Retrieved 22 January, 2022, from https://www.sida.se/English/how-we-work/our-fields-of-work/health/sidas-response-to-covid-19

Shaw, D. J. (2007). *World food security: A history since 1945*. Palgrave Macmillan.

Su, Z., McDonnell, D., Li, X., Bennett, B., Segalo, S., Abbas, J., Chesmehzangi, A., Xiang, Y. T. (2021). COVID-19 vaccine donations—Vaccine empathy or vaccine diplomacy? A narrative literature review. *Vaccines: Uptake and Equity in Times of the COVID-19 Pandemic* Special Issue. https://doi.org/10.3390/vaccines9091024

Tendall, D. M., Joerin, J., Kopainsky, B., Edwards, P., Shreck, A., Le, Q. B., Kruetli, P., Grant, M., & Six, J. (2015). Food system resilience: Defining the concept. *Global Food Security, 6*, 17–23.

Torres, I., Lopez-Cevallos, D., Artaza, O., Profeta, B., Kang, J., & Machado, C. V. (2021). Vaccine scarcity in LMICs is a failure of global solidarity and multilateral instruments. *The Lancet, 397*(10287), 1804. https://doi.org/10.1016/S0140-6736(21)00893-X

Tudge, C. (1977). *The famine business*. Faber and Faber.

Twigg, J. (2007). *Characteristics of a disaster-resilient community*. Department for International Development, DFID DRR Interagency Coordination Group.

UN. (2020). *Policy brief: The impact of COVID-19 on food security and nutrition*. United Nations. Retrieved 22 January, 2022, from https://www.un.org/sites/un2.un.org/files/sg_policy_brief_on_covid_impact_on_food_security.pdf

UNEP. (2020). *The global biodiversity outlook 5 (GBO-5)*. United Nation's Environment Programme. Retrieved 22 January, 2022, from https://www.cbd.int/gbo5

UNICEF. (Undated). *Resilience, humanitarian assistance and social protection for children in Europe and Central Asia*. United Nation's Children's Fund. Retrieved 22 January, 2022, from https://www.unicef.org/eca/media/2671/file/Social_Protection2.pdf

UNICEF. (2020a). *Build resilience in children to help them stay safe on social media*. United Nation's Children's Fund. Retrieved 22 January, 2022, from https://www.unicef.org/eap/press-releases/build-resilience-children-help-them-stay-safe-social-media

UNICEF. (2020b). *Building food and nutrition resilience in Quezon City: A case study on integrated food systems*. United Nation's Children's Fund. Retrieved 22 January, 2022, from https://www.unicef.org/media/89406/file/Building-food-nutrition-resilience-Quezon-City.pdf

UNICEF. (2020c). *Lives Upended: How COVID-19 threatens the futures of 600 million South Asian children*. United Nation's Children's Fund. Retrieved 22 January, 2022, from https://www.unicef.org/rosa/reports/lives-upended

UNICEF. (2020d). *UNICEF-WHO-World Bank: Joint child malnutrition estimates —Evels and trends* (2020d ed.). United Nations and World Bank Group. Retrieved 22 January, 2022, from https://data.unicef.org/resources/jme-report-2020d/

UNICEF. (2021). *UNICEF integrated budget, 2022–2025*. United Nation's Children's Fund. Retrieved 22 January, 2022, from https://www.unicef.org/executiveboard/media/7546/file/2021_SRS-Item_12-Integrated_budget-T.Asare-Presentation-EN-2021.09.02.pdf

UNSCEB. (2017). *Adopting an analytical framework on risk and resilience: a proposal for a more proactive, coordinated and effective United Nations action. Prepared by a task team led by the World Food Programme*. United Nations System Chief Executives Board for Coordination. https://unsceb.org/sites/default/files/imported_files/RnR_0.pdf

Upton, J., Cissé, J., & Barrett, C. (2016). Food security as resilience: Reconciling definition and measurement. *Agricultural Economics, 47*, 135–147.

USAID. (Undated). *The resilience agenda: Measuring resilience in USAID*. USAID. Retrieved 22 January, 2022, from https://pdf.usaid.gov/pdf_docs/pdacx975.pdf

USAID. (2014). *Multi-sectoral nutrition strategy 2014–2025*. USAID. Retrieved 22 January, 2022, from https://www.usaid.gov/sites/default/files/documents/1867/USAID_Nutrition_Strategy_5-09_508.pdf

USAID. (2017). *Global fod security strategy technical guidance objective 2: Strengthened resilience among people and systems*. USAID. Retrieved 22 January, 2022, from https://cg-281711fb-71ea-422c-b02c-ef79f539e9d2.s3.us-gov-west-1.amazonaws.com/uploads/2018/03/GFSS_TechnicalGuidance_Resilience.pdf

USAID. (2020). *Bureau for resilience and food security*. USAID. Retrieved 22 January, 2022, from https://www.usaid.gov/who-we-are/organization/bureaus/bureau-resilience-and-food-security

USAID. (2021a). *Responding to COVID-19's impact on resilience and food security*. USAID. Retrieved 22 January, 2022, from https://www.usaid.gov/who-we-are/organization/bureaus/bureau-resilience-and-food-security/responding-to-covid-19-impact-on-resilience-and-food-security

USAID. (2021b). *U.S. government global food security strategy implementation report of 2020*. USAID. Retrieved 22 January, 2022, from https://cg-281711fb-71ea-422c-b02c-ef79f539e9d2.s3.us-gov-west-1.amazonaws.com/uploads/2021b/02/Final-GFSS-Implementation-Report-021721.pdf

USAID. (2021c). *U.S. Government global food security strategy (fiscal year 2022–2026)*. USAID. Retrieved 22 January, 2022, from https://cg-281 711fb-71ea-422c-b02c-ef79f539e9d2.s3.us-gov-west-1.amazonaws.com/uploads/2021c/10/Global-Food-Security-Strategy-FY22-26_508C.pdf

Watkins, K. (2001). *Rigged rules, double standards*. Oxfam Publications.

Willett, W., Rockström, J., Loken, B., Springmann, M., Lang, T., Vermeulen, S., Garnett, T., et al. (2019). Food in the anthropocene: The EAT–Lancet Commission on healthy diets from sustainable food systems. *The Lancet, 393*(10170), 447–492.

WFP. (2015). *Policy on building resilience for food security and nutrition*. United Nations World Food Programme. Retrieved 22 January, 2022, from https://www.wfp.org/publications/policy-building-resilience-food-security-and-nutrition

WFP. (2016). *Building resilience for zero hunger—Zimbabwe*. United Nations World Food Programme. Retrieved 22 January, 2022, from https://www.wfp.org/operations/200944-building-resilience-zero-hunger

WFP. (2017). *The three-pronged approach (3PA) factsheet*. United Nations World Food Programme. Retrieved 22 January, 2022, from https://www.wfp.org/publications/2017-three-pronged-approach-3pa-factsheet

WFP. (2019a). *WFP Regional Resilience Framework: North Africa, Middle East, Central Asia and Eastern European Region*. Rome: United Nations World Food Programme. Retrieved 22 January, 2022, from https://docs.wfp.org/api/documents/WFP-0000112565/download/

WFP. (2019b). *Rome-based agencies—Canada resilience initiative—2018 Report* United Nations World Food Programme. Retrieved 22 January, 2022, from https://www.wfp.org/publications/rome-based-agencies-canada-resilience-initiative-2018-annual-report

WFP. (2019c). *Livelihoods—Building resilience*. United Nations World Food Programme. Retrieved 22 January, 2022, from https://www.wfp.org/publications/2019c-livelihoods-building-resilience

WFP. (2020). *Resilience programming*. United Nation's World Food Programme. Retrieved 22 January, 2022, from https://www.wfp.org/resilience-programming

WFP. (2021a). *COVID-19 pandemic*. United Nation's World Food Programme. Retrieved 22 January, 2022, from https://www.wfp.org/emergencies/covid-19-pandemic

WFP. (2021b). *Hunger pandemic*. United Nation's World Food Programme. Retrieved 22 January, 2022, from https://www.wfp.org/stories/hunger-pandemic-food-security-report-confirms-wfps-worst-fears

World Bank. (2013). *Building resilience: Integrating climate and disaster risk into development: The World Bank group experience*. World Bank Group.

Retrieved 22 January, 2022, from https://openknowledge.worldbank.org/handle/10986/16639

World Bank. (2016). *On the road to resilience*. World Bank Group. Retrieved 22 January, 2022, from https://www.worldbank.org/en/programs/road-to-resilience

World Bank. (2017). *Operational guidance for monitoring and evaluation (M&E) in climate and disaster resilience-building operations*. Washington, D.C.: World Bank Group. Retrieved 22 January, 2022, from https://documents1.worldbank.org/curated/en/692091513937457908/pdf/122226-ReME-Operational-Guidance-Note-External-FINAL.pdf

World Bank. (2020). *Building resilience to food and nutrition insecurity shocks*. World Bank Group. Retrieved 22 January, 2022, from https://projects.worldbank.org/en/projects-operations/project-detail/P155475

World Bank. (2021a). *Opportunities for climate finance in the livestock sector: Removing obstacles and realizing potential report*. World Bank Group. Retrieved 22 January, 2022, from https://openknowledge.worldbank.org/handle/10986/35495

World Bank. (2021b). *Food security and COVID-19 brief*. World Bank Group. Retrieved 22 January, 2022, from https://www.worldbank.org/en/topic/agriculture/brief/food-security-and-covid-19

World Bank. (2021c). *Resilience rating system: A methodology for building and tracking resilience to climate change*. World Bank Group. Retrieved 22 January, 2022, from https://openknowledge.worldbank.org/handle/10986/35039

World Bank. (2021d). *The World Bank group's response to the COVID-19 (coronavirus) pandemic*. World Bank Group. Retrieved 22 January, 2022, from https://www.worldbank.org/en/who-we-are/news/coronavirus-covid19

World Bank, (2022). *Stagflation risk rises amid sharp slowdown in growth*. Press Release. Retrieved 4 January, 2023, from https://www.worldbank.org/en/news/press-release/2022/06/07/stagflation-riskrises-amid-sharp-slowdown-in-growth-energy-markets

**Open Access** This chapter is licensed under the terms of the Creative Commons Attribution 4.0 International License (http://creativecommons.org/licenses/by/4.0/), which permits use, sharing, adaptation, distribution and reproduction in any medium or format, as long as you give appropriate credit to the original author(s) and the source, provide a link to the Creative Commons license and indicate if changes were made.

The images or other third party material in this chapter are included in the chapter's Creative Commons license, unless indicated otherwise in a credit line to the material. If material is not included in the chapter's Creative Commons license and your intended use is not permitted by statutory regulation or exceeds the permitted use, you will need to obtain permission directly from the copyright holder.

CHAPTER 5

# Food Security and Resilience: The Potential for Coherence and the Reality of Fragmented Applications in Policy and Research

*Mark A. Constas*

## INTRODUCTION

Climate change dynamics, stressed agro-ecological systems, political conflicts, mass migrations, and volatile economic conditions are commonly viewed as risks that threaten food security and damage the elements of a given food system on which food security depends (Bernard de Raymond et al., 2021; Hasegawa et al., 2018; Martin-Shields & Stojetz, 2019; Von Braun, 2009). In studies that emphasize

---

M. A. Constas (✉)
Charles H. Dyson School of Applied Economics and Management, Cornell University, Ithaca, NY, USA
e-mail: mark.constas@cornell.edu

Tata-Cornell Institute for Agriculture and Nutrition, Cornell University, Ithaca, NY, USA

Cornell Atkinson Center for Sustainability, Cornell University, Ithaca, NY, USA

© The Author(s) 2023
C. Béné and S. Devereux (eds.), *Resilience and Food Security in a Food Systems Context*, Palgrave Studies in Agricultural Economics and Food Policy, https://doi.org/10.1007/978-3-031-23535-1_5

capacity-building to enable households and communities recover from risk-exposure events, the concept of resilience has frequently been paired with that of food security. This is evidenced by large-scale food resilience-oriented food-security initiatives led by major donors, including the United States Agency for International Development (USAID), the United Kingdom's Department for International Development (DFID),[1] and the European Union (EU).[2] As a topic of research, the surge of activity occurring at the intersection of food security and resilience has been marked by the recent publication of a series of review articles in the peer-reviewed literature (e.g., Ansah et al., 2019; Béné, 2020; Bullock et al., 2017; Serfilippi & Ranmath 2018; Tendall, 2015). As an indicator of broad adoption, the appearance of review articles suggests a high level of sustained activity for given concept (Keathley-Herring et al., 2016). This appears to be the case for resilience, suggesting that a considerable body of work has been amassed.

Although resilience has been cited as having a long and varied history (see Alexander, 2013; Cicchetti & Garmezy, 1993; Vernon, 2004), that history was not initially connected to development. While seminal discussions of food security are based in agricultural production (Botero, 2012/1588; Malthus, 1798), interest in resilience has separate origins. Early work on resilience focused on subjects such as optics and acoustics in connection with reflection and echoes, respectively (Bacon, 1625). Other examples of early work on resilience by physicists focused on the elasticity of gases (Gott, 1670) while mechanical engineers modeled the rigidity and ductility of steel beams (Rankine, 1858).[3] What some have called the renaissance of resilience (Bahadur et al, 2010; Béné et al., 2016) is grounded in the frequently cited work of Holling (1973), who studied the resilience of ecological systems.[4] Not surprisingly, the initial uptake of

---

[1] In 2020, DFID was merged with the Foreign and Commonwealth Office to create the Foreign, Commonwealth and Development Office (FCDO).

[2] Examples of early initiatives that helped to build this connection include USAID's Partnership for Resilience and Economic Growth (PREG) and Resilient, Inclusive, and Sustainable Environments (RISE) Challenge, DFID's Building Resilience and Adaptation to Climate Extremes and Disasters (BRACED), and the EU's Global Alliance for Resilience Initiative and Supporting the Horn of Africa's Resilience.

[3] For a detailed history of resilience as a concept, see Gößling-Reisemann et al. (2018).

[4] While focused mainly on resistance, earlier work by Lewontin (1969) on the stability of ecological systems may also be viewed as a seminal reference for resilience.

resilience in the period following Holling's seminal work is most evident among those working on ecological problems (Pimm, 1984; Walker, 1992) or environmental problems (Timmerman, 1981; Tobin, 1999). While physicists and engineers have lfong made use of resilience, applications to humans are relatively recent. As noted by Gößling-Reisemann et al. (2018), psychological ideas such as "mental elasticity" (Miles, 1935) and "psychological equilibration" (Bentley, 1938) first appeared well in the twentieth century.

The application of resilience to food security is even more recent than its application in psychology, with the work of Pingali et al. (2005) cited frequently as one of the earlier contributions. The record of empirical research is extraordinarily young in comparison with that devoted to food security. In a recent review of the resilience of local food systems (Béné, 2020), for example, the earliest resilience-related study cited was Fafchamps and Lund (2003), who examined how households in rural Philippines use risk-sharing networks to cope with income and expenditure shocks. One empirical study that is regularly cited as marking the early stages of resilience work in development is Alinovi et al. (2008). A paper by Ansah et al. (2019) that reviewed resilience also cited Alinovi et al. (2008) but included Keil et al. (2008) as another resilience-focused paper that was published in the same year.

While the concept of food security has circulated far longer than the concept of resilience, differences are apparent in how each is featured in development work. As a vital indicator of well-being, food security is at the center of work in development and humanitarian assistance. This is evidenced, for instance, by the fact that Zero Hunger is one of 17 goals that comprise the Sustainable Development Goals of the 2030 Agenda. Whereas food security is an outcome, the focus on resilience capacities helps us understand how threats to and losses of food security can be managed (Constas et al., 2014). From a modeling perspective, the resilience of an outcome like food security can be presented as a variable to be predicted. Resilience can also be treated as a dependent variable in cases where the goal of an intervention is to build resilience capacity (Béné et al., 2012; d'Errico et al., 2020).

Assumptions about the benefits of combining food security and resilience provide no guarantee of complete application or coherent integration. This uncertainty is tied in part to the fact that each concept is inherently multidimensional with multi-level manifestations. The conception of food security that has thus far been most broadly accepted is structured around four components or pillars (FAO, 1996): availability, access, utilization, and stability. This conceptualization, which presents its own challenges, does not make explicit reference to food sovereignty or agency. As an expansion of and reaction to conventional notions of food security, food sovereignty emphasizes the right to food. Moving beyond the 1996 World Food Summit (WFS), the Nyéléni Declaration for Food Sovereignty introduced six pillars that could be considered as central to food security (Via Campesina, 2007).[5] This suggests that describing food security in terms of the original four pillars or dimensions provides at best a partial account of food security.

By focusing on absorptive capacities, adaptive capacities, and transformative capacities, resilience also exhibits multidimensionality (Béné et al., 2012, 2014, 2016; Walker, et al., 2004). Other sources have included anticipatory capacity (Weingärtner et al., 2020) or resistance and adaptive preference (Béné & Doyen, 2018) as fourth and fifth capacities, adding further complexity to the concept of resilience. Like the call for food sovereignty in connection with food security, questions about social inclusivity and equity have also been raised when contemplating resilience (Forsyth, 2018). Examples of such work, which might be viewed as extensions of Sen's (1999) capability approach, can be found in Bohle et al. (2009) and Coulthard (2012).[6]

Additional dimensions of resilience can be added if one considers the scale of (idiosyncratic or covariate) shocks and their origins (e.g., weather, social unrest or conflict, failed government, and weak institutions). Furthermore, both food security and resilience may be observed and studied at multiple levels/scales. Policies, programs, and units of analysis for measurement and research may focus on individuals, households, communities, or higher-level units. Extending questions about scale beyond households or communities, the concept of food systems has steadily gained momentum. While reviews of the literature (Pingali & Sunder, 2017; Reardon & Timmer, 2012) have demonstrated that a focus on food systems does not represent a particularly new strand of discussion

---

[5] As an expansion of and reaction to conventional notions of food security, the concept of food sovereignty emphasizes the right to food in virtue of the six pillars of the Nyéléni Declaration for Food Sovereignty (see Via Campesina, 2007; HLPE, 2017). These pillars are discussed in a subsequent section of this chapter.

[6] A concise discussion of early work that considers the overlap between resilience and power may be found in Béné et al. (2014).

in food-security, the United Nations' Food Systems Summit in September 2021 has elevated interest and spurred activity.

Underlining the importance of commensurability (Kuhn, 1982) and the consistency of associated propositions (Thagard, 2000, 2007), coherence is fundamental to advancing knowledge within a given domain of understanding. As separate ambitions, the desire to achieve coherence *within* food security or resilience presents a pair of significant challenges. Yet, the desire to achieve coherence *between* food security and resilience requires an even more ambitious effort. Despite the challenges involved in building food security–resilience connections, it appears that applications in policy, programming, and research have taken hold (Brown, 2016) and continue to expand (UNDRR, 2021).

The conceptual complexity that exists *within* and *between* the concepts of food security and resilience raises questions about the nature of what might be possible. With this in mind, in the present xhapter I sought to: (1) *Explore the potential for conceptual coherence*—how might the concepts of food security and resilience be integrated into a coherent form? (2) *Explore applications*—how have the concepts of food security and resilience appeared together in policy and in research? To examine how food security and resilience might be integrated, a conceptual model is offered that suggests points of intersection. The model is then used to explore applications through examination of a high-profile policy document and a sample of peer-reviewed articles, using lexical analysis and content-analysis methods. The lexical analysis portion of the study, which counted joint occurrences of food security and resilience, revealed simple usage trends over time. The content analysis examined how the concept of resilience has been applied to that of food security.

## Toward an Integrated Conceptual Model of Food Security Resilience

To motivate the effort to build a conceptual model that promotes integration between food security and resilience, two questions are introduced: (1) *How are distinct aspects of food security—based on distinct perspectives of food security—affected by shocks and stressors at varying scales?* (2) *How do distinct resilience capacities help households or higher-level units (e.g., communities, regions, countries) anticipate, manage, and recover from exposure to shocks and stressors?* Considered together, these two questions help in identifying intersections where integration between food security and resilience is possible. To add substance to these questions, the concepts of food security and resilience are considered. The aim of this brief discussion is to identify the dimensions around which a model can be constructed.

### *Food Security: From Pillars to Systems and the Importance of Agency*

The emergence and evolution of food security as a focal point for development work is noteworthy in its own right (see Shaw, 2007). The objective here is to briefly summarize three perspectives on food security—the 1996 WFS perspective, the food sovereignty perspective, and the food systems perspective. These three perspectives represent, respectively, the dominant approach, a counterpoint to the dominant approach that introduces agency as an important feature of food security, and an approach that captures a broad array of elements that contribute to food security.

As noted above, the 1996 WFS definition of food security (i.e., availability, access, utilization, and stability) (FAO, 2008a, 2008b) is perhaps the most familiar. Availability is a function of agricultural production, food trade, and net food stocks when considering import–export dynamics. Food accessibility reflects resources provided by food that is produced for self-consumption, market access, and the ability to purchase food at the household level (also known as affordability). Food utilization highlights the importance of feeding practices, including preparation and consumption behaviors. As a cross-cutting pillar, stability draws attention to periodic fluctuations in availability, access, and utilization.

Although the four pillars of food security have been used widely, agency and power are not explicit parts of the 1996 WFS conception. A more recent FAO discussion document points out this gap (Gordillo, 2013, p. vi), stating that "[t]he concept of food security—adopted by FAO member states—is somehow a neutral concept in terms of power relations. It does not prejudge the concentration of economic power in the different links of the food chain and in the international food trade, or the ownership of key means of production such as land, or more contemporarily, access to information." Gordillo then added that "the concept of food sovereignty begins precisely with noting the asymmetry of power" (Gordillo, 2013, p. vi). Attention to food sovereignty as an element of food security has been discussed on many occasions. More than 20 years before the 1996 WFS, for example, the 1974 World

Conference on food insisted that "every man, woman and child has the inalienable right to be free from hunger and malnutrition" (FAO, 1974). By 1996, the version of food security put forth by WFS did not highlight the importance of food sovereignty or power as an element of food security. Food sovereignty was emphasized again in 2006 (FAO, 2006) in a report titled *The Right to Food Guidelines: Information Papers and Case Studies*. It is unclear why interest in food sovereignty and power relations as an element of food security within FAO's writings has been inconsistent.[7]

As an overarching framework, the food systems perspective can serve as a heuristic device that may help in unifying the various elements on which food security depends. A conceptual framework of food systems offered by HLPE describes how the food security pillars of availability, access, and utilization are supported by three main food systems components—*food supply chains*, *food environments*, and *consumer behaviors* (HLPE, 2017). The *food supply chain* includes production systems, storage and distribution, processing and packaging, and retail trade and markets. The *food environment* reflects the role of physical access to food (e.g., distance to markets), economic access (e.g., affordability), promotion (e.g., food messaging and advertising), and food quality and safety. Acknowledging the importance of context, the HLPE also describes drivers of food systems. Food systems drivers, which can influence one or more components of a food system, include biophysical and environmental drivers, innovation and technology drivers, political and economic factors, socio-cultural drivers, and demographic drivers.

While not featured in Fig. 5.1, a more recent model offered by the HLPE (2020) includes the right to food. As pointed out in a review of food systems frameworks (Brouwer et al., 2020), a wide selection of food systems representations can be found. In the same review, Brouwer et al. (2020) summarize the drivers, components, and outcomes associated with 32 reports and studies that feature food systems. While considerably diversity can be found across the reviewed systems, most representations are similar (or close) to the HLPE's conception that focuses attention on the inputs to and processes involved in food supply chains, food environments, and consumption (consumer behavior). Introducing greater

---

[7] The addition of sustainability and agency as core components of food security, noted in a report by the High-Level of Panel of Experts (HLPE), may indicate that food security is a concept that is undergoing a transition (HLPE, 2020).

complexity, some have called for an expanded conception of food systems. Caron et al. (2018, p. 38), for example, note that a food system should be conceived broadly as a "nexus that links food security, nutrition, and human health, the viability of ecosystems, climate change, and social justice." One of the compelling qualities of a food systems perspective is that it approaches the problem of food insecurity in a comprehensive manner.

To summarize the above discussion, food security can be conceptualized according to one or some combination of three versions. The first, and perhaps most widely subscribed to, is based on the 1996 WFS. This version (henceforth referred to as "*WFS Food Security*") is represented by the four dimensions of availability, accessibility, utilization, and stability. A second version of food security is based in the notion of food sovereignty. With its emphasis on power and agency, the food sovereignty perspective on food security emphasizes the right to food, the role of local food providers, and tension between community needs and corporatist approaches to food production. Finally, the food systems perspective represents the aspiration to integrate all factors and processes that contribute to food security, including everything from ecological inputs to production methods, post-harvest food processing, supply chains, markets, and consumption.

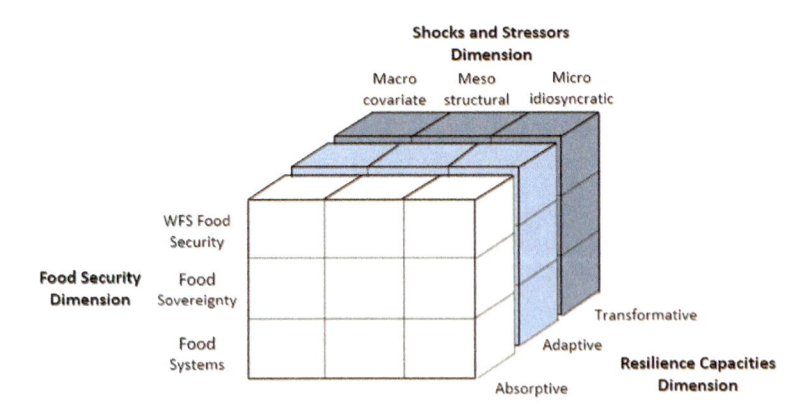

**Fig. 5.1** Integrated food security and resilience model (*Source* Author)

### *Resilience: Shocks, Stressors, and Resilience Capacities*

As noted earlier, awareness of the complex risk landscape faced by the world's most vulnerable populations explains largely why resilience has captured the attention of the community of donors, policymakers, and nongovernmental organizations working in development assistance or humanitarian aid. Drawing attention to this awareness highlights the chief defining quality of resilience work—increased sensitivity to shocks and stressors (Choularton et al., 2015; Zseleczky & Yosef, 2014).

While the phrase *"shocks and stressors"* is sometimes invoked without definition, we must distinguish between shocks and stressors as well as between types of shocks. Shocks, which are thought to be more pronounced and more conspicuous than stressors, may arise from various sources (e.g., weather, political conflicts, earthquakes, economic crises, and health shocks—including epidemics and pandemics). Shocks may have widespread or macro-level effects that threaten the welfare of large geographic areas or a significant proportion of a population (these are known as covariate shocks). Shocks at the micro-level are typically highly localized (these are known as idiosyncratic shocks), affecting individual households (e.g., the death of a family member). While the effects of stressors are often viewed as less pronounced than those of shocks, their negative impacts may be just as corrosive to household and community welfare. Stressors include events or factors such as a family member's prolonged illness, poorly functioning governments and weak institutions, lack of physical infrastructure, inadequate provision of education, sociopolitical unrest, ethnic tensions, and gender inequality. Such stressors frequently undermine food security and general well-being.

Although it is a long-standing practice to describe shocks and stressors as disturbances that are categorized as local covariate or idiosyncratic (e.g., Deaton, 1997; Dercon, 2002), it is also useful to understand that shocks and stressors can be characterized as meso-level disturbances. Such shocks and stressors may, for example, affect villages or districts. It is important to highlight here that shocks and stressors can exert negative impacts at one or more levels, from microscale to mesoscale to macroscale effects.

Awareness of a more complex configuration of shocks and stressors has spurred an array of funded programs designed to help build households and help communities manage and recover from various risk exposure events and/ conditions that undermine well-being.[8] Such programs and interventions have been designed to build absorptive, transformative capacities. Consistent with earlier work by Walker et al. (2004), Béné et al., (2012) described absorptive capacity as a factor that enables households (or other entities) to persist in the face of shocks and stressors, while adaptive capacity makes it easier to adjust one's livelihood or to maintain an acceptable level of food security. Béné et al., (2014, 2016, 2012) also highlighted the importance of transformative capacity as a factor that enables significant shifts in governance, policies, systems of social protection, and/or systemic change. Béné et al. (2014) noted that resilience capacities may overlap in a temporal sense; they may be drawn upon to manage a given shock or stressor or a collection of shocks or stressors.

### *Toward an Integrated Model of Food Security and Resilience*

The above discussion suggests that an integrated approach to food security and resilience can be expressed using a three-dimensional model that illustrates potential points of intersection between food security, shocks and stressors, and resilience capacities. The *food security dimension* highlights the WFS approach (FAO, 1996), the food sovereignty approach (via Campesina, 2007), and the food systems approach (HLPE, 2017) as perspectives that constitute—either separately or in some combination—how food security is conceptualized. The *shocks and stressors dimension* highlights three categories of disturbances that may undermine food security at three different scales. The *resilience capacities dimension* lists three types of capacities which may be drawn upon to deal with shocks and stressors. The *integrated food security resilience model,* presented in Fig. 5.1, illustrates how food security, shocks and stressors, and resilience capacities intersect.

With 27 "cells" as points of intersection, the model illustrates the complexity involved in bringing food security and resilience into conversation with each other. Bearing in mind that each of the elements within each dimension is also multidimensional, the model offers a simplified representation of how food security and resilience may be integrated. This

---

[8] Much of the work on resilience has been concerned with household- or community-level dynamics. Work on country-level resilience, focused on macroeconomic indicators, can be found in Boorman et al. (2013), Briguglio et al. (2009), and Kose and Prasad (2010).

model, like any model, is necessarily incomplete and somewhat reductionist. It is presented here to convey what might be involved in such an integration. The model is offered as a heuristic on which further expanded conceptualizations may be based. The main point here is that considering food security and resilience jointly introduces a certain set of demands, demands that could plausibly give shape to policy and define the focus of empirical work. In the next section, the ways in which these demands are enacted in policy discourse and in research are considered.

## A Case Study: Lexical Analysis and Content Analysis of Food Security and Resilience

An investigation of connections between food security and resilience that focus on the State of Food Security Insecurity (SOFI) and two peer-reviewed journals (details provided below) was conducted by combining elements of a lexical analysis (Laver & Benoit, 2003; Tausczik & Pennebaker, 2010) with a scoping review (Arksey & O'Malley, 2005). The lexical analysis was used to examine the occurrence of the term *"resilience"* in a sample of food security-focused publications. Based on examining documents over an extended period of time, the lexical analysis provided some form of reconnaissance view of food security–resilience connections over time. Moving beyond basic frequency counts that record the appearance of *"resilience"* in food security publications, scoping methods were applied to investigate the specific ways in which food security and resilience are connected to one another. Drawing on these methods, two questions drove the analysis: (1) *Simple trend analysis*—What is the broad trend in the inclusion of resilience in food security discussions? (2) *Content analysis*—How has resilience been used in and integrated into policy and research work on food security? Applying these two questions to both the policy literature and the research literature generates four questions. Table 5.1 summarizes the analytical focus of the study.

### *Methods*

Two procedures jointly comprise the methods for the present study. (i) *Document selection* describes the rationale that was used to decide which documents would be examined. (ii) *Analytic methods* describe how lexical analysis and content analysis methods were performed.

**Table 5.1** Logic of the review

*Analysis of food security and resilience*
*Focus on trends and applications*

| Type of analysis | Object of analysis | |
|---|---|---|
| | Policy | Research |
| Trend Analysis | 1. To what extent has resilience been featured in policy discourse on food security? | 2. To what extent has resilience been featured in the research literature on food security? |
| Content Analysis | 3. In what ways have resilience and food security been applied in the policy discourse? | 4. In what ways have resilience and food security been applied in the research literature? |

*Document Selection*

A single policy-oriented document and a sample of research articles focused on food security were the objects of analysis. The policy document selected was the *State of Food Insecurity* (SOFI), for two reasons. First, SOFI is a highly visible publication on food security. Second, SOFI is published under the auspices of multiple United Nations agencies, most of which (the FAO, the World Food Programme, and the International Fund for Agriculture) place food security at the center of their work.[9] For the simple trend analysis, the entire set of SOFI reports was considered (1999 through 2020).[10]

For the research literature, two leading peer-reviewed journals with a food security focus were selected—*Global Food Security*, and *Food Policy*. The time-period reviewed for these journals was shorter than that for the trend analysis. Using a high-profile policy statement from USAID (USAID, 2012) as a temporal marker, 2012 was treated as the baseline year for the trend analysis of the peer-reviewed literature. To analyze the conceptual integration of food security and resilience in the peer-reviewed literature, articles from just one of the two journals, *Global Food Security*, were considered. As will be reported, *Global Food Security* published a

---

[9] SOFI was published from 1999 through 2008 by the FAO only. In 2009, the WFP and the IFAD joined as partners in SOFI. This arrangement continued through 2015. From 2016 forward, the United Nations Children's Fund and the World Health Organization joined as partners in producing SOFI.

[10] No SOFI reports were published in 2007 or 2016.

higher proportion of articles featuring the concept of resilience. The other justification for selecting *Global Food Security* was its comparatively higher impact factor and cite score.[11]

*Procedures for the Lexical Trends Analysis*
The lexical trend analysis documented any occurrence of *"resilience"* in SOFI reports and in the two peer-reviewed journals included in the study as the unit of analysis. For SOFI, this involved a search for *"resilience"* and its cognates (i.e., *"resilient," "resilience," "resiliency"*) in any part of a report for a given year. This search excluded reference listings, figures and table headings, report titles, and titles of sections and subsections. Occurrences of the unit of analysis in a given year and in sentences were treated as data points. This meant that repeated use of *"resilience"* within a single sentence was not counted as multiple instances of use. The relative prevalence of use was computed by simply dividing the number of pages that included *"resilience"* by the total number of pages (excluding front matter, references, and annexes) and was also recorded for a given year.

The search for *"resilience"* in journal articles was limited to titles, abstracts/keywords, highlights, and the main texts of the articles. With the intent of maximizing inclusion, only one condition needed to be met for a given article to be recorded as an instance that included the use of *"resilience."* For SOFI, where each report covered a broad range of issues, assessing the relative prevalence of use made sense. The selection criteria for the peer-reviewed articles implied that any item included in the frequency count involved some kind of discussion (however passing or superficial) of resilience. The unit of analysis was one journal year with occurrences of *"resilience"* aggregated across all volumes to obtain a frequency count of articles for a given year. Unlike with SOFI, here multiple occurrences within a given article were not considered as part of the frequency count.

*Procedures for Content Analysis*
Content analysis of both SOFI volumes and *Global Food Security* articles was organized to document how *"resilience"* was deployed in the context of food security. This stance was taken because food security was

---

[11] While *Food Policy* outperforms *Global Food Security* in the H-Index, the H-index is upwardly biased by the age of the journal. As of 2021, *Food Policy* had been published for 19 years compared with 10 years for *Global Food Security*.

viewed as the "*incumbent concept*" in reference to resilience. This seems like a reasonable assumption to make for SOFI and for the peer-reviewed journal included in the review; indeed, food security is, and has been, at the center of SOFI and a central focus of research published in *Global Food Security*.

Applying the relevant part of the above conceptual model, the content analysis of SOFI was structured around h two questions: (1) Which version of food security (following the above discussion) was featured? (2) How frequently and in what ways was resilience connected to food security? Each *Global Food Security* article that was identified as having a resilience focus was carefully read to document: (1) the version(s) of food security that was (were) used, (2) the "*shocks and stressors*" context that was highlighted, (3) the definition of and/or the topic to which "*resilience*" was applied, and (4) the extent to which resilience was included in the substantive conclusions that were offered. In addition to these four points, details regarding the country(ies) of study and the type of study conducted (quantitative, qualitative, mixed methods, or policy paper) were recorded.

### *Results of the Lexical Trend Analysis of SOFI Reports*

For the first eight years of SOFI (1999–2008), with the exception of 2004, "*resilience*" did not appear at all in the texts of the reports. In 2005 and 2006, occurrences returned to zero and SOFI was not published in 2007. From 2009 through 2015, the use of "*resilience*" remained consistently low with occurrences in each report ranging from three in 2009 and 2013 to a high of 14 in 2010. In 2017, the theme of the SOFI report was *Building Resilience for Peace and Security*. Not surprisingly, a notable increase in the use of "*resilience*" was evident for that report, with 51 occurrences. This represented an approximate tripling over the previous high in 2014 and a nearly fivefold increase over the most recent year of publication, 2015.[12] In 2018, there were 119 occurrences of "*resilience.*" Here again, the spike in the use of "*resilience*" reflected a thematic focus of SOFI for that year's report, *Building Climate Resilience for Food Security and Nutrition*. Figure 5.2 displays the frequency with which "*resilience*" occurred in SOFI reports (the solid line) and the linear

---

[12] As noted earlier, no SOFI reports for 2007 and 2016 were published.

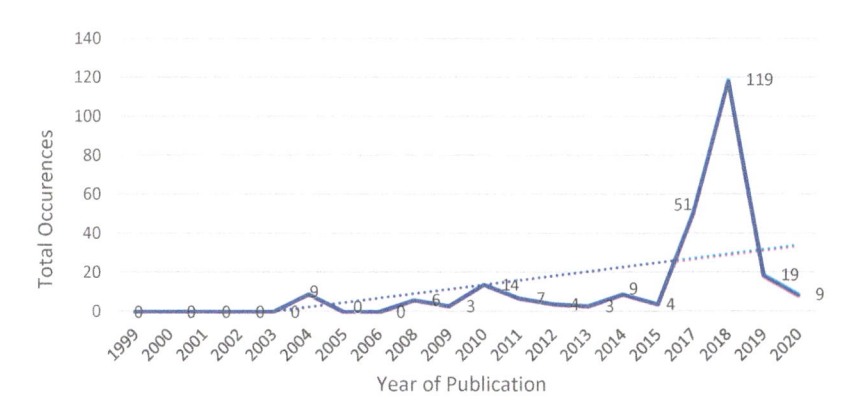

Fig. 5.2 Appearance of *"resilience"*: trends in SOFI, 1999–2020 (*Source* Author)

trend (the dotted line) for the entire period covered by the lexical trend analysis.

The graph lines in Fig. 5.2 show clearly that 2017 and 2018 did not mark the beginning of a trend but rather were idiosyncratic events. This is evidenced by the return to a low level of occurrence in 2019 and a yet lower level in 2020, with occurrences falling to 19 and nine, respectively, in those years.

The second stage of the lexical trend analysis focused on the relative prevalence of *"resilience"* within a given report. Looking beyond the absolute value of occurrences for a given year of SOFI, prevalence indicates whether *"resilience"* occurred on only a few pages or appeared across many sections of a SOFI report. In this way, data on prevalence revealed the consistency with which "resilience" appeared throughout a given SOFI volume. Figure 5.3 graphically depicts the relative prevalence of "resilience" within a given report and across years. The solid line traces occurrences for a given year and the dotted line represents the linear trend across the 17 years of observations.

As reflected in Fig. 5.3, the prevalence-based results paint a different picture from that represented by the absolute number of occurrences shown in Fig. 5.2. While the trend lines suggest an overall increase in relative occurrence, the change over time is erratic in comparison with the

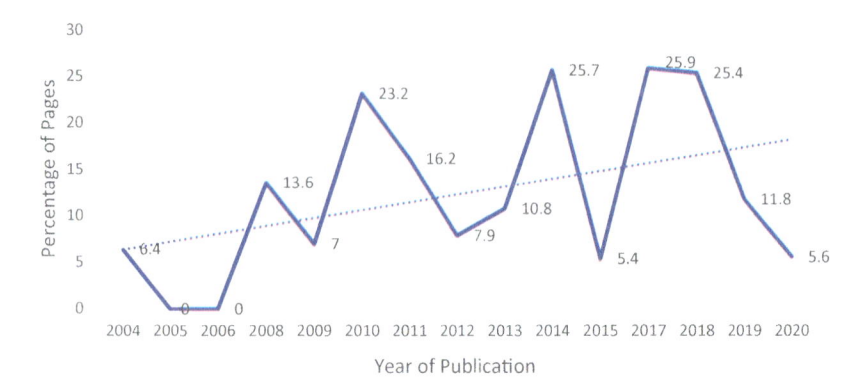

**Fig. 5.3** Appearance of *"resilience"*: trends in SOFI (2004–2020) (*Source* Author)

results obtained from simple frequency counts of articles within a given year.

The increase observed for the lexical presence of "resilience," followed by a marked descent, raises questions about the durability of resilience as a concept around which ongoing policy discussions on food security might be structured. In principle, resilience is a cross-cutting topic. It has, for example, been suggested as a perspective that could support efforts to build the humanitarian-development-peace (HDP) nexus (e.g., Béné et al., 2018; EU, 2021; Hilhorst, 2018; Howe, 2019). The original motivation that drove researchers and policymakers to focus on resilience was grounded in shared recognition of a more complex risk landscape. As noted above, shocks and stressors have become more pronounced and less predictable. With COVID-19 as a massive global shock, the motivation to draw on resilience should be amplified. If this is the case, one would expect resilience to exhibit a certain durability in discussions of food security. In the two periods following the SOFI that was thematically oriented toward resilience, dramatic decline in the appearance of "resilience" occurred. This decline may simply reflect the fact that SOFI is organized around themes that change from year to year. The designation of SOFI themes for each year may, in and of itself, generate discontinuities or discursive shifts in what is emphasized. In its first three years (1999–2002), SOFI was published under a single title, *Food Security When People Live with Hunger and Fear Starvation*. For the next two years (2003

Global Food Security and Food Policy (2011–2020)

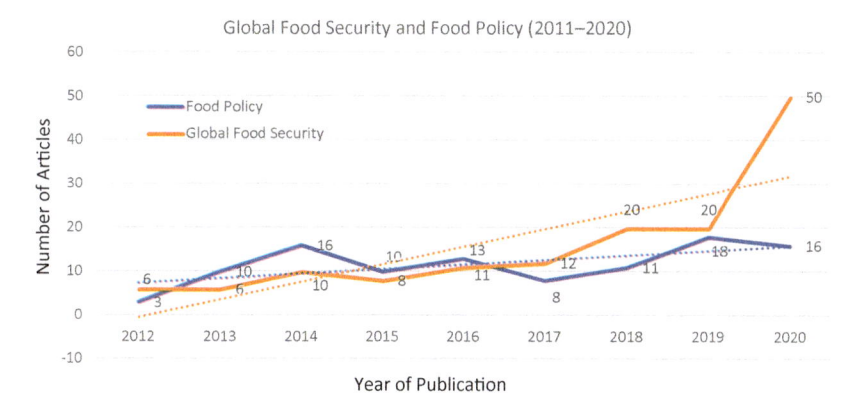

**Fig. 5.4** Appearance of *"resilience"* in the research literature (*Source* Author)

and 2004), another title was used: *Monitoring Progress Towards the WFS and the Millennium Development Goals*. From 2005 onward, SOFI was published under a different title or theme each year.[13]

## *Results of Lexical Trends in Representative Peer-Reviewed Journals*

The lexical search, covering nine years of articles in *Global Food Security* and *Food Policy*, revealed an overall pattern of increase in the appearance of the term "resilience." The graph lines shown in Fig. 5.4 display the relative frequency of the appearance of "resilience" in the two peer-reviewed journals from 2012 through 2020. Following the same format as above, the solid lines depict actual occurrences in a given year and the dotted lines show linear trends in use over time.

For *Global Food Security*, the pattern of increase was consistent, starting with a low of six articles in 2012 and increasing to 50 articles in 2020. The rate of increase varied, with the most dramatic increase occurring between 2019 and 2020, where the number of articles containing "resilience" more than doubled, from 20 to 50. The period with the next largest increase occurred between 2017 and 2018, when the number of articles increased from 12 to 20. The pattern of increase for *Food Policy* was less consistent and less dramatic. From 2012 through 2020, the number of

---

[13] A complete listing of SOFI titles is provided in Appendix 5.1.

articles with the term "resilience" increased from six to 16. The largest increase occurred between 2012 and 2014, when the number of articles increased from three to 10. The doubling observed is largely inconsequential in light of the small base number and the absence of an overall consistent increase.

The observed trends are shown as simple frequencies (absolute values) of journal articles over time, so it is important to consider the possibility that the results become distorted by year-to-year fluctuations in the number of articles published. To account for differences in articles published per year, the number of articles containing "resilience" in any given year was divided by the overall number of articles published in that same year. Figure 5.5 graphs trends in the appearance of "resilience" that are not biased by differences in the number of articles published in a given year by each journal. The figure also allows for a fair comparison between journals for any given year.

The trends displayed in Fig. 5.5, based on prevalence, are not as pronounced as the trends based on absolute occurrences shown in Fig. 5.4. The relative differences between journals regarding prevalence is, however, roughly consistent with results of the simple frequency count. While the presence of an overall pattern of increase for *Global Food Security* is reflected in both Figs. 5.4 and 5.5, the absence of clear evidence of a pattern of any kind is suggested for *Food Policy*. A comparison of the

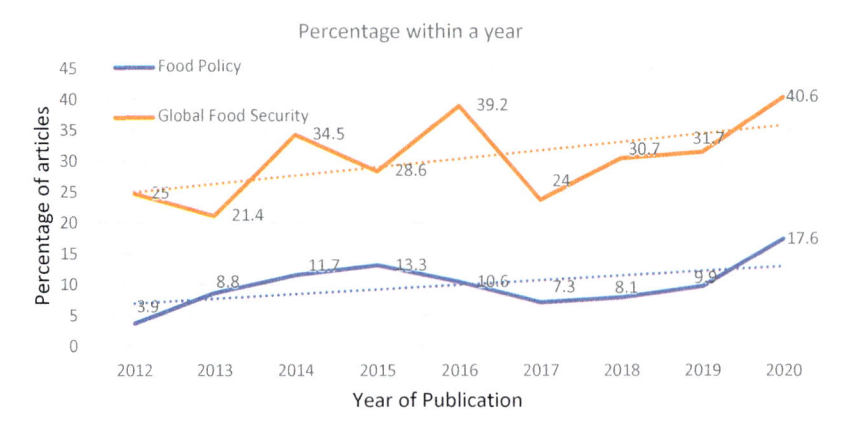

**Fig. 5.5** Relative prevalence of "resilience" in the research literature: Food Policy and Global Food Security (*Source* Author)

occurrences of "resilience" in the initial time period (2012–2014) with such occurrences in the most recent time period (2018–2019) suggests a modest pattern of increase in both *Food Policy* and *Global Food Security*. While these findings are less dramatic than those based on the simple frequency count of articles, the trend lines (the dotted lines) shown in Fig. 5.5 reveal a modest pattern of increase in relative frequency.

### *Results from the Content Analysis of the Use of "Resilience" in SOFI*

Using the results of the lexical trend analysis as a point of departure, the content analysis offers a more detailed view of how the concepts of food security and resilience were discussed in SOFI. Following the categories included in the food security dimension of the conceptual model (Fig. 5.1), each occurrence of "resilience" was reviewed to see if and how the concept was discussed in connection with food security. Any mention of "food security" was examined to categorize it as a general reference (i.e., "food security" mentioned but not differentiated), as a specific reference to one or more of the 1996 WFS pillars (availability, access, utilization, or stability), as a reference to food systems, or as a reference to food sovereignty (including references to power dynamics and the right to food). In those cases where resilience was connected to something other than food security, food systems, or food sovereignty, the connection was recorded as "Other or General." The year-by-year results are displayed in Table 5.2 with the most dominant pairing with "resilience" for any given year highlighted in yellow.

In addition to a year-by-year analysis, the column totals shown in Table 5.2 provide an aggregate picture of combinations for the 13 years of SOFI reports that were analyzed. On those occasions when resilience was discussed in a SOFI report, 11 of 13 years were categorized as referring to resilience under the heading "Other or General." Examples of occurrences that fall into this category include phrasing such as "build on a foundation of resilience" (FAO, 2004, p. 26), "many countries remain on the list for several years owing to the lingering effects of drought and/or conflict and low resilience" (FAO, 2008b, p. 19), "-; reduce risk and increase the resilience of the most vulnerable" (FAO, 2012, p. 33), and "build lasting climate resilience" (FAO, 2018, p. 94). When "resilience" was mentioned, it was paired with "food security" or "food systems" but rarely with both. Of the 110 instances where "resilience" was mentioned, only two involved a pairing with "food security" and "food systems" at

**Table 5.2** SOFI—resilience, food security, and food systems

| | Co-occurrences among "Resilience," "Food Security," and "Food Systems"[1] SOFI reports with resilience occurrences (1999–2020) | | | |
|---|---|---|---|---|
| | Connection between Resilience, and Food Security–Related Concepts | | | |
| Year of SOFI | Food and Nutrition Security | Food Systems | Food Sovereignty* | Other or General |
| 2004 (9) | 2 | 1 | 0 | 6 |
| 2008 (6) | 1 | 3 | 0 | 2 |
| 2009 (3) | 0 | 0 | 0 | 3 |
| 2010 (14) | 2 | 0 | 0 | 12 |
| 2011 (7) | 1 | 1 | 0 | 5 |
| 2012 (4) | 1 | 1 | 0 | 2 |
| 2013 (3) | 0 | 0 | 0 | 3 |
| 2014 (9) | 4 | 0 | 0 | 5 |
| 2015 (4) | 0 | 1 | 0 | 3 |
| 2017 (51) | 19 | 1 | 0 | 31 |
| 2018* (119) | 23 | 29 | 0 | 70 |
| 2019 (19) | 2 | 7 | 0 | 11 |
| 2020 (9) | 0 | 9 | 0 | 0 |
| Total (257) | 55 | 50 | 0 | 153 |

*Note* (1) Numbers indicate whether occurrences were discussed in connection with food and nutrition security, food systems, or food sovereignty (*Source* Author)

the same time. Both instances were found in the 2018 SOFI report when the theme was *Building Climate Resilience for Food Security and Nutrition*. Somewhat surprisingly, no explicit pairing was found in the 2020 SOFI report, when the title and theme for that year was *Transforming Food Systems for Affordable Healthy Diets*. Most noticeable in the results is the complete inattention to food sovereignty in SOFI reports for any year.

One notable observation that emerged from the SOFI content analysis was the lack of conceptual clarity. In not a single case where resilience was mentioned did such a mention include a definition or conceptual discussion of the meaning of "resilience." No clear distinction was made between resilience as a capacity and resilience as an outcome or property of an outcome, such as food security. It follows from this point that no distinctions were made between distinct types of resilience capacities and the relationship between resilience capacities and food security was not discussed.

Overall, the results of the SOFI content analysis reveal that connections between resilience and food security and between resilience and food systems are typically low. SOFI 2017 featured the greatest co-occurrence of "food security" and "resilience," with 23 co-occurrences recorded. This, however, was to be expected when considering the theme and title of SOFI 2017—*Building Resilience for Peace and Security*. The same lack of connection found between "resilience" and "food security" was observed between "resilience" and "food systems." While SOFI 2008 included more references to food systems in connection with resilience, only three co-occurrences of the terms were identified in that period. Only once did SOFI (FAO, 2018) connect "resilience" and "food systems" at an appreciable level, with 23 occurrences. Interestingly, this occurred in a year when the food systems concept was not an explicit focus of the 2018 report. That report, titled *Building Climate Resilience for Food Security and Nutrition*, did not highlight food systems in its forward or key messages. It appears, however, that challenges associated with climate change provide a compelling case for exploring points of intersection between the concepts of resilience and food systems.[14] This is a plausible explanation, as climate change is frequently identified as a driver of many kinds of shocks to which the development of more resilient food systems is a needed response. The complete lack of connection between resilience and food sovereignty presents a starkly contrasting picture. This finding is somewhat surprising when one considers that the FAO, as one of the organizations leading SOFI, has a history of drawing attention to food sovereignty and the right to food (Gordillo, 2013).

### Content Analysis Results: The Use of "Resilience" in Global Food Security Articles

Because the lexical analysis of the journal articles included articles in which "resilience" occurred in any part of an article in the frequency counts for a given year, the first stage of the content analysis required a culling of articles from the 143 articles that were selected from 2012 through 2020. One criterion for narrowing this set of articles was based on article type. The focus was on research articles, while reviews were excluded. To ensure that the concept of resilience was an actual focus in any given article, only articles that included "resilience" in the title or the abstract were included.

Of the 143 articles in the inclusive set for the lexical search, 12 included "resilience" in the title, the abstract, or keywords/highlights.

---

[14] Connections among climate change, resilience, and food security are examined in detail by De Pinto, Islam, and Katic (Chapter 7 in this volume) in an analysis of 12 Adaptation Fund projects.

It should be reiterated here that the original set of 143 articles was based on "generous" inclusion criteria. For an article to be selected, "resilience" needed to appear at least once anywhere in the article. At this stage, no additional criteria were applied to determine whether an article actually focused on resilience in any part. Table 5.3 lists the titles of the 12 articles from *Global Food Security* that were included in the content analysis.

As shown in Table 5.3, 11 of the 12 selected articles were published in 2020 and one was published in 2019. Thus, prior to 2019 no articles included "resilience" in the title, abstract, or article highlights. While one might argue that this temporal bunching is an artifact of the inclusion criteria, the criteria were applied consistently over the nine-year search period (2012–2020). With 12 articles as a sample of peer-reviewed research articles with a focus at the intersection of food security and

**Table 5.3** *Global Food Security* articles examined for content analysis

| "Resilience" Appeared in Titles, Abstracts, or Highlights |
| --- |
| 1. Small farms' resilience strategies to face economic, social, and environmental disturbances in selected regions in Poland and Latvia (Czekaj et al., 2020) |
| 2. Interplay of trade and food system resilience: Gains on supply diversity over time at the cost of trade independency (Kummu, et al., 2020) |
| 3. A brighter future: Complementary goals of diversity and multifunctionality to build resilient agricultural landscapes (Frei et al., 2020) |
| 4. Choosing awareness over fear: Risk analysis and free trade support global food security (Adamchick & Perez, 2020) |
| 5. Policy options for mitigating impacts of COVID-19 on domestic rice value chains and food security in West Africa (Arouna et al., 2020) |
| 6. Filling knowledge gaps to strengthen livestock policies in low-income countries (Serra et al., 2020) |
| 7. Global changes in crop diversity: Trade rather than production enriches supply (Aguiar et al., 2020) |
| 8. Food policy and the unruliness of consumption: An intergenerational social practice approach to uncover transforming food consumption in modernizing Hanoi, Vietnam (Wertheim-Heck & Raneri, 2020) |
| 9. Perspective article: Actions to reconfigure food systems (Loboguerrero et al., 2020) |
| 10. Food securers or invasive aliens? Trends and consequences of non-native livestock introgression in developing countries (Leroy et al., 2020) |
| 11. Alternative discourses around the governance of food security: A case study from Ethiopia (Jiren et al., 2019) |
| 12. Using local initiatives to envision sustainable and resilient food systems in the Stockholm city region (Sellberg et al., 2020) |

*Source* Compiled by the author

resilience, the content analysis proceeded to a stage that entailed a more carefully detailed reading.

Before discussing the main findings from the content analysis, several basic qualities of the papers should be discussed. Regarding study types, four of the studies (Adamchick & Perez, 2020; Arouna et al., 2020; Loboguerrero et al., 2020; Serra et al., 2020) categorized as policy papers were non-empirical studies. Of the remaining eight studies, four were quantitative, three were qualitative, and one was a mixed methods study. The question regarding how resilience is defined in research articles revealed that the majority of the articles did not provide explicit definitions or conceptions of resilience. Of the 12 articles examined, only two (Adamchick & Perez, 2020; Czekaj et al., 2020) offer explicit definitions or conceptions of resilience. Czekaj et al., (2020, p. 2) state that resilience "refers to the capacity and ability of physical or socio-ecological systems to recover from a disturbance of any type and maintain the original function." As part of their definition, Czekaj et al. (2020) also note that resilience includes elements of persistence, adaptability, and transformability. Adamchick and Perez (2020, p. 3) state that "[r]esilience includes the readiness to anticipate and mitigate the impact of epidemic events that are expected to happen without knowledge of when or where they will occur." Four out of 12 articles (Frei et al., 2020; Jiren et al., 2019; Loboguerrero et al., 2020; Sellberg et al., 2020) offer what can be interpreted as implied definitions that reveal varying perspectives on resilience. In these cases, one can deduce the intended meaning of resilience even though no explicit definition was provided. Sellberg et al. (2020), for example, note that "enhancing resilience requires substantial innovation, experimentation and transformation." In work that was based on the stakeholder view of food security in Ethiopia, Jiren et al. (2019) state resilience "typically takes a complex adaptive systems perspective, emphasizing feedbacks, slow drivers of systems behavior, and emergent system dynamics resulting from self-organization."

None of the remaining articles among the selected 12 provide either an explicit or an implied definition of resilience (Aguiar et al., 2020; Arouna et al., 2020; Kummu, et al., 2020; Leroy et al., 2020; Serra et al., 2020; Wertheim-Heck & Raneri, 2020). Arouna et al. (2020) state that the aim of their study was "to reduce the current and potential impact of the COVID-19 pandemic on domestic rice value chains' resilience and their capacity to sustain food security in West Africa." Loboguerrero et al. note that "[f]ood systems need to shift towards more sustainable, inclusive, healthy and climate-resilient futures." In these cases, and others where no definitions are offered, resilience as a concept is mentioned but not developed. Such usage suggests that, as a concept, resilience stands in need of definition rather than being used simply as a "*buzzword*" (Béné et al., 2017; Staal, 2016).

Regarding instances where resilience might be connected to food security and/or food systems, eight of the 12 selected articles focus on food security alone. In several articles that highlight food systems, references are general and specific definitions or conceptions of food systems are not included. This is true for five of the eight articles that make reference to food systems (Czejak et al., 2020; Kummu, et al., 2020; Adamchick & Perez, 2020; Leroy et al., 2020; Loboguerrero et al., 2020).

Discussions of the shocks-and-stressors context are similarly mixed and typically general. It appears that most of the work was motivated by shocks and stressors, but the references are not accompanied by associated empirical work that involved corresponding metrics or analyses. This is true for all 12 articles examined. The conclusions that are offered highlight resilience only occasionally. The three articles that do reference resilience in their conclusions (Adamchick & Perez, 2020; Arouna, et al., 2020; Czekaj et al., 2020) do so on a conceptual level with limited detail. Table 5.4 synopsizes the content analysis findings from the 12 resilience-oriented articles that were examined in *Global Food Security*.

Perhaps, the most basic finding to highlight is that, over a nine-year period, only 12 of 143 articles met the criteria for inclusion (i.e., "resilience" or "resilient" in the title, abstract, or keywords/highlights). While the results of the lexical analysis suggest increasing interest in resilience over time, there is also little evidence of evolution. Referring back to the three dimensions that were part of the conceptual model presented earlier, a lack of detailed conceptions was found for food security, for the discussion of shocks and stressors, and for the description of resilience capacities. Most concerning was a marked tendency to use "resilience" without defining the concept.[15] On the whole, there is no basis for concluding that progress has been made regarding how resilience is being used research or in policy. The 2020 surge in resilience-related publications in *Global Food Security* may suggest that an inflection point has been reached. Time will tell if such an inflection point will be marked

---

[15] This finding aligns with the results obtained by De Pinto, Islam and Katic (Chapter 7 in this volume).

**Table 5.4** Summary of findings of the content analysis of *Global Food Security* Research Articles from Global Food Security, 2012–2020

| Food Security Focus, Location & Study Type | Shocks and Stressor Context | Definition and/or Application of Resilience |
|---|---|---|
| *Czekaj et al. (2020) Small farms' resilience strategies to face economic, social, and environmental disturbances in selected regions in Poland and Latvia [empirical]* ||||
| Food systems (GR)[1]<br>• Poland & Latvia<br>• Qualitative | Economic, social, and environmental disturbances | **Definition**: capacity and ability to recover and bounce back, with specific references to different re capacities<br>**Application**: farms' resilience, resilience strategies of small-scale farmers (SSFs) in relation to disturbances |
| **Conclusion**: "[The study] illustrates the overall spectrum of strategies employed by SSFs in these two countries, thereby providing a basis for further analysis of the differences in, and prevalence of, specific resilience strategies of SSFs in different countries" |||
| *Kummu et al. (2020) Interplay of trade and food system resilience: Gains on supply diversity over time at the cost of trade independency [empirical]* |||
| Food security, WFS with nutritional focus and food systems (GR)<br>• Multi-country, global<br>• Quantitative | Trade shocks, food shocks, unanticipated shocks, production shocks, import shocks | **Definition**: None provided<br>**Application**: Applied resilience principles to food production diversity, food supply diversity, independence of food imports, import connections |
| **Conclusion**: "Our findings thus highlight the interconnected trade-offs between trade-related aspects of food system resilience, and provide important information for global actors, as well as national policy makers" |||
| *Frei et al. (2020) A brighter future: Complementary goals of diversity and multifunctionality to build resilient agricultural landscapes* |||
| Food security (production) emphasis<br>• Quebec, Canada<br>• Quantitative | Environmental and socioeconomic stressors, unpredictable stressors and change | **Definition**: None provided<br>**Application**: Resilience of socio-ecological systems and agrodiversity |
| **Conclusion**: "Managing agricultural landscapes for ES [ecosystem services] multifunctionality, including multiple facets of food production as well as regulating and cultural services, enables the dual goals of feeding the world and conserving a diversity of ecosystem functions" |||

(continued)

**Table 5.4** (continued)

*Research Articles from* Global Food Security, *2012–2020*

| Food Security Focus, Location & Study Type | Shocks and Stressor Context | Definition and/or Application of Resilience |
|---|---|---|

*Adamchick and Perez (2020) Choosing awareness over fear: Risk analysis and free trade support global food security [non-empirical]*

| Food security WFS<br>Food systems (GR)[1]<br>• Global<br>• Policy paper | Population growth, public health risks, global spread of pathogens, reference to COVID-19 | **Definition**: Defined in terms of readiness to anticipate and mitigate<br>**Application**: Risk analysis capacity and use increases local and global food system resilience |
|---|---|---|

**Conclusion**: "The capacity and use of risk analysis coupled with sound understanding of underlying system dynamics will contribute to resilient and enduring food systems"

*Arouna et al. (2020) Policy options for mitigating impacts of COVID-19 on domestic rice value chains and food security in West Africa [non-empirical]*

| Food security WFS<br>• West Africa<br>• Policy paper | Trade disruptions, zoonotic pathogens, rice prices | **Definition**: None provided<br>**Application**: Resilience of domestic rice value and rice value chains |
|---|---|---|

**Conclusion**: "[The study] assess[ed] the potential impact of the COVID-19 pandemic on domestic rice value chains' resilience and their capacity to sustain food security in the region"

*Serra et al. (2020) Filling knowledge gaps to strengthen livestock policies in low-income countries [non-empirical]*

| Food security -WFS<br>• Global<br>• Policy paper | Market shocks, climate, disease, conflict, demand and pricing of animal source foods | **Definition**: None provided<br>**Application**: The use of data-based models to assess livestock's contribution to the economy, trade, food security and resilience |
|---|---|---|

**Conclusion:** "It is imperative to consider which type of data collection and modeling ought to be prioritized in low-income, livestock-rich countries to strengthen livestock policies and enhance the positive impact on household incomes and dietary diversity"

(continued)

**Table 5.4** (continued)

| Research Articles from Global Food Security, 2012–2020 | | |
|---|---|---|
| Food Security Focus, Location & Study Type | Shocks and Stressor Context | Definition and/or Application of Resilience |

Aguiar et al. (2020) *Global changes in crop diversity: Trade rather than production enriches supply*

| | | |
|---|---|---|
| Food system -GR<br>• Multiple countries (N=152)<br>• Quantitative | Crop production shocks linked to droughts, floods, pests, and wars | **Definition**: None provided<br>**Application**: Resilience of global food systems and the diversity of crop production globally |

**Conclusion**: "Our results indicate that the expansion and diversification of crop trade was the main driver of the global diversification of supply since within-country production slightly increased"

Wertheim-Heck and Raneri (2020) *Food policy and the unruliness of consumption: An intergenerational social practice approach to uncover transforming food consumption in modernizing Hanoi, Vietnam*

| | | |
|---|---|---|
| Food security (ND)<br>• Vietnam<br>• Qualitative | Market transformation away from traditional healthy diets, coping with altered food retail market | **Definition**: None provided<br>**Application**: No references to resilience in body of the article |

**Conclusion:** "…Traditional shopping and food preferences is strong; creative agency results in food security resilience; and pester power is driving food preparation and subsequently dietary changes at the home dinner table"

Loboguerrero et al. (2020) *Perspective article: Actions to reconfigure food systems* [non-empirical]

| | | |
|---|---|---|
| Food systems<br>• Global<br>• Policy paper | Climate change, global warming, extreme events | **Definition**: None provided<br>**Application**: Reconfiguring food systems; resilience of food systems' agents under rapid change and |

**Conclusion**: "Food systems need to shift towards more sustainable, inclusive, healthy and climate-resilient futures".

Leroy et al. (2020) *Food securers or invasive aliens? Trends and consequences of non-native livestock introgression in developing countries*

| | | |
|---|---|---|
| Food Security & Food systems (ND) reference to sustainable<br>• Multi-country (N = 83)<br>• Quantitative | Resilience to climate change, resistance to endemic diseases, consequences of genetic erosion | **Definition**: None provided<br>**Application**: Resilience of livestock production systems (i.e., resilience of livestock from varied production systems) |

(continued)

**Table 5.4** (continued)

*Research Articles from* Global Food Security, *2012–2020*

| Food Security Focus, Location & Study Type | Shocks and Stressor Context | Definition and/or Application of Resilience |
|---|---|---|

**Conclusion:** "Animal genetic resources can be regarded as the centre of a complex social, environmental and economic system, so policies need to address the challenges related to sustainability in a holistic manner, accepting trade-offs where necessary, and considering, at different scales, the relationships and dynamics between the animals, their herders, the production systems, agroecosystems, and the market"

*Sellberg et al. (2020) Using local initiatives to envision sustainable and resilient food systems in the Stockholm city-region*

| Food systems-socio-ecological systems<br>• Sweden<br>• Qualitative case study | Uncertain change | **Definition**: None provided<br>**Application**: Food system resilience and transformation |
|---|---|---|

**Conclusion**: "We found that the Seeds of Good Anthropocene scenario methodology helped to understand more of the dynamics and divergent views in a transformation process in a specific social-ecological context"

*Jiren et al. (2019) Alternative discourses around the governance of food security: A case study from Ethiopia*

| Food security, food system & biodiversity (GR)<br>• Ethiopia<br>• Mixed methods | Resilience to shocks and uncertainties | **Definition**: None provided<br>**Application**: Resilience as an approach to food security |
|---|---|---|

**Conclusion**: "Adaptive co-management of food security—that is, collaboration among stakeholders with diverse interests across governance levels … could be one way to harmonize contradictions, integrate divergent discourses and interests, bridge current gaps and incorporate multiple framings to open a new pathway for sustainability"

*Notes* (1) "GR" denotes general reference only; (2) "ND" denotes not defined

as only a shallow rhetorical shift that fails to apply the concept of resilience in a meaningful way, or as a shift that draws on foundational work on resilience (Holling, 1973), resilience thinking (Walker & Salt, 2006), resilience theory for development (Barrett & Constas, 2014), or resilience for food security (Béné et al., 2016).

## Conclusion

Ongoing applications of resilience to food security make it clear that the initial interest in resilience is not ephemeral. The effort to integrate food security and resilience, however, represents a new challenge. In response to this challenge, the present chapter offers a conceptual model, the *integrated food security resilience model*, that highlights possibilities for integration. The model can be used as a heuristic to explore how resilience and food security may be combined. It may also support work toward a data architecture to develop a more comprehensive set of indicators for resilience analysis that is concerned with food security. Guided by the model, the review of a sample of policy documents revealed uneven application of the resilience concept over time, with very little evidence of the adoption of an integrated perspective at any given point in time. The peer-reviewed articles that were reviewed exhibited greater consistency in the use of resilience, but with little evidence of integration. In both reviews, neither food security nor resilience was conceptualized effectively. The findings compel us to consider a more basic question of how food security and resilience, as standalone concepts, are defined and used in research and policy.

In closing, the findings presented in this chapter suggest that there is room for improvement in work that aims to integrate resilience and food security in policy discourse and in research. If resilience is to be a meaningful addition to our understanding of food security, the specific requirements introduced by resilience need to be explored (see Constas et al., 2020). Growing attention to food systems as a means of understanding the challenge of food security (HLPE, 2017; Pingali & Sunder, 2017; Reardon & Timmer, 2012) must be given closer attention. Building on both the early work of Via Campesina (2007) and the more recent work of the HLPE (2020), greater attention must also be paid to agency and food sovereignty as emerging elements of food security. The ways in which resilience and food security may be integrated to reflect the full complexity of each concept need further attention. Future work will reveal whether the aspiration for integration will remain an undeveloped effort that lacks precision or will be realized as a true substantive shift that leads to greater conceptual coherence.

## Appendix

See Table 5.5.

Table 5.5 SOFI titles/themes 1999–2020

| Year | Title/theme |
| --- | --- |
| 1999 | Food Insecurity: When People Live with Hunger and Fear Starvation |
| 2000 | Food Insecurity: When People Live with Hunger and Fear Starvation |
| 2001 | Food Insecurity: When People Live with Hunger and Fear Starvation |
| 2002 | Food Insecurity: When People Live with Hunger and Fear Starvation |
| 2003 | Monitoring progress towards the World Food Summit and Millennium Development Goals |
| 2004 | Monitoring progress towards the World Food Summit and Millennium Development Goals |
| 2005 | Eradicating World Hunger—Key to Achieving the Millennium Development Goals |
| 2006 | Eradicating World Hunger—Taking Stock Ten Years After the World Food Summit |
| 2007 | No SOFI report published |
| 2008 | High Food prices and Food Security—Threats and Opportunities |
| 2009 | Economic Crises—Impacts and Lessons Learned |
| 2010 | Addressing Food Insecurity in Protracted Crises |
| 2011 | How Does International Price Volatility Affect Domestic Economies and Food insecurity? |
| 2012 | Economic Growth is Necessary but Not Sufficient to Accelerate Reduction of Hunger and Malnutrition |
| 2013 | The Multiple Dimensions of Food Security |
| 2014 | Strengthening the Enabling Environment for Food Security and Nutrition |
| 2015 | Meeting the 2015 International Hunger Targets: Taking Stock of Uneven Progress |
| 2016 | No SOFI report published |
| 2017 | Building Resilience for Peace and Security |
| 2018 | Building Climate Resilience for Food Security and Nutrition |
| 2019 | Safeguarding Against Economic Slowdowns |
| 2020 | Transforming Food systems for Affordable Healthy Diets |

## REFERENCES

Adamchick, J., & Perez, A. (2020). Choosing awareness over fear: Risk analysis and free trade support global food security. *Global Food Security, 26*. https://doi.org/10.1016/j.gfs.2020.100445

Aguiar, S., Texiera, M., Garibaldi, L., & Jobaggy, E. (2020). Global changes in crop diversity: Trade rather than production enriches supply. *Global Food Security, 26*. https://doi.org/10.1016/j.gfs.2020.100385

Alexander, D. E. (2013). Resilience and disaster risk reduction: An etymological journey. *Natural Hazards Earth Systems Science, 13*, 2707–2713.

Alinovi, L., Mane, E., & Romano, D. (2008). Towards the measurement of household resilience to food insecurity: Applying a model to Palestinian household data. In *Deriving food security information from the national household budget surveys. Experiences, achievement, challenges* (pp. 137–152). FAO.

Ansah, I., Gardebroek, C., & Ihle, R. (2019). Resilience and household food security: A review of concepts, methodological approaches and empirical evidence. *Food Security, 11*, 1187–1203.

Arksey, H., & O'Malley, L. (2005). Scoping studies: Towards a methodological framework. *International Journal of Social Research Methodology, 8*, 19–32. https://doi.org/10.1080/1364557032000119616

Arouna, A., Soullier, G., Mendez del Villar, P., & Demont, M. (2020). Policy options for mitigating impacts of COVID-19 on domestic rice value chains and food security in West Africa. *Global Food Security, 6*. https://doi.org/10.1016/j.gfs.2020.100405

Bacon, F. (1625). *Sylva Sylvarum, or of Natural History in ten Centuries*. W. Lee.

Bahadur, A. V., Ibrahim, M., & Tanner, T. (2010). The resilience renaissance? Unpacking of resilience for tackling climate change and disasters. CSR Discussion Paper No. 1, Institute of Development Studies, Strengthening Climate Resilience programme, Brighton, 45.

Barrett, C., & Constas, M. (2014). Toward a theory of resilience for international development applications. *Proceedings of the National Academy of Sciences, 111*(40), 14625–14630.

Béné, C. (2020). Resilience of local food systems and links to food security—A review of some important concepts in the context of COVID-19 and other shocks. *Food Security, 12*, 805–822. https://doi.org/10.1007/s12571-020-01076-1

Béné, C., Chowdhury, F. S., Rashid, M., Mamun, R., Dhali, S., Sabbir, A., & Jahan, F. (2017). Squaring the circle: Reconciling the need for rigor with the reality on the ground in resilience impact assessment. *World Development, 97*, 212–223. https://doi.org/10.1080/17565529.2017.1301868

Béné C., Cornelius, A., & Howland, F. (2018). Bridging humanitarian responses and long-term development through transformative changes—Some initial reflections from the World Bank's Adaptive Social Protection Programme in the Sahel. *Sustainability, 10*, 1697, https://doi.org/10.3390/su10061697

Béné, C., & Doyen, L. (2018). From resistance to transformation: A generic metric of resilience through viability. *Earth's Future, 6*. https://doi.org/10.1002/2017EF000660

Béné, C., Godfrey-Wood, R., Newsham, A. & Davies, M. (2012). Resilience: New utopia or new tyranny?—Reflection about the potentials and limits of the concept of resilience in relation to vulnerability reduction programmes. IDS Working Paper 405. Institute of Development Studies.

Béné, C., Headey, D., Haddad, L., & von Grebmer, K. (2016). Is resilience a useful concept in the context of food security and nutrition programmes? Some conceptual and practical considerations. *Food Security, 8*(1), 123–138.

Béné, C., Newsham, A., Davies, M., Ulrichs, M., & Godfrey-Wood, R. (2014). Review article: Resilience, poverty and development. *Journal of International Development, 26*, 598–623.

Bentley, M. (1938). Retrospect and prospect. *The American Journal of Psychology, 51*(2), 357–360.

Bernard de Raymond, A., Alpha, A., Ben Ari, T., Daviron, B, Nesme, T., & Tétart, G. (2021). Systemic risks a food security. Emerging trends and future avenues for research. *Global Food Security, 29*, https://doi.org/10.1016/j.gfs.2021.100547

Bohle, H., Etzold, B., & Keck, M. (2009). Resilience as agency. *International Human Dimensions Programme, Update, 2*, 8–13.

Botero, G. (2012). *On the causes of the greatness and magnificence of cities*. Geoffrey Symcox (Trans and intr.). University of Toronto Press: Toronto (Original work published in 1588).

Boorman, J. Faigenbaum, J. Ferhai H. Bhaskaran, D. & Kohli, A. (2013). The centennial resilience index: Measuring countries' resilience to shock. *Global Journal of Emerging Market Economies, 2*, 57–98, https://doi.org/10.1177/0974910113494539

Briguglio, L., Cordina, G., Farrugia, N., & Vella, S. (2009). Economics vulnerability and resilience: Concepts and measurements. *Oxford Development Studies, 37*, 229–347.

Brown, K. (2016). *Resilience, development and global change*. Routledge. https://doi.org/10.4324/9780203498095

Brouwer, I., McDermott, J., & Ruben, R. (2020). Food systems everywhere: Improving relevance in practice. *Global Food Security, 26*. https://doi.org/10.1016/j.gfs.2020.100398

Bullock, K., Dhanjal-Adams, A., Milne, T., Oliver, L., Todman, A., Whitmore, & Pywell R. (2017). Resilience and food security: Rethinking an ecological concept. *Journal of Ecology, 105,* 880–884.

Caron, P., de Ferrero, Y., & Loma-Osario, G. (2018). Food systems for sustainable development: Proposals for a profound four-part transformation. *Agronomy and Sustainable Development, 38.* https://doi.org/10.1007/s13 593-018-0519-1

Cicchetti, D., & Garmezy, N. (1993). Milestones in the development of resilience [Special issue]. *Development and Psychopathology, 5*(4), 497–774.

Constas, M., Frankenberger, T., Hoddinott, J., Mock, N., Romano, D., Béné C., & Maxwell, D. (2014). *A proposed common analytical model for resilience measurement: A general causal framework and some methodological options.* Published by the Food and Agriculture Organization and the World Food Program under the Food Security and Information Network.

Constas, M., Mattioli, L. & Russo, L. (2020). What does resilience imply for development practices? Tools for more coherent programming and evaluation of resilience. *Development Policy Review,* https://doi.org/10.1111/dpr.12518

Choularton, R., Frankenberger, T., Kurtz J., & Nelson, S. (2015). *Measuring shocks and stressors as part of resilience Measurement.* Resilience Measurement Technical Working Group (Technical Series No. 5). Food Security Information Network, www.fsincop.net/fileadmin/user_upload/fsin/docs/resour ces/FSIN_TechnicalSeries_5.pdf

Coulthard, S. (2012). Can we be both resilient and well, and what choices do people have? Incorporating agency into the resilience debate from a fisheries perspective. *Ecology and Society, 17*(1), 4.

Czekaj, M., Adamsone-Fiskovica, A., Tyran, E. & Kilis, E. (2020). Small farms' resilience strategies to face economic, social, and environmental disturbances in selected regions in Poland and Latvia. *Global Food Security, 26.* https://doi.org/10.1016/j.gfs.2020.100416

Deaton, A. (1997). *The analysis of household surveys: A microeconomic approach.* Johns Hopkins University Press.

De Pinto, A., Islam, M. M, & Katic, P. (2023). Food security under a changing climate: Exploring the integration of resilience in research and practice. [citation to be completed by editors]

Dercon, S. (2002). Income risk, coping strategies, and safety nets. *The World Bank Research Observer, 17,* 141–166.

d'Errico, M., Garbero, A., Letta, M., & Winters, P. (2020). Evaluating program impact on resilience: Evidence from Lesotho's Child Grants Programme. *Journal of Development Studies, 56*(12), 2212–2234.

European Union. (2017). Operationalizing the humanitarian-development nexus—Council conclusions. *Outcome of Proceedings of the European Union,*

*19 May 2017*, Brussels. Retrieved from www.consilium.europa.eu/media/24010/nexus-st09383en17.pdf

Fafchamps, M., & Lund, S. (2003). Risk-sharing networks in rural Philippines. *Journal of Development Economics, 71*(2), 261–287.

Food and Agriculture Organization (FAO). (1974). World Food Conference, Rome, November 5 and 16.

FAO. (1996). World food summit: Rome declaration on world food security and world food summit plan of action, November 13–17, Rome, Italy.

FAO. (2002). *Report of the World Food Summit five years after*. FAO.

FAO. (2004). *The state of food insecurity in the world 2004: Monitoring progress towards the World Food Summit and Millennium Development Goals*. FAO.

FAO. (2006). *The Right to Food Guidelines: Information Papers and Case Studies*. FAO.

FAO. (2008a). An introduction to the basic concepts of food security. Rome: FAO. www.fao.org/3/a-al936e.pdf

FAO. (2008b). *The state of food insecurity in the world 2008: Monitoring progress towards the World Food Summit and Millennium Development Goals*. FAO.

FAO. (2012). *The state of food insecurity in the world 2012: Monitoring progress towards the World Food Summit and Millennium Development Goals*. FAO.

FAO. (2018). *The state of food insecurity in the world 2018: Monitoring progress towards the World Food Summit and Millennium Development Goals*. FAO.

Forsyth, T. (2018). Is resilience to climate change socially inclusive? Investigating theories of change processes in Myanmar. *World Development, 111*, 13–26.

Frei, B., Queiroz, C., Chaplin-Kramer, B., Andersson, E., Renard, D., Rhemtulla, J., & Bennet, E. (2020). A brighter future: Complementary goals of diversity and multifunctionality to build resilient agricultural landscapes. *Global Food Security, 26*. https://doi.org/10.1016/j.gfs.2020.100407

Gößling-Reisemann, S., Hellige, H. D., & Thier, P. (2018). The Resilience Concept: from its historical roots to theoretical framework for critical infrastructure design. (artec-paper, 217). Bremen: Universität Bremen, Forschungszentrum Nachhaltigkeit (artec). https://nbn-resolving.org/urn:nbn:de:0168-ssoar-59351-6

Gordillo, G. (2013). *Food security and sovereignty (a base document for discussion)*. FAO.

Gott, S. (1670). *The divine history of the genesis of the world explicated & illustrated*. E. C. & A. C. for Henry Eversden.

Hasegawa, T., Fujimori, S., Havlík, P., et al. (2018). Risk of increased food insecurity under stringent global climate change mitigation policy. *Nature Climate Change, 8*, 699–703.

Hilhorst, D. (2018). Classical humanitarianism and resilience humanitarianism: Making sense of two brands of humanitarian action. *Journal of International Humanitarian Action*, 3(15), https://doi.org/10.1186/s41018-018-0043-6

HLPE. (2017). Nutrition and food systems. A report by the High-Level Panel of Experts on Food Security and Nutrition of the Committee on World Food Security. HLPE report 12. HLPE.

HLPE. (2020). Food security and nutrition: Building a global narrative towards 2030. A report by the High-Level Panel of Experts on Food Security and Nutrition of the Committee on World Food Security. HLPE.

Holling, C. S. (1973). Resilience and stability of ecological systems. *Annual Review of Ecology and Systematics*, 4, 1–23.

Howe, P. (2019). The triple nexus: A potential approach to supporting the achievement of the sustainable development goals? *World Development*, 124. https://doi.org/10.1016/j.worlddev.2019.104629

Jiren, T., Dorresteijn, I., Hanspach, J., Shultner, J., Bergsten, A., Manlosa, A., Jager, N., Senbata, F., & Fischer, J. (2019). Alternative discourses around the governance of food security: A case study from Ethiopia. *Global Food Security*, 24. https://doi.org/10.1016/j.gfs.2019.100338

Keathley-Herring, H., Van Aken, E., Gonzalez-Aleu, F., Deschamps, F., Letens, G., & Cardenas Orlandini, P. (2016). Assessing the maturity of a research area: Bibliometric review and a proposed framework. *Scientometrics*, 109, 927–951.

Keil, A., Zeller, M., Wida, A., Sanim, B., & Birner, R. (2008). What determines farmers' resilience towards ENSO-related drought? An empirical assessment in Central Sulawesi, Indonesia. *Climatic Change*, 86, 291–307.

Kose, M. & Prasad, E. (2010). *Emerging markets: Resilience and growth amid global turmoil*. Brookings Institution Press.

Kuhn, T. (1982). Commensurability, comparability, and communicability. *Proceedings of the Biennial Meeting of the Philosophy of Science Association*, 2, 669–688.

Kummu, M., Kinnunen, P., Lehikoinen, E., Porkka, M., Querioz, C., Röös, E., Troell, M., & Weil, C. (2020). Interplay of trade and food system resilience: Gains on supply diversity over time at the cost of trade independency. *Global Food Security*, 24. https://doi.org/10.1016/j.gfs.2020.100360

Laver, M., & Benoit, K. (2003). Extracting policy positions from political texts using words as data. *Americana Political Science Review*, 97, 211–331.

La Via Campesina. (2007). Nyéléni Declaration of the Forum for Food Sovereignty. https://nyeleni.org/spip.php?article290

Leroy, G., Boettcher, P., Besbes, B., Raul Pena, C., Jaffrezic, F., & Baumung, R. (2020). Food securers or invasive aliens? Trends and consequences of non-native livestock introgression in developing countries. *Global Food Security*, 26. https://doi.org/10.1016/j.gfs.2020.100420

Lewontin, R. (1969). The meaning of stability. *Brookhaven Symposia in Biology*, 22, 12–24.

Loboguerrero, A., Thornton, P., Wadsworth, J., Campbell, B., Herrero, M., Mason-D'Croz, D., Dinesh, D., Huyer, S., Jarvis, A., Millan, A., Wollenberg, W., & Zebiak, S. (2020). Perspective article: Actions to reconfigure food systems. *Global Food Security, 26*. https://doi.org/10.1016/j.gfs.2020.100432

Malthus, T. (1798). An essay on the principle of population as it affects the future improvement of society, with remarks on the speculations of Mr. Godwin, M. Condorcet, and other writers. J. Johnson in St. Paul's Churchyard, www.esp.org/books/malthus/population/malthus.pdf

Martin-Shields, C., & Stojetz, W. (2019). Food security and conflict: Empirical challenges and future opportunities for research and policy making on food security and conflict. *World Development, 119*, 150–164.

Miles, W. (1935). The material as dealt with by the psychologist. *The American Journal of Psychiatry, 92*(2), 285–295.

Pimm, S. (1984). The complexity and stability of ecosystems. *Nature, 307*, 321–326.

Pingali, P., & Sunder, N. (2017). Transitioning toward nutrition-sensitive food systems in developing countries. *Annual Review of Resource Economics, 9*, 439–459.

Pingali, P., Alinovi, L., & Sutton, J. (2005). Food Security in complex emergencies: Enhancing food system resilience. *Disasters*, 29(1), Food security in complex emergencies, S5–S24. https://doi.org/10.1111/j.0361-3666.2005.00282.x

Rankine, W. J. M. (1858). *A manual of applied mechanics*. Richard Griffin.

Reardon, T., & Timmer, P. (2012). The economics of the food system revolution. *Annual Review of Resource Economics, 4*, 225–264.

Sellberg, M., Norström, A., Peterson, G., & Gordon, L. (2020). Using local initiatives to envision sustainable and resilient food systems in the Stockholm city region. *Global Food Security, 24*. https://doi.org/10.1016/j.gfs.2019.100334

Sen, A. (1999). *Development as freedom*. Oxford University Press.

Serfilippi, E., & Ramnath, G. (2018). Resilience measurement and conceptual frameworks: A review of the literature. *Annals of Public and Cooperative Economics, 89*, 645–664.

Serra, R., Kiker, G., Minten, B., Valerio, V., Varijakshapanicker, P, & Wane, A. (2020). Filling knowledge gaps to strengthen livestock policies in low-income countries. *Global Food Security, 26*, www.sciencedirect.com/science/article/abs/pii/S2211912420300821

Shaw, D. J. (2007). *World food security: A history since 1945*. Palgrave Macmillan.

Staal, T. (2016). Resilience is today's buzzword in international development, but what does it really mean? *USAID Frontlines*, January 13, 2016. https://medium.com/usaid-frontlines/insights-tom-staal-5a7ab307d818

Tausczik, Y., & Pennebaker. (2010). The psychological meaning of words: LIWC and computerized text analysis methods. *Journal of Language and Social Psychology, 28*, 24–54.

Tendall, D., Joerin, M., Kopainsky, J., Edwards, B., Shreck, P., Le, A., Q. B., Kruetli, P., Grant, M., & Six, J. (2015). Food system resilience: Defining the concept. *Global Food Security, 6*, 17–23.

Thagard, P. (2000). *Coherence in thought and action*. MIT Press.

Thagard, P. (2007). Coherence, truth and the development of scientific knowledge. *Philosophy of Science, 74*, 26–47.

Timmerman, P. (1981). Vulnerability, resilience and the collapse of society. Environmental Monograph 1, Institute for Environmental Studies, Toronto University, Toronto.

Tobin, G. A. (1999). Sustainability and community resilience: The holy grail of hazards planning? *Environmental Hazards, 1*(1), 13–25.

United Nations Office of Disaster Risk Reduction (UNDRR). (2021). *From risk to resilience: Accelerating adaptation action at COP26*. United Nations. www.undrr.org/event/risk-resilience-accelerating-adaptation-action-cop26

United States Agency for International Development (USAID). (2012). Building resilience to recurrent crisis: USAID policy and program guidance. www.usaid.gov/policy/resilience

Vernon, R. F. (2004). A brief history of resilience. In Clauss-Ehlers C. S. & Weist M. D. (Eds.), *Community planning to foster resilience in children*. Springer, Boston, MA. https://doi.org/10.1007/978-0-306-48544-2_2

Von Braun, J. (2009). *Food security risks must be comprehensively addressed: Annual report essay, 2008–2009*. International Food Policy Research Institute.

Walker, B. (1992). Biological diversity and ecological redundancy. *Conservation Biology, 6*(1), 18–23.

Walker, B., Holling, C. S., Carpenter, S., & Kinzig A. (2004). Resilience, adaptability, transformability in socio-ecological systems. *Ecology and Society, 9*(2), 5. www.ecologyandsociety.org/vol9/iss2/art5/

Walker, B., & Salt, D. (2006). *Resilience thinking: Sustaining ecosystems and people in a changing world*. Island Press.

Wertheim-Heck, & Raneri, J. (2020). Food policy and the unruliness of consumption: An intergenerational social practice approach to uncover transforming food consumption in modernizing Hanoi, Vietnam. *Global Food Policy, 26*, https://doi.org/10.1016/j.gfs.2020.100418

Weingärtner, L., Pforr, T., & Wilkinson, E. (2020). *The evidence-base on anticipatory action*. World Food Programme.

Zseleczky, L. & Yosef, S. (2014). Are shocks really increasing? A selective review of the global frequency, severity, scope, and impact of five types of shocks. 2020 Conference Paper 5, May. IFPRI

**Open Access** This chapter is licensed under the terms of the Creative Commons Attribution 4.0 International License (http://creativecommons.org/licenses/by/4.0/), which permits use, sharing, adaptation, distribution and reproduction in any medium or format, as long as you give appropriate credit to the original author(s) and the source, provide a link to the Creative Commons license and indicate if changes were made.

The images or other third party material in this chapter are included in the chapter's Creative Commons license, unless indicated otherwise in a credit line to the material. If material is not included in the chapter's Creative Commons license and your intended use is not permitted by statutory regulation or exceeds the permitted use, you will need to obtain permission directly from the copyright holder.

CHAPTER 6

# Food Systems, Resilience, and Their Implications for Public Action

*John Hoddinott*

## INTRODUCTION

The last 15 years has seen global food systems subject to two major shocks—the 2008 food prices crisis and the COVID-19 pandemic that began in early 2020. Country-level shocks such as drought and floods continue—in 2020 alone, nearly 34 million people in China and India were affected by flooding (CRED, 2021). While global per capita food production continues to increase, the multi-decade trend of falling numbers of persons considered to be undernourished appears to be coming to an end. There is also increased concern over the quality of

---

J. Hoddinott (✉)
Division of Nutritional Sciences, Charles H. Dyson School of Applied Economics and Management and Department of Global Development, Cornell University, Ithaca, NY, USA
e-mail: jfh246@cornell.edu

© The Author(s) 2023
C. Béné and S. Devereux (eds.), *Resilience and Food Security in a Food Systems Context*, Palgrave Studies in Agricultural Economics and Food Policy, https://doi.org/10.1007/978-3-031-23535-1_6

diets that global food systems provide, and the environmental impacts of the farming practices that provide the food for these diets.[1]

These stylized facts have informed increased use of two concepts: food systems and resilience. This chapter seeks to contribute to efforts that bring these two notions together with a view towards understanding how a resilience lens can improve our understanding of food systems at local and global levels, how resilience can be better measured and assessed, and how this, in turn, contributes to improving food security interventions and policy. It consists of three sections: Building blocks; Linking resilience to food systems; and Implications for public action.

## BUILDING BLOCKS

The last ten years has seen an outpouring of work on the concept of food systems, based on a recognition that because of the multiplicity of factors that underpin food security, a too heavy emphasis on a single component (say production or markets) is insufficient (Reardon & Timmer, 2012). Examples of this work include HLPE (2017), Béné (2020), FAO, IFAD, UNICEF, WFP and WHO (2020), FAO (2021), and Herrero et al. (2021). While these approaches differ in detail, common to many are the following actors—producers; processors; distributors; and consumers—and a view that the objective of a resilient food system is to meet dietary needs in a sustainable fashion (HLPE, 2017). These approaches to food systems are careful to note that there are multiple food products within a food system, each characterized by its own value chain and that there are heterogeneities within each element. Further, distinctions between these elements are not always clear cut—a notable example being households in low-income countries who are food producers, processors, and consumers of their own production.

Just as there are many definitions of food systems, there are many definitions of resilience. These include: (1) Resilience as the *capacity* to withstand or absorb sudden or chronic shock (Béné, 2020; Constas et al., 2014; FAO, 2016); (2) Resilience as *recovery*; the extent to which food

---

[1] I have benefitted from ongoing discussions about food systems and resilience with Chris Barrett, Andrea Cattaneo, Mark Constas, Marco d'Errico, Rebecca Pietrelli, and Maximo Torero. This chapter has been made substantially better by the detailed comments provided by Chris Béné and Stephen Devereux. Errors are mine.

security (or other measure) returns to its pre-shock state. This conceptualization of resilience hews closely to the use of the concept in ecology and engineering, and to the word's etymological roots (Hoddinott, 2014; Hoddinott and Knippenberg, 2017; Béné, 2020); and (3) Resilience as a *normative condition*; the capacity to avoid adverse well-being states (or achieve a desirable state) in the face of exposure to shocks and stressors (Barrett & Constas, 2014; Cissé & Barrett, 2018).

Central to discussions of resilience are the concepts of shocks and stressors to different actors within the food system. Shocks are events, either positive or negative, the timing and severity of which cannot be precisely predicted in advance. A stressor is a long-term trend that adversely affects a system and increases the vulnerability of actors within that system to shocks. Shocks emanate from the settings (Hoddinott & Quisumbing, 2010)[2]—physical, social, political, legal, and economic—in which actors operate (a covariant shock), or they could be restricted to only one person or household (an idiosyncratic shock) (Dercon et al., 2005). The distinction between covariant and idiosyncratic shocks is not always clear-cut. A drought in only one locality might result in poor, rainfall-dependent households selling assets to richer, non-rainfall dependent households so, although the event was common to both, it adversely affected only the poor. Further, shocks can vary in terms of speed of onset, duration, and intensity. These shocks can affect the settings themselves, household assets, or the processes by which these assets are used to generate income (Hoddinott & Quisumbing, 2010).

Building on these ideas, food systems resilience can be defined as the capacity over time of a food system to sustainably provide sufficient, appropriate, and accessible food to all, in the face of shocks and stressors.

---

[2] The physical setting refers to natural phenomena such as the level and variability of rainfall, the natural fertility of soils, distances to markets, and quality of infrastructure. The social setting captures such factors as the existence of certain norms of behaviour, of social cohesion and strife. The legal setting can be thought of as the general "rules of the game" in which exchange takes place, which, in turn, is partly a function of the political setting that captures the mechanisms by which these rules are set. Finally, there is an economic setting that captures policies that affect the level, returns, and variability of returns on assets (Hoddinott & Quisumbing, 2010).

## Linking Resilience to Food Systems

Two themes underpin this section. First, more is known about the resilience of some (but not all) components of food systems than is commonly recognized. Second, there is relatively little understanding of how the resilience of these components fit together.

Some additional comments about the first theme are warranted. There is a long-standing literature within agricultural economics that examines how (and how successfully) food producers manage risk (see for example, Moschini and Hennessy [2001] for a summary of the older literature on this topic and Tack and Yu [2021] for a more recent review, defined as the possibility of different types of shocks (covariate or idiosyncratic) being realized. Characterizing these ex ante actions is consistent with the notion of resilience as the *capacity* to absorb shocks. Within the literature on the functioning of food markets (distributors, both wholesalers and retailers), there has been a considerable body of research into the extent to which these markets are integrated and why some markets are more integrated than others. As discussed below, such approaches are consistent with the notion of resilience as *recovery*. Lastly, there is a large literature on household resilience, much of which will be discussed elsewhere in this book. This literature will be very briefly summarized. By contrast, there appears to be much less work on food systems resilience (or related literatures such as risk management), the focus of this chapter.

### *Food Production Resilience*

Food production is seen as a core component of food systems (FAO, 2021; HLPE, 2017). However, with the recent exceptions of FAO (2021) and Constas et al. (2021), there is much less explicit work on measurement of food production resilience. In part, this reflects enormous variations in agro-ecological conditions and in how food is produced around the world. As described by Savary et al., "These production units include the large-scale commercial farms of the global North, with their high level of mechanisation and inputs (synthetic and also biological, with highly selected and specialised seed) as well as the small-scale, smallholder farms of the global South, with their large labour force, their crop diversity, the frequent inclusion of livestock in agriculture, and their limited reliance on external inputs" (Savary et al., 2020, p. 695). That noted, consider the notion of resilience as the capacity to withstand or absorb sudden shocks or chronic stressors. Measures of the resilience of food production can either be a summary statistic that captures this ability or as characteristics of the organization, structure, and process of food production that are believed to be associated with the summary statistics. With the important caveat that the ideas below are exploratory, we describe both approaches below.

Before doing so, we briefly set out a conceptual model of food production. The farming unit has endowments of capital and labour. Capital includes physical capital (agricultural tools, livestock), natural capital (land), human capital (in the form of knowledge, skills, and health), financial capital, and social capital (Scoones, 1998). The farmer allocates these endowments, along with purchased inputs, across a series of agricultural activities. (For simplicity, we ignore allocations to non-agricultural activities.) These allocations are based on perceptions of the level and variability of activities returns, as well as their covariance. For example, farmers may decide to grow a mix of crops that embody differing levels of susceptibility to climatic shocks and returns. Crops may be grown in different locations, may be temporally diverse (that is, grown at different times or different crops may grow to maturity at different speeds) or may be intercropped. Once these allocations are made, shocks that threaten crop or livestock production (covariant or idiosyncratic) occur; these are outside the direct control of the farmer. It may respond to these shocks through undertaking compensating or reinforcing actions (for example, undertaking additional weeding in fields affected by a weed infestation; spending more time harvesting a field where production had been atypically high) (Hoddinott & Quisumbing, 2010).

A summary statistic of food production resilience at the national level—or to phrase more precisely, a summary statistic of domestic food production resilience—captures the outcome of these allocations, the shocks and the compensating or reinforcing actions can be measured in physical or monetary terms; it does not attempt to disaggregate or disentangle how the outcome has come about. One such summary statistic is outlined in Zampieri et al. (2020); a simplified version of their approach goes as follows.

Consider circumstances where the shocks adversely affecting food production are so severe, or the ability of farmers to respond to shocks so limited, that the consequence of the shocks is total crop failure. The probability that this occurs is given by $F$ where $0 < F < 1$. We can think of a farm (or locality or country) with resilient food production as the reciprocal of the probability of total crop failure.

$$R = 1/F \tag{6.1}$$

As $F \to 0$, $R$ rises in value. As $F \to 1$, $R$ approaches 1 and so an increase in the value of $R$ captures the notion of greater resilience of food production. Next, make the strong assumptions that mean production over time and variations in production over time are both trendless. Following Zampieri et al. (2020), define $P$ as the level of production that occurs when conditions are optimal and allow only two states of the world: one where crop production is optimal and one where crop production fails totally. With these strong assumptions in mind, over time, the mean and variance of food production are given by:

$$\mu = P(1-F) \tag{6.2}$$

and

$$\sigma^2 = P^2(1-F)(F) \tag{6.3}$$

Manipulating these expressions yields:

$$R = \mu^2/\sigma^2 \tag{6.4}$$

Equation (6.4) is the summary statistic. Food production resilience is the inverse of the coefficient of variation of production squared. A higher value for $R$ corresponds to greater production resilience. Note that this is consistent with intuition. Variability in production (the denominator) increases when farmers are less able to minimize the effects of adverse shocks (for example, where farmers lack access to irrigation, there will be greater year-to-year fluctuations in output because of differences in rainfall over time; this increased variability increases the magnitude of the denominator and thus lowers $R$. Conditional on $\sigma^2$, increased production is associated with greater resilience; the intuition here being that at higher levels of production, a given level of variability represents a small fraction of total output.[3]

---

[3] Zampieri et al. (2020) show how to adapt this approach to circumstances where production is non-stationary or where more than one crop is produced.

We end this section noting the following. First, elements of the conceptual model described above contain characteristics believed to be associated with summary statistics for food production resilience. For example, we should expect that measures of endowments and actions that reduce variability in production will be correlated with increased production resilience. Examples include measures of farming practices that reduce the likelihood of crop failure (for example, intercropping and crop rotation), the availability of irrigation (or more generally, resources that reduce reliance on rainfall and improve water control), and improved availability of inputs (captured, for example, through measure of the thickness of input markets.) Second, as Savary et al. (2020), FAO (2021), and many others have noted, diversification—the choices farmers make about what they grow, where, and when—is seen as one way in which farming can be made more resilient. Diversification indices for what is grown include the Shannon and Simpson diversity indices; as the literature on farm fragmentation shows, these can also be adapted to capture spatial diversity in production (see Knippenberg et al., 2020, for references and FAO (2021) for more recent work on measuring production diversity). However, there is a risk that "too much" diversification might come at the cost of reduced efficiency in production, for example, because of a loss of economies of scale or in comparative advantage.

Second, this measure pertains to resilience of domestic food production. It does not account for international trade. In principle, low domestic production resilience can be offset through food imports, though this does expose domestic food supply to shocks that emanate outside the country. We return to this point when we discuss implications for public action.

### *Resilience in the Food Processing Sector*

Globally, the food processing sector is enormously heterogeneous, ranging from the large meat processing plants employing hundreds of workers to women grinding grain harvested from their own fields. Unlike food security, there are no well-developed, validated metrics for resilience in the food processing sector. Nor, unlike food production or food markets, are there measures that can be adapted to capture aspects of resilience. That said, literatures on supply chains and on recent experiences arising from the COVID-19 pandemic suggest several possible metrics that could be developed to capture resilience within the food processing sector.

The supply chain literature: Aboah et al. (2019) argue that flexibility is a key attribute in the resilience of value chains. Applying their approach specifically to the processing sector, flexibility includes the ability to: re-organize production/processing in response to a shock; obtain raw foods from other sources should disruptions affecting existing suppliers; and tap alternative distribution channels. Relatedly, stock holdings can also play a role by reducing processors' vulnerability to transitory shocks in the supply of inputs.

COVID-19 experiences: Taylor et al. (2020) document the spread of COVID-19 among workers in US livestock plants. They note that such operations are susceptible to the transmission of coronaviruses for several reasons, including that their employees work long shifts in close proximity to coworkers. They also note that in the United States, 12 plants produce more than 50 per cent of the country's beef and 12 other plants are responsible for more than 50 per cent of pork production. Rotz and Fraser (2015) also document increased concentration within the North American food processing sectors. Three examples from their paper illustrate this:

- As early as 1962, the largest 50 processing firms in the United States controlled 70% of market sales
- As of 2015, the four largest processors in the United States milled more than half of all wheat flour
- The three largest US meat packers control 80 per cent of the American beef market.

Given all this, Savary et al.'s (2020) description of how COVID-19 affected the North American food processing sector is not surprising. "Labour shortages have also been an issue for large-scale food processors and suppliers. A growing number of workers are taken ill in food processing facilities where the operational model is not conducive to safe physical distancing. Consequently, a large number of food processing plants temporarily suspended production in Europe and North America" (Savary et al., 2020, p. 704).

Putting these disparate studies together suggests that resilience within the processing sector reflects three considerations: (a) the extent of market concentration within the sector. Countries where food processing is dominated by a small number of firms may be less resilient to shocks that affect their workforces; (b) the availability of substitutes, for example, through imports. While the shocks described by Savary et al. (2020) were disruptive—particularly for meat processing and packaging—the availability of other sources of animal source foods lessened their impacts on consumers; and (c) more speculatively, the degree of labour intensity within the processing sector with greater intensity associated with lower resilience (Reardon & Swinnen, 2020). We do not have good candidate summary statistics that capture all of this.

## *Resilient Food Markets*

There is a vast academic literature on the structure, conduct, and performance of food markets in low-, middle- and high-income countries. This literature rarely speaks directly to the notion of resilient food markets. However, the literature on spatial market integration provides relevant insights.

We begin with an adaptation of the Takayama—Judge model described by Fackler and Goodwin (2001) and Fackler and Tastan (2008) as a point-location model. Geographically separated locations are represented by points or nodes. Assume that no node is connected to another. Within each node, the price of food is determined by local production (supply) and local demand. An adverse shock to supply—say a drought or flood occurs—causing supply to fall. With no means of offsetting this, food prices rise and remain persistently high until supply is restored. In extreme versions of this (and where there are no offsetting increases in wages or income), the result could be famine; see Devereux (1988) and Ravallion (1987, 1997). Seen in this way, these unconnected geographically separated food markets are not resilient—they lack the *capacity* to withstand or absorb sudden or chronic shock and their *recovery*—the extent to which prices return to their pre-shock state—is slow.

Next, we relax the strong assumption that no node is connected to another by introducing a set of transportation routes or links. Links and nodes together constitute a trade (market) network (Fackler & Tastan, 2008). Again, consider a supply shock in one node. The initial effect is to raise prices in that node but, by so doing, prices differ across the two nodes. Traders can exploit this through arbitrage, buying food in the node not affected by the supply shock, then transporting it to and selling it in the node where the supply shock occurred. This has the effect of slowing the rise in food prices in the affected node, allowing them to return to their pre-shock state more quickly. But it also, potentially, causes prices to change in the non-affected node. The extent of the transmission of the exogenous price shock in the affected node to prices in the non-affected node is captured by measures and methods of assessing market integration. These include error correction models, cointegration analysis, and parity bounds models (von Cramon-Taubadel, 2017; Kabbiri et al., 2016; Varela et al., 2012).

With caveats that we return to below, we assert that more integrated food markets are more resilient food markets. In turn, this takes us to the question as to what features influence the extent to which markets are integrated. These fall into four categories: (1) Information flows; (2) Transactions costs; (3) Government regulations on trade; and (4) Market structure.

Knowing that prices differ across markets is necessary for arbitrage to take place (Jensen, 2010). Quantifying these information flows is challenging. A proxy measure used in a handful of studies is some measure of access to communications technology. An older study by Goletti et al. (1995) examining rice market integration in Bangladesh between 1989 and 1992 found that the number of telephones per capita was associated with reduced market integration, a somewhat counterintuitive finding. By contrast, Aker (2010) finds that the introduction of mobile phones reduces dispersion in prices in rural Niger, and Jensen (2007) shows how arbitrage in south Indian fish markets increased after the introduction of mobile telephony.

Transaction costs may also affect market integration. *Ceteris paribus*, these will be higher the farther markets are away from each other, and several studies show this (for example, see Varela et al., 2012). Direct measures of transaction costs are rare, however with Zant (2013) being an exception. Instead, road density and quality can be used as proxies for transaction costs—higher quality roads can be travelled more quickly and can support larger vehicles, allowing for greater economies of scale in transport. FAO (2021) extends the measurement of road networks to encompass two additional ideas: route redundancy (the availability of alternative routes when a road link is broken); and detour costs, the extra costs incurred when a route is closed and the shortest alternative route needs to be taken. Scale economies may also arise when markets are larger (put differently, per unit transportation costs are an inverse function of volume, Jensen, 2010). Some studies capture this idea of scale economies by including measures of the size of the market, population or population density or incomes per capita.

Arbitraging across spatially separated markets will be affected by government regulations on trade as well as the broader legal and policy environment in which trade takes place. The strongest version of these are prohibitions on the movement of food products across administrative borders. Requirements that marketed surpluses be sold to state-owned entities accompanied by the use of fixed, below market procurement prices are another form of intervention as is the use of government buffer stocks (both purchases for and sales to). Dercon (1995) and Rozelle et al. (1997) document that reductions in government involvement in grain markets in Ethiopia and China, respectively, improved market integration. That said, Ismet, Barkley, and Llewelyn (1998) argue that government intervention in Indonesian rice markets enhanced market integration but with the caveat that procurement prices were relatively high. Martin and Anderson (2012) argue that restrictions on movement of food products, specifically exports, was a significant factor in contributing to the rise in global food prices in 2008; noting that once a few countries started to do so, others quickly followed suit resulting in a cascade of export bans and subsequently panic buying by other countries that were dependent on food imports to meet domestic food security needs. Possibly having learned from the policy mistakes made in 2007–08, fewer government-imposed restrictions on food exports; by late 2020, only 13 countries had done so. These affected only a minimal amount of the volume of food traded globally, around one per cent of global-traded calories (Martin & Glauber, 2020) and many of these were subsequently rescinded.

Market structure could contribute to either enhancing or detracting from market integration. As Kabbiri et al. (2016) note, market concentration may allow for economies of scale in the collection of information on prices and on transport, thus allowing such traders to respond more quickly to price differentials. But they also note that traders may have an incentive to sustain market segmentation to keep prices artificially high. Evidence on the impact of market structure on market integration appears to be lacking.

We end with two interrelated caveats. First, the description provided here focuses on domestic markets. Integration into regional and/or global markets can also provide resilience to domestic food markets as well as potentially reducing prices. Just as with domestic food markets, the quality of infrastructure linking markets in different countries along with the regulatory environment—specifically rules and tariffs governing cross-border trade—will affect the extent of market integration; see Brenton et al. (2014) for an example. Second, tensions exist between open and closed food systems. Food markets that are more regionally and globally integrated are more likely to be affected by shocks that occur elsewhere, see Bekkers et al. (2017). Put differently, while integration into global markets that creates *dependence* on imports may result in a less resilient food system whereas *diversification* of food supplies so that they include imports (or the ability to import as needed) may increase food system resilience.[4]

## *Food Security Resilience*[5]

Up to this point, our focus has been on the supply side of the food system. Analyses of food systems, however, also include the demand side; more specifically food security outcomes (see, for example, HLPE, 2017). In contrast to other elements of the food system, extensive attention has been paid to resilience at the level of households as consumers, often described as food security resilience or more generally as development resilience. We summarize four approaches here. Three of these are based on the concept of "resilience as ex ante capacity", the fourth uses the idea of "resilience as a normative condition".

Resilience as ex ante capacity can be thought of as "the capacity to withstand or absorb sudden or chronic shock; cope with temporary disruption while minimizing the damages and costs from hazard; restore after an event; manage or maintain basic functions and structures to become suitable for future situation" (Birhanu et al., 2017, p. 2). In the existing literature, this is operationalized in several ways.

One approach is to build on the Sustainable Livelihoods framework that conceptualizes well-being as a function of five asset categories: financial, human, natural, physical, and social capitals (Scoones, 1998; Quandt et al., 2019). For example, Ranjan (2014) focuses on the roles that social and financial capital play in dealing with drought. Quandt et al. (2019) constructs a series of composite indices across the five asset categories

---

[4] My thanks to Stephen Devereux for suggesting this phrasing.

[5] This material draws heavily on joint work found in Barrett et al. (2021).

using principal components (Browne et al., 2014). Composite indicators implicitly allow for substitution across asset categories (that is, two households could score at the same aggregate level but hold different types of assets in different quantities). An alternative approach used, for example, by Stanford et al. (2017) argue that households who score highly on each asset category (as well as a sixth that they call institutional capital) are best placed to be resilient. In contrast to the approach taken by Browne et al. (2014) and Quandt et al. (2019), Stanford et al. (2017) make the strong assumption that assets in different categories do not substitute for each other.

A limitation of the asset approach is that it does not account for resources beyond the household, such as infrastructure and social services that might also contribute to resilience (Birhanu et al., 2017, Stanford et al., 2017). A second approach to "resilience as ex ante capacity" seeks to remedy this weakness, the Resilience Indicators for Measurement and Analysis (RIMA), developed by the Food and Agriculture Organization (FAO). An updated version, RIMA-II (FAO, 2016), uses factor analysis to estimate four latent variables, labelled "pillars". One is assets (AST) and as such, is similar in spirit to the approaches based on the Sustainable Livelihoods framework. RIMA-II adds three additional pillars—Access to Basic Services (ABS), Social Safety Nets (SSN), and Adaptive Capacity (AC). Data for all pillars are typically found in standard household and community surveys. These are combined into an overall resilience capacity index (RCI). For example, in d'Errico et al.'s (2018) study of resilience in Tanzania and Uganda, AST has four elements: an agricultural asset index, a wealth index, tropical livestock units, and land. ABS is captured through consideration of an infrastructure index and distances to schools and markets. SSN includes public and private transfers. AC is based on income diversification, education, and income earners' share.

A third approach to "resilience as ex ante capacity" is found in Smith and Frankenberger (2018). They also conceptualize resilience as a latent capacity but use different pillars than those found in RIMA-II. Smith and Frankenberger consider three capacities. Absorptive capacities seek to mitigate the impact of shocks and include the availability of cash savings, access to informal safety nets, assets, bonding social capital (local norms of trust and reciprocity), and the availability of disaster mitigation. Adaptive capacities show the extent to which households can alter their livelihood strategies in the face of changing circumstances—diversity of livelihoods, assets, education, access to information, and bridging and linking social capital all contribute to these adaptive capacities, as does confidence in one's abilities to adapt. Finally, transformative capacity refers to "enabling conditions that foster more lasting resilience. It relates to governance mechanisms, access to markets, services and infrastructure, community networks, and formal safety nets that are part of the wider system in which households and communities are embedded" Smith and Frankenberger (2018, p. 366).

The fourth approach, "resilience as a normative condition" is developed in Barrett and Constas (2014) and Cissé and Barrett (2018). Here, resilience reflects the capacity to avoid adverse well-being states, rather than a capacity itself. Cissé and Barrett (2018) translate this conceptualization into an econometric method, estimating resilience as a conditional probability of satisfying some normative standard of living; for example, a food consumption score.

## Implications for Public Action

We begin by noting significant knowledge gaps in our understanding of food systems resilience.

First, although this paper is situated within the literature on food systems, it is important to recognize that there is no one "food system". Food systems differ by commodity and country's income level. There is also likely to be path dependence. Historical circumstances, legal traditions, and the like will all have influenced how food systems have developed. A consequence of all this is that households may be simultaneously exposed to very different food systems. For example, a smallholder Ethiopian farmer might grow maize and coffee. The former being a crop that is produced, processed, and consumed without ever leaving the household, while production of the latter is the first step in a complex value chain that encompasses multiple actors across multiple continents.

Second, we know much more—both conceptually and empirically—about some components of the food system, notably production and consumption, than we do about processing. Even in those sectors where the knowledge base is substantial, there are significant evidence gaps—a notable example being that of limited information on market structure in the distribution and processing sectors.

Third, language used in discourse surrounding food systems and resilience (including that found in this paper) often defaults to words such as households or actors. But households differ by place, race, ethnicity, and socioeconomic status and this can have implications for how different households interact with the food system. Further, as Barrett et al. (2021) note, much of the discourse surrounding food security resilience ignores the crucial role of gender. The chapter by Bryan et al. (Chapter 8 in this volume) discusses this at length.

Fourth, our understanding of how these components fit together is limited. Do all elements within a food system need to be resilient for the system to be resilient? Or can greater resilience in some elements compensate for more limited resilience in others? Questions such as these are relevant to how we prioritize public investments.

Given these knowledge gaps, it would be unwise to make strong statements regarding the implications of applying a resilience lens to food systems for the purposes of contributing to improved food security interventions and policy. Mindful of this important caveat, we note the following.

Adapting the language of United Nations (2020), governments can enhance the resilience of food systems through improved anticipation, prevention, absorption, adaptation, and transformation. While, by definition, the timing of shocks cannot be precisely predicted in advance, it is possible to improve knowledge of when and where they are most likely to occur. The famines of the 1980s spurred investment in early warning systems such as FEWSNET now encompassed within the Integrated Phase Classification (see IPC Global Partners, 2021) are a good example of strengthening resilience through improved anticipation. However, access to this information remains uneven—in many cases, researchers in high-income countries may have better knowledge about when these shocks are likely to occur than do poor people in low-income countries. Investments in information dissemination will better equip all actors within food systems to anticipate and pre-emptively react to shocks before they occur. Better information flows will also enable food markets to function more efficiently. Investments in infrastructure that also facilitate to integrate market infrastructure such as roads are also likely to enhance resilience, improving both absorptive and adaptative capacities. Social safety nets and deepening of financial markets (improving the availability of savings products and insurance) can improve the ability of households to cope or absorb shocks. On the former, there is growing evidence that safety nets that were in place prior to the start of the COVID-19 pandemic were effective in mitigating the pandemic's adverse effects on food security (see Bottan et al., 2021 and Abay et al., forthcoming for evidence on this from Bolivia and Ethiopia); it is less clear that, at least in low-income countries, safety nets that were hurriedly implemented in response to the crisis were effective. This suggests that in a world that is increasingly shock-prone, governments and their development partners need to invest in social protection measures that are shock-responsive. This should include mechanisms that permit the rapid implementation of increased benefit levels to existing clients when needed (vertical expansion) and the ability to incorporate new beneficiaries into existing programmes (horizontal expansion) (Devereux, 2021). It also suggests that moving towards standing or rights-based social safety nets might be advantageous. Finally, governments can undertake investments in adaptation and transformation and/or undertake actions that incentivize the private sector to make such investments. That all said, it is less clear what government actions should be prioritized. Should they focus on protecting individuals and households from the likelihood that shocks will occur, on minimizing the impacts on income generation or maintaining consumption levels through, for example, social safety nets.

A theme in much of the work on resilience is the beneficial effects of diversification. But there is a tension between diversification and gains from specialization. How can public action reduce this tension, for example, through improving access to financial services such as savings and insurance (both public and private)? Relatedly, what role does diversification within food systems play in making them more resilient. For example, are systems with multiple value chains more resilient? As noted above, in high-income countries, certain parts of the food system (such as processing) are highly concentrated. Anti-trust policy and interventions can be used to limit excessive concentration but at the potential cost of loss of economies of scale.

Tensions exist between open and closed food systems. The last 40 years have seen horrific famines in Ethiopia and North Korea. While both had multiple and complex causes, both were exacerbated by limited intra- and international market integration. For example, during the 1972–74 famine in Wollo, Ethiopia, limited road networks outside the major cities were a significant factor in preventing food from reaching drought affected areas (Devereux, 1988); during the 1984 famine, grain prices in drought affected areas rose by 2.5 times their pre-drought levels (Cutler, 1991). By contrast, severe flooding in 1998 covered, at one point, 75 per cent of Bangladesh but famine was averted because of policy changes that allowed food to be imported from India (del Ninno & Dorosh, 2001).

That said, as noted above, open food systems expose countries to shocks that occur elsewhere. So here, the policy trade-off is one between an open system that reduces the likelihood that really bad outcomes such as famines occur but potentially increases susceptibility to more frequent, but possibly less severe, shocks.

Finally, it is important to note that public action can be a means of enhancing resilience, but it can also be a source of shocks; for example, where governments unexpectedly intervene in food markets. Such actions can be a source of shocks in themselves; further, such actions may create disincentives to investment in food system resilience.

## References

Abay, K., Berhane, G., Hoddinott, J. and Tafere, K. (Forthcoming). COVID-19 and food security in Ethiopia: Do social protection programs protect? *Economic Development and Cultural Change*. https://doi.org/10.1086/71583

Aboah, J., Wilson, M. M., Rich, K. M., & Lyne, M. C. (2019). Operationalising resilience in tropical agricultural value chains. *Supply Chain Management: An International Journal, 24*(2), 271–300. https://doi.org/10.1108/SCM-05-2018-0204

Aker, J. (2010). Information from markets near and far: Mobile phones and agricultural markets in Niger. *American Economic Journal: Applied Economics, 2*(3), 46–59. https://doi.org/10.1257/app.2.3.46

Barrett, C. B., & Constas, M. (2014). Toward a theory of resilience for international development applications. *Proceedings of the National Academy of Sciences of the United States of America, 111*(40), 14625–14630. https://doi.org/10.1073/pnas.1320880111

Barrett, C., Ghezzi-Kopel, K., Hoddinott, J., Homami, N., Tennant, E., Upton, J., & Wu, T. (2021). A scoping review of the development resilience literature: Theory, methods and evidence. *World Development, 146*, 105612. https://doi.org/10.1016/j.worlddev.2021.105612

Bekkers, E., Brockmeier, M., Francois, J., & Yang, F. (2017). Local food prices and international price transmission. *World Development, 96*, 216–230. https://doi.org/10.1016/j.worlddev.2017.03.008

Béné, C. (2020). Resilience of local food systems and links to food security—A review of some important concepts in the context of COVID-19 and other shocks. *Food Security, 12*, 805–822. https://doi.org/10.1007/s12571-020-01076-1

Birhanu, Z., Ambelu, A., Berhanu, N., Tesfaye, A., & Woldemichael, K. (2017). Understanding resilience dimensions and adaptive strategies to the impact of recurrent droughts in Borana Zone, Oromia Region, Ethiopia: A grounded theory approach. *International Journal of Environmental Research and Public Health, 14*(2), 118. https://doi.org/10.3390/ijerph14020118

Bottan, N., Hoffmann, B., & Vera-Cossio, D. (2021). Stepping up during a crisis: The unintended effects of a noncontributory pension program during the COVID-19 pandemic. *Journal of Development Economics, 150*, 102635. https://doi.org/10.1016/j.jdeveco.2021.102635

Brenton, P., Portugal-Perez, A., & Régolo, J. (2014). *Food prices, road infrastructure, and market integration in Central and Eastern Africa*. Policy Research Working Paper 7003. World Bank.

Browne, M., Ortmann, G. F., & Hendriks, S. L. (2014). Household food security monitoring and evaluation using a resilience indicator: an application of categorical principal component analysis and simple sum of assets in five African countries. *Agrekon, 53*(2), 25–46. https://doi.org/10.1080/03031853.2014.915477

Cissé, J. D., & Barrett, C. B. (2018). Estimating development resilience: A conditional moments-based approach. *Journal of Development Economics, 135*, 272–284. https://doi.org/10.1016/j.jdeveco.2018.04.002

Center for Research on the Epidemiology of Disasters (CRED). (2021) *The Non-COVID Year in Disasters*. CRED.

Constas, M., Frankenberger, T. R., & Hoddinott, J. (2014). Resilience measurement principles: Towards an agenda for measurement design. *Food Security Information Network, Resilience Measurement Technical Working Group, Technical Series*, 1.

Constas, M., d'Errico, M., Hoddinott, J., & Pietrelli, R. (2021). Resilient food systems: A proposed analytical strategy for empirical applications. Background paper for *The State of Food and Agriculture 2021*. FAO Agricultural Development Economics Working Paper 21-10. Rome. https://doi.org/10.4060/cb7508en

Cutler, P. (1991). The political economy of famine in Ethiopia and Sudan. *Ambio, 20*(5), 176–178. https://www.jstor.org/stable/4313816

Del Ninno, C., & Dorosh, P. (2001). Averting a food crisis: private imports and public targeted distribution in Bangladesh after the 1998 flood. *Agricultural Economics, 25*(2–3), 337–346. https://doi.org/10.1111/j.1574-0862.2001.tb00213

Dercon, S. (1995). On market integration and liberalisation: Method and application to Ethiopia. *Journal of Development Studies, 32*(1), 112–143. https://doi.org/10.1080/00220389508422404

Dercon, S., Hoddinott, J., & Woldehanna, T. (2005). Shocks and consumption in 15 Ethiopian villages, 1999–2004. *Journal of African economies, 14*(4), 559–585. https://doi.org/10.1093/jae/eji022

d'Errico, M., Romano, D., & Pietrelli, R. (2018). Household resilience to food insecurity: Evidence from Tanzania and Uganda. *Food Security, 10*(4), 1033–1054. https://doi.org/10.1007/s12571-018-0820-5

Devereux, S. (1988). Entitlements, availability, and famine. *Food Policy, 13*, 270–282. https://doi.org/10.1016/0306-9192(88)90049-8

Devereux, S. (2021). Social protection responses to COVID-19 in Africa. *Global Social Policy*, 1–27. https://doi.org/10.1177/14680181211021260

Fackler, P., & Goodwin, B. K. (2001). Spatial price analysis. In R. Evenson & P. Pingali (Eds.), *Handbook of agricultural economics* (pp. 971–1024). North Holland.

Fackler, P., & Tastan, H. (2008). Estimating the degree of market integration. *American Journal of Agricultural Economics, 90*(1), 69–85. https://doi.org/10.1111/j.1467-8276.2007.01058

FAO. (2021). *The state of food and agriculture 2021. Making agri-food systems more resilient to shocks and stresses.* FAO. https://doi.org/10.4060/cb4476en

FAO, IFAD, UNICEF, and WFP. (2017). *The state of food security and nutrition in the world 2017: Building resilience for peace and food security.*

FAO, IFAD, UNICEF, WFP & WHO. (2020). *The state of food security and nutrition in the World 2020. Transforming food systems for affordable healthy diets.* https://doi.org/10.4060/ca9692en

Goletti, F., Ahmed, R., & Farid, N. (1995). Structural determinants of market integration: The case of rice markets in Bangladesh. *The Developing Economies, 33*(2), 185–202.

Herrero, M., Hugas, M., Lele, U., Wira, A. & Torero, M. (2021). *Shift to healthy and sustainable consumption patterns. A paper from the Scientific Group of the UN Food Systems Summit.* https://scfss2021.org/wpcontent/uploads/2021/04/Action_Track_2_paper_Shift_to_Healthy_Consumption.pdf

HLPE. (2017). *Nutrition and food systems. A report by the high-level panel of experts on food security and nutrition of the committee on World Food Security, Rome.*

Hoddinott, J. (2014). Looking at development through a resilience lens. *Resilience for food and nutrition security, 19.*

Hoddinott, J., & Knippenberg, E. (2017). Shocks, social protection and resilience: Evidence from Ethiopia. *ESSP-IFPRI Discussion paper 109.* Addis Ababa.

Hoddinott, J. & Quisumbing, A. (2010). Methods for microeconometric risk and vulnerability assessment. In R. Fuentes-Nieva & P. Seck (Eds). *Risk, vulnerability and human development: On the brink.*

Ismet, M., Barkley, A., & Llewelyn, R. (1998). Government intervention and market integration in Indonesian rice markets. *Agricultural Economics, 19*(3), 283–295. https://doi.org/10.1111/j.1574-0862.1998.tb00532

IPC Global Partners. (2021). *Integrated food security phase classification technical manual version 3.1. Evidence and standards for better food security and nutrition decisions.*

Jensen, R. (2010). Information, efficiency, and welfare in agricultural markets. *Agricultural Economics, 41*, 203–216. https://doi.org/10.1111/j.1574-0862.2010.00501

Jensen, R. (2007). The digital provide: Information (technology), market performance and welfare in the south Indian fisheries sector. *Quarterly Journal of Economics, 122*(3), 879–924. https://doi.org/10.1162/qjec.122.3.879

Kabbiri, R., Dora, M., Elepu, E., & Gellynck, X. (2016). A global perspective of food market integration: A review. *Agrekon, 55*(1–2), 62–80. https://doi.org/10.1080/03031853.2016.1159589

Knippenberg, E., Jolliffe, D., & Hoddinott, J. (2020). Land fragmentation and food insecurity in Ethiopia. *American Journal of Agricultural Economics, 102*(5), 1557–1577. https://doi.org/10.1002/ajae.12081

Martin, W., & Anderson, K. (2012). Export restrictions and price insulation during commodity price booms. *American Journal of Agricultural Economics, 94*(2), 422–427. https://doi.org/10.1093/ajae/aar105

Martin, W., & Glauber, J. (2020) Trade policy and food security. COVID-19 and trade policy: Why turning inward won't work. In R. Baldwin & S. Evenett (Eds.), *COVID-19 and trade policy: Why turning inward won't work*. CEPR Press.

Moschini, G., & Hennessy, D. A. (2001). Uncertainty, risk aversion, and risk management for agricultural producers. In R. Evenson & P. Pingali (Eds.), *Handbook of agricultural economics* (pp. 87–153). North Holland.

Quandt, A., Neufeldt, H., & McCabe, J. T. (2019). Building livelihood resilience: What role does agroforestry play? *Climate and Development, 11*(6), 485–500. https://doi.org/10.1080/17565529.2018.1447903

Ranjan, R. (2014). Multi-dimensional resilience in water-scarce agriculture. *Journal of Natural Resources Policy Research, 6*(2–3), 151–172. https://doi.org/10.1080/19390459.2014.898872

Ravallion, M. (1997). Famines and economics. *Journal of Economic Literature, 35*(3), 1205–1242.

Ravallion, M. (1987). *Markets and famines*. Oxford University Press.

Reardon, T., & Swinnen, J. (2020). COVID-19 and resilience innovations in food supply chains. In *COVID-19 and global food security*, International Food Policy Research Institute (IFPRI).

Reardon, T., & Timmer, P. (2012). The economics of the food system revolution. *Annual Review of Resource Economics, 4*, 225–264. https://doi.org/10.1146/annurev.resource.050708.144147

Rotz, S., & Fraser, E. D. (2015). Resilience and the industrial food system: Analyzing the impacts of agricultural industrialization on food system vulnerability. *Journal of Environmental Studies and Sciences, 5*(3), 459–473. https://doi.org/10.1007/s13412-015-0277-1

Rozelle, S., Park, A., Huang, J., & Jin, H. (1997). Liberalization and rural market integration in China. *American Journal of Agricultural Economics, 79*(2), 635–642. https://www.jstor.org/stable/1244163

Savary, S., Akter, S., Almekinders, C., et al. (2020). Mapping disruption and resilience mechanisms in food systems. *Food Security, 12*, 695–717. https://doi.org/10.1007/s12571-020-01093-0

Scoones, I. (1998). *Sustainable rural livelihoods: A framework for analysis*. IDS Working Paper 72. IDS Sussex.

Smith, L. C., & Frankenberger, T. R. (2018). Does resilience capacity reduce the negative impact of shocks on household food security? Evidence from the 2014 floods in northern Bangladesh. *World Development, 102*, 358–376. https://doi.org/10.1016/j.worlddev.2017.07.003

Stanford, R. J., Wiryawan, B., Bengen, D. G., Febriamansyah, R., & Haluan, J. (2017). The fisheries livelihoods resilience check (FLIRES check): A tool for evaluating resilience in fisher communities. *Fish and fisheries, 18*(6), 1011–1025. https://doi.org/10.1111/faf.12220

Taylor, C., Boulos, C., & Almond, D. (2020). Livestock plants and COVID-19 transmission. *Proceedings of the National Academy of Sciences, 117*(50), 31706–31715. https://doi.org/10.1073/pnas.2010115117

Tack, J., & Yu, J. (2021). Chapter 78: Risk management in agricultural production. In C. B. Barrett & D. R. Just (Eds.), *Handbook of agricultural economics 5* (pp. 4135–4231). Elsevier.

United Nations. (2020). *United Nations common guidance on helping build resilient societies*.

Varela, G., Aldaz-Carroll, E., & Iacovone, L. (2012). *Determinants of market integration and price transmission in Indonesia*. Policy Research Working Paper 6098. World Bank.

von Cramon-Taubadel, S. (2017). The analysis of market integration and price transmission—Results and implications in an African context. *Agrekon, 56*(2), 83–96. https://doi.org/10.1080/03031853.2017.1295655

Zant, W. (2013). How is the liberalization of food markets progressing? Market integration and transaction costs in subsistence economies. *World Bank Economic Review, 27*(1), 28–54. https://doi.org/10.1093/wber/lhs017

Zampieri, M., Weissteiner, C. J., Grizzetti, B., Toreti, A., van den Berg, M., & Dentener, F. (2020). Estimating resilience of crop production systems: From theory to practice. *Science of The Total Environment, 735*, 139378. https://doi.org/10.1016/j.scitotenv.2020.139378

**Open Access** This chapter is licensed under the terms of the Creative Commons Attribution 4.0 International License (http://creativecommons.org/licenses/by/4.0/), which permits use, sharing, adaptation, distribution and reproduction in any medium or format, as long as you give appropriate credit to the original author(s) and the source, provide a link to the Creative Commons license and indicate if changes were made.

The images or other third party material in this chapter are included in the chapter's Creative Commons license, unless indicated otherwise in a credit line to the material. If material is not included in the chapter's Creative Commons license and your intended use is not permitted by statutory regulation or exceeds the permitted use, you will need to obtain permission directly from the copyright holder.

# CHAPTER 7

# Food Security Under a Changing Climate: Exploring the Integration of Resilience in Research and Practice

*Alessandro De Pinto, Md Mofakkarul Islam, and Pamela Katic*

## INTRODUCTION

Climate change already affects vulnerable populations in many low- and middle-income countries and is expected to alter the lives of many more people in even more areas of the world in the future (IPCC, 2018). The nutritional status of these people is of particular importance because climate change will exacerbate the incidence of malnutrition in these

---

A. De Pinto (✉) · M. M. Islam · P. Katic
Natural Resources Institute, University of Greenwich, London, UK
e-mail: A.DePinto@greenwich.ac.uk

M. M. Islam
e-mail: M.M.Islam@greenwich.ac.uk

P. Katic
e-mail: P.G.Katic@greenwich.ac.uk

© The Author(s) 2023
C. Béné and S. Devereux (eds.), *Resilience and Food Security in a Food Systems Context*, Palgrave Studies in Agricultural Economics and Food Policy, https://doi.org/10.1007/978-3-031-23535-1_7

areas (Fanzo et al., 2018; Myers et al., 2017; Phalkey et al., 2015). Some projections show an increase of 4.8 million undernourished children worldwide by 2050 due to climate change, and 97% of the people at risk of hunger will be in low-income countries (IFPRI, 2017). Climate change has a direct influence on food availability because it affects habitats and crop productivity. Furthermore, ensuing increases in prices are expected to reduce accessibility to healthy foods such as vegetables, fruits, and animal-source foods with repercussions on people's diets (Springmann et al., 2016; Wiebe et al., 2015). Despite the uncertainty in these projections, regional differences in agricultural production are expected to widen the gap between the rich and the poor (Nelson et al., 2010; Parry et al., 2004; Stevanovic et al., 2016). Trade (including food aid) might not be able to fully buffer localized food shortages and ease these problems (Elbehri et al., 2015; Nelson et al., 2009; Stevanovic et al., 2016).

Climate change will not only affect food production and sourcing; its effects are expected to ripple throughout value chains and food systems. Studies indicate that higher temperatures and prolonged exposure to high levels of $CO_2$ concentration could lead to losses in nutrient content (e.g., zinc, iron, proteins) of key food crops, induce changes in important quality parameters (e.g., dry matter, sugar content, citric and malic acid, organic acids, antioxidant compounds) (Dong et al., 2018; Högy & Fangmeier, 2009; Moretti et al., 2010; Myers et al., 2014), and increase the incidence of foodborne pathogens and mycotoxins (Battilani et al., 2016; Tirado et al., 2010). Storage, marketing, and retail systems will need to adapt as areas become hotter and transportation will have to negotiate with less durable and more frequently flooded roads as well as damaged port infrastructure (Nicholls & Cazenave, 2010; Shi et al., 2015). Electrical grid failures caused by hydroelectric dams that run dry or by an overburdening demand will affect retail shops as well as consumers (Portier et al., 2013). The utilization of food will also be affected. Studies suggest that increased contamination of drinking water supplies and increases in the prevalence of respiratory diseases and diarrhoea are possible, particularly in semiarid areas (Signorelli et al., 2016).

Concerns for vulnerable populations have increased not only because of the likely impact of climate change and a better understanding of the long-lasting implications of malnutrition (Alderman et al., 2006; Martins et al., 2011), but also because researchers, practitioners, and development agencies have developed a greater appreciation for the complexities of the issues surrounding food security. For example, discussions about food security immediately after World War II were mostly related to commodity trade, tariffs, barriers, food processing, and calorie availability with little primary concern regarding malnutrition (FAO, 1946). By the beginning of the 1980s, new and richer ideas had entered into the debate. Notably, Amartya Sen pointed out how, during twentieth-century famines, it was not the lack of food (total calories output from agriculture) that caused problems of hunger, but rather the inability of the poor to access it. Today, there are numerous definitions of food security, but most recent ones share common traits and describe it as a multifaceted problem linked to the availability, accessibility, utilization, and stability of food over time that affects people's physical, social and economic development (see Mark Constas' Chapter 5 in this volume). These multidimensional definitions are better suited to comprehend and address the complexity of the threats posed by climate change.

Given the magnitude and the broad reach of the challenges that the world is facing, it is not surprising that the scope of development interventions has broadened from food production to integrated approaches that target entire food systems (FAO-WHO, 2014; Mbow et al., 2019; Oliver et al., 2018). One obvious consequence of this new paradigm is that interventions require approaches of increasing complexity and collaboration among experts from different fields.

A relatively new concept supporting contemporary development thinking and interventions in climate change, humanitarian, and food security contexts is that of resilience. Resilience is now regularly used in the academic literature and within the international development community as an approach to deal with adverse shocks and to promote sustainable development (Serfilippi & Ramnath, 2018). Prosperi et al. (2016) and Vonthron et al. (2016) note that research on resilience and vulnerability could provide support when framing the principles of sustainable food systems and that the concept of resilience can be useful to rethink food emergencies and development. Because it is integrative by construction, resilience has been recognized to link together research areas that have often been considered in isolation (e.g., gender, social protection, health and nutrition, climate change, energy, infrastructure) (Béné et al., 2016), and it is expected to provide support for a systems approach with recognition for the relations among human capabilities and natural systems (Xu et al., 2015). Given the multifaceted impacts of climate change that can affect virtually all dimensions of food security, resilience has the potential to be a useful concept to help develop a coherent and inclusive method to support decision-making and plan for interventions that require work at the intersection of multiple disciplines (Grafton et al., 2019; Quandt et al., 2017; Wilson, 2010).

In this chapter, we explore how the concept of resilience has been integrated into the work on climate change and food security and whether its use has helped researchers and practitioners advance their agenda. First, we review the academic literature to determine how academics have engaged with the concept of climate resilience in a food security context. Then, we provide a case study of the way resilience is used by implementers on the ground. Finally, we draw key conclusions and suggestions on how to move the climate resilience agenda for improved food security forward.

## Resilience in the Academic Literature on Food Security and Climate Change

Two parallel processes should be considered when analysing the influence of resilience on the work on climate change and food security. First, during the last two decades, published research became increasingly receptive of the progress made in the field of food security research and broadened its scope from a distinct focus on agricultural production and agricultural policies (Nelson et al., 2009, Rosegrant et al., 2014) to considering instruments associated with social inclusion and protection as well as a vision for entire food systems (Nelson et al., 2018; Rosenzweig et al., 2020; Schwan & Yu, 2018). Second, at approximately the same time, the interpretation and use of the concept of resilience went through changes that made it more usable for researchers and practitioners. The concept evolved from one describing ecosystem stability (Holling, 1973) to one illustrating the ability of social systems to absorb shocks and stressors and, through adaptive processes, to reorganize into fully functioning entities. This conceptual broadening is also how resilience became a prominent concept in the literature on food security and disaster and risk management during the first decade of the twenty-first century (Alinovi et al., 2008; Pingali et al., 2005). Furthermore, after an initial focus on resilience as an end in itself and the ensuing efforts to quantify and measure it, researchers moved to an interpretation of resilience as a means to achieving an ultimate end such as food security (Ansah et al., 2019).

A search in the Web of Science Core Collection shows that researchers working on climate change have steadily and increasingly included the concept of resilience in their work. The search of the published literature reveals that the words 'climate change' and 'resilience' appeared together in the published literature a total of 48 times during the period 1996–2000 and 8,626 times during the period 2016–2020 (Fig. 7.1).

A similar pattern is apparent when a comparable search is carried out for articles in which the words 'food security' and 'resilience' or 'climate change', 'food security', and 'resilience' appear together. In all combinations, the incidence of the word 'resilience' increases significantly after the year 2005 and even more pronouncedly after 2015 (Fig. 7.1).

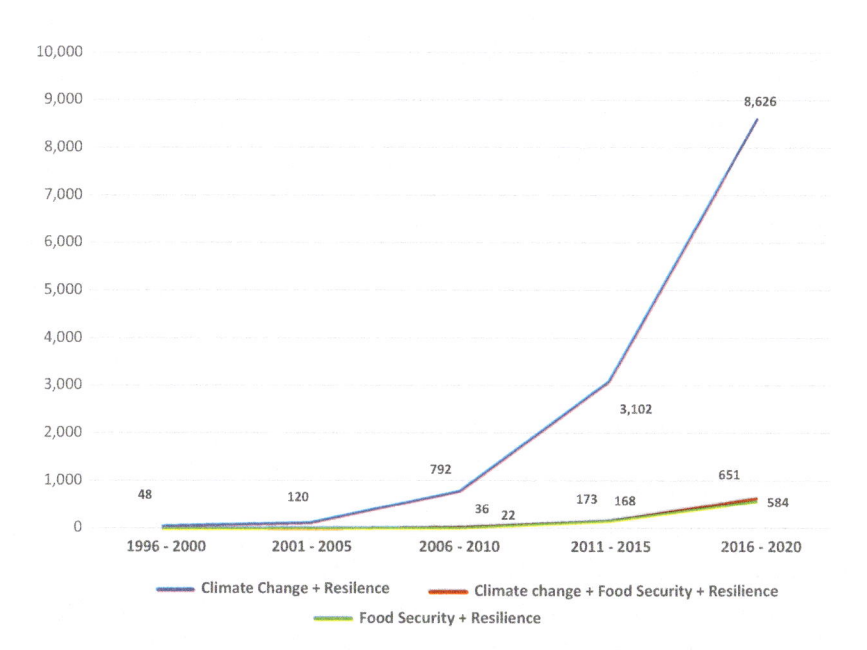

**Fig. 7.1** Incidence of the words 'climate change and resilience'; 'food security and resilience'; 'climate change, food security and resilience' in selected academic literature, years 1996–2020[1] (*Note* [1] Search strings used in the Web of Science Core Collection literature search: TS = [climate change] AND TS = [resilience]; TS = [food security or food insecurity] AND TS = [resilience]; TS = [climate change] AND S = [Food Security or Food Insecurity] AND TS = [Resilience]. *Source* Authors)

What follows is a brief review of selected research at the nexus of climate change, resilience, and food security. This review reveals the wide range of responses covering all the major components of food systems that have been investigated and the wealth of information on actions and interventions that have the potential to increase people's resilience to climate shocks and stressors and improve people's food security under a changing climate.

### *Agricultural Food Production*

While discussions on agricultural risk management have overwhelmingly concentrated on the use of crop insurance and index-based insurance (Cole et al., 2013; Giné & Yang, 2009; Giné et al., 2008; Hill et al., 2016, 2019), a range of viable strategies beyond financial instruments are available to lower climate-related risks by reducing vulnerability or by making food production systems better able to cope with and recover from shocks (Hallegatte et al., 2017; Lipper et al., 2018). Improved agricultural and pest management practices have also been widely demonstrated to have positive impacts on crop production (Arefi et al., 2017; Deb et al., 2018; Garnett et al., 2013; Midega et al., 2018). They are expected to increase and stabilize yields in the face of long-term changes in temperature, precipitation, and of the frequency and severity of extreme weather events (De Pinto, Cenacchi et al., 2020; Rosegrant et al., 2014). Gains in productivity and reductions in yield volatility are possible in both rainfed and irrigated systems through the expansion of soil and water management practices that increase water availability for crops in rainfed systems (Rockström & Barron, 2007) and the introduction of water-saving technologies in irrigated systems (Molden et al., 2010). Increasing the availability of diverse genetic material (FAO, 2011), advances in crop breeding and new genome editing systems such as CRISPR/Cas (Mojica et al., 2009) can protect crop production from a deteriorating climate (Rosegrant et al., 2014). A closer integration of crops, trees, and livestock into more complex systems is expected to stabilize or increase productivity and protect agricultural production from extreme weather events (Altieri et al., 2015; Asfaw et al., 2019; Lin, 2011; Weindl et al., 2015) and to increase the resilience of agricultural livelihoods (Quandt et al., 2017). Climate-related advisory services are considered essential to protect production and reduce output volatility by facilitating preparedness and timely responses (Aker & Mbiti, 2010; Goddard, 2016; Schimmelpfennig, 2016; Woodfine, 2009).

### *Distribution, Processing, and Marketing*

Investment in processor and distributor networks can make supply chains less vulnerable to climate change and better suited to withstand shocks from extreme weather events (Zilberman et al., 2012). Innovations in packaging, processing, and storage practices improve efficiency, reduce waste, and increase the availability of nutritious albeit perishable foods (James & James, 2010). Expanding rural electrification is expected to increase the availability and reduce the cost of nutrient-rich, highly perishable foods such as vegetables and fruits, not only by facilitating their production using irrigation, but also by providing more cold-storage options (Arndt, 2019). Investments in processing and cold-storage facilities, feeder roads, and cooled transportation have the additional benefit of smoothing income shocks that small producers face from seasonality, market volatility, and weather shocks (da Silva & Fan, 2017). Risks of food poisoning and food spoilage can be abated with the development of quality assurance and control tools and methods that prevent or control microbiological risks (Jacxsens et al., 2010; Tirado et al., 2010). Processing foods (e.g., drying and salting meat and fish, processing milk into yogurt and cheese) is shown to reduce the need for cold storage and prevent the spoilage of nutritious foods, thus increasing its availability to consumers (Berlin et al., 2008; GLOPAN, 2016).

The general consensus in the literature is that trade will play an important role in adjusting to the shifts in agricultural and food production patterns resulting from climate change (Brenton et al., 2022; Nelson et al., 2009, 2010), improving household food access by moderating price increases, and reducing shocks to food availability (Brown & Kshirsagar, 2015; Brown et al., 2017; Lybbert & Sumner, 2012). However, in order to ensure that the stabilizing power of trade is realized, investments in maintaining, expanding, and climate-proofing existing infrastructure are considered necessary, particularly in order to reach geographically isolated, poor, and/or socially marginalized communities (Thacker et al., 2019).

## Food Preparation and Consumption

The consequences of hunger and malnutrition on people's health greatly affect already vulnerable people's capacity to respond, cope, and adapt to the negative consequences of climate change (Tirado et al., 2015; Wheeler & von Braun, 2013). Addressing undernutrition and micronutrient deficiencies under a changing climate requires that micronutrient-rich foods such as vegetables, fruits, nuts, seeds, and pulses are made widely available and affordable despite unfavourable production conditions (Headey et al., 2018; Nelson et al., 2018; Ruel et al., 2017). Efforts to develop fortified food, biofortified crop varieties, and the supplementation of targeted micronutrients can help reduce people's nutritional deficiencies (Chakrabarti et al., 2018; Martorell et al., 2015) and mitigate the reduced nutrient quality in crops caused by climate change (Beach et al., 2019). In low-income countries, food processing and preservation techniques can increase food safety and preserve nutritional value of foods while minimizing the need for cold storage (FAO, 2016). Communication and participatory approaches are essential in both rural and urban settings to change consumers' behaviour and promote a dietary shift from carbohydrate-rich staples to a more diverse and healthier diet that addresses micronutrient deficiencies (Leroy & Frongillo, 2007; Ruel, 2001; Ruel et al., 2017). Nutrition labelling, advertising restrictions, taxes on unhealthy foods such as sugar-rich sodas, and nutrition education in schools and health centres are suggested policy levers that can be used to encourage behaviour change that favours resilience to climate change (Fanzo et al., 2018; Hawkes et al., 2017).

## Capacity Building in Institutions and Governance

The role of national and regional governments, community organizations, and market institutions is now recognized as essential to provide information and encourage the types of innovation and investments necessary to manage climate risks (Meinzen-Dick et al., 2013). In particular, investments in improving governments and local organizations' capacities to provide effective leadership are considered necessary for the coordination of responses that span across numerous stakeholders and economic sectors (Babu et al., 2019). Many new technologies that are expected to improve people and systems' resilience require that coherent policies and regulatory procedures be in place (England et al., 2018; Saito, 2013) and that at times, these must be globally harmonized (such as in the case of genetic engineering) to avoid trade bans and implementation bottlenecks (Duensing et al., 2018).

Informal institutions, for example social networks, are recognized to promote cooperation in resource management and income diversification, and thereby contribute to livelihood and ecological resilience (Kristjanson et al., 2017). They can also help reduce the gender gap in information about climate risks and in decision-making power which are recognized as detrimental to develop efficient responses to climate change and to achieve better nutrition and health outcomes (Bryan et al., 2017; De Pinto, Seymour et al., 2020; Peterman et al., 2014). Stronger governance is essential to reduce investment gaps that penalize vulnerable groups and to ensure that resilience-enhancing investments are spread across economic, social, and environmental dimensions (McGregor et al., 2020; Meinzen-Dick et al., 2013).

### *An Operational Problem*

Notwithstanding the breadth and depth of the research at the nexus of climate change, food security, and resilience, our review reveals some of its limits. We found that, overwhelmingly, studies use the word 'resilience' to give an intuitive depiction of a desirable trait of food systems, without connecting it with a specific theory or framework to back up their claims about increasing the resilience of households or communities to climate stresses or shocks. We also found that despite the complexity of the system analysed, the characterization of resilience (resilience of what and to what) often lacks specificity. Only a small portion of the literature actually attempts to model or measure resilience or how resilience contributes to food security (Bene et al., 2017). Furthermore, most studies linking resilience and food security track multiple commonly available production-oriented proxies (e.g., yields, production, and revenues) but very few consider multiyear resilience-related outcomes. These observations mirror what other authors have found in their reviews of the literature. Hogeboom et al. review the nexus water, energy, and food (Hogeboom et al., 2021) and find that only a few studies model (20%) or measure (13%) resilience. In a thorough analysis of papers that look at resilience in agri-food supply chains, Stone and Rahimifard (2018) find that there is a poor consensus on what elements are the most important for resilience. Ansah et al. (2019) suggest that since studies do not resort to a common resilience framework, the number of different indicators used, the length of time that they were tracked, the different units of analysis, and the different approaches (statistical vs modelling) all make comparing the results difficult, if possible at all. Constas et al. (2014) note how variable selection tends to be context-specific and driven by data availability rather than theory.

Operational issues might be at the root of these shortcomings. Largely guided by the work of the WFP/FAO 'Resilience Measurement Technical Working Group', important conceptual progress has been made in the measurement of resilience in the context of food security and humanitarian interventions (Constas et al., 2014). Efforts to increase the operational viability of the concept of resilience can be found in studies that attempt to connect more formally agricultural and household activities with resilience and human well-being (for example, Quandt et al., 2019; Robinson et al., 2015; Rockström, 2003; Silici et al., 2011; Verchot et al., 2007). However, despite these valuable contributions, difficulties remain in connecting concepts that are both intuitive and complex. This is in part due to the structural complexities of working with systems made of many interconnected parts in which a shock affects the functioning or behaviour of one component of the system or of a group of people and then ripples through the system to reach other components or groups of individuals (Béné, 2020). Predicting the effects of investments and interventions that aim to abate the negative effects of climate change is also difficult because of positive and negative feedback loops (Bryan et al., 2017), non-linear relationships among factors that determine the system functioning, and because of the existence of thresholds below which a change in one component does not result in some appreciable difference in the performance of the whole system (Levine, 2014).

These conceptual difficulties might also explain why the literature provides limited evidence on the causal relationship between actions that purportedly increase resilience to climate stresses and shocks and improved food security. For example, Wilson (2010) points to the limits of multifunctional agricultural systems in building resilience while Cochrane and Cafer (2018) find limits and exceptions to the expected positive effect of diversification in agricultural livelihoods and smallholder production on community resilience. Gil et al. (2017) review the literature on integrated farming systems and find that studies generally claim a positive association between integrated farm systems and enhanced resilience (by virtue of increased yields, reduced yield variance, or increased incomes), but they also stress that very few studies identify the causal pathways that lead to increased climate resilience. Rosenstock et al. (2019), perform an extensive review of the published literature on climate smart agriculture (CSA), an approach to agriculture that includes among its objectives improving resilience, and find that less than a fifth of all articles attempt to connect CSA practices with resilience. They find that most of the articles that do so focus on a few indicators (e.g., soil quality and input-use efficiency) and assume that improvements in these indicators signify an increased resilience to climate change. The authors also state that the general disagreement among researchers on what to measure and what indicators to use, might explain why the literature on CSA provides so little information on one of its foundational pillars. The few cases where resilience per se has been measured directly are through a self-assessed recovery index, estimated through series of recall questions and psychometric techniques (Béné & Haque, 2021). However, these tools have been developed to analyse resilience at the household level rather than at a broader food system level.

Therefore, it appears that despite the amount of research on how to respond to various climate threats, our knowledge of how these actions translate into resilience is still limited, and an identification of a sequence of investments that can generate climate-resilient pathways to food security still appears elusive.

## Resilience in Practice—The Case of Projects Implemented Through the Adaptation Fund (AF)

The international development community has embraced the concept of resilience as a proxy for long-term growth, with a clear recognition that resilience to stressors and adverse shocks is essential for individuals and communities to achieve sustainable development. As a result, the role of resilience is codified in several targets of the United Nations' 2015 Sustainable Development Goals (e.g., SDG targets 1.5, 2.4, 13.1). In order to understand how, up to now, practitioners on the ground have engaged with the concept of resilience in a climate change and food security context, we analysed a series of projects which were implemented through the Adaptation Fund (AF). The AF was chosen for two reasons. Firstly, the AF was created specifically to finance adaptation projects in low- and middle-income countries that are parties to the Kyoto Protocol and are particularly vulnerable to the adverse effects of climate change

(Adaptation Fund, 2021a), and secondly, agriculture and food security are two of the key areas in which the AF provides support (Adaptation Fund, 2021b).

As of April 2021, the AF had financed over 160 projects since its inception in 2010 (Adaptation Fund, 2021c). Of these, we selected twelve projects (Table 7.1) for in-depth analysis based on the following criteria: (i) the projects had to have a strong 'food security' component, (ii) they had to have 'resilience' building as a key objective, and (iii) they had to be at the implementation stage or completed. We selected cases from diverse countries and regions to account for variations in climatic, ecological, and socioeconomic contexts. We excluded projects that were in high-income countries (e.g., Moldova), funded primarily for readiness building of recipients, or were of very short duration, e.g., one year or less.

Within each project, we explored: (i) whether resilience was conceptualized as a means to achieving food security and/or other developmental outcomes or was conceptualized as an end in itself; (ii) whether a characterization of resilience (resilience of what and to what) was provided; (iii) the type of interventions proposed for enhancing resilience; and (iv) what indicators and methods were used to evaluate whether resilience was achieved. From a methodological viewpoint, our approach is 'descriptive-exploratory' in nature and uses 'typical' or 'illustrative' cases (Yin, 2009).

With few exceptions (e.g., PN10), the projects we analysed commonly framed resilience as an 'end' goal for the project. However, only two projects (PN7 and PN8) provided a formal definition of resilience. In all the projects, the term was often framed in contrast to the 'vulnerability' of certain entities against various climate changes and stresses. This narrative indicates that resilience was considered as an 'antidote' to vulnerability. Another term, 'adaptive capacity', which was conceptualized as a mediatory variable between vulnerability and resilience, was commonly found. The implicit Theory of Change (ToC) common to all those projects implied that certain interventions would increase the adaptive capacity of vulnerable entities, reduce loss and damages, and eventually reduce vulnerability, which, in turn, would improve resilience (as an end-goal) (Fig. 7.2).

Depending on the project, the targeted beneficiaries included smallholder farmers, farm households, agropastoralists, rural communities, small pond-based aquaculture systems, agricultural sector, livestock systems, or even 'natural systems' such as forests (Fig. 7.2). Project narratives about specific entities of interest were not obvious and often involved

**Table 7.1** Case study projects

| Code | Project title | Sector | Country | Start date | Duration |
|---|---|---|---|---|---|
| PN1 | Agricultural Climate Resilience Enhancement Initiative (ACREI) | Food Security | Ethiopia, Kenya, Uganda | 30/08/2018 | 3 years |
| PN2 | Building Adaptive Capacities of Small Inland Fishermen Community for Climate Resilience and Livelihood Security, Madhya Pradesh, India | Food Security | India | 18/11/2018 | 3 years |
| PN3 | Reducing the Vulnerability by Focusing on Critical Sectors (Agriculture, Water Resources and Coastlines) in order to Reduce the Negative Impacts of Climate Change and Improve the Resilience of these Sectors | Multisector | Costa Rica | 10/07/2015 | 5 years |
| PN4 | Enhancing Adaptive Capacity and Increasing Resilience of Small and Marginal Farmers in Purulia and Bankura Districts of West Bengal | Agriculture | India | 28/05/2015 | 4 years |
| PN5 | Enhancing the Resilience of the Agricultural Sector and Coastal Areas to Protect Livelihoods and Improve Food Security | Multisector | Jamaica | 02/11/2012 | 3.5 years |

(continued)

Table 7.1 (continued)

| Code | Project title | Sector | Country | Start date | Duration |
|---|---|---|---|---|---|
| PN6 | Enhancing Climate Resilience of Rural Communities Living in Protected Areas of Cambodia | Ecosystem-based Adaptation | Cambodia | 21/05/2013 | 5 years |
| PN7 | Building resilience to climate change and variability in vulnerable smallholders | Agriculture | Uruguay | 22/10/2012 | 5 years |
| PN8 | Adapting to Climate Change Through Integrated Risk Management Strategies and Enhanced Market Opportunities for Resilient Food Security and Livelihoods | Food Security | Malawi | 06/11/2020 | 5 years |
| PN9 | Improving adaptive capacity of vulnerable and food-insecure populations in Lesotho | Food Security | Lesotho | 10/08/2020 | 4 years |
| PN10 | Adapting to climate induced threats to food production and food security in the Karnali Region of Nepal | Food Security | Nepal | 10/26/2018 | 4 years |
| PN11 | Enhancing Resilience of Communities to the Adverse Effects of Climate Change on Food Security in Mauritania | Food Security | Mauritania | 08/14/2014 | 4 years |
| PN12 | Building Resilient Food Security Systems to Benefit the Southern Egypt Region (Phase-I) | Food Security | Egypt | 03/31/2013 | 4 years |

*Source* Compiled by the authors

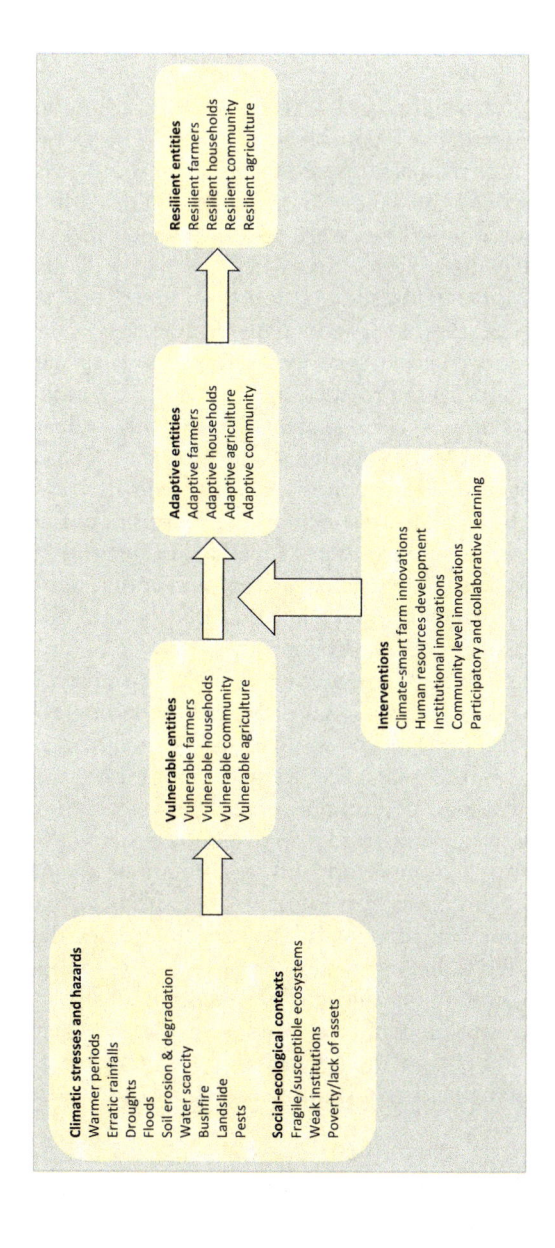

**Fig. 7.2** The generic resilience intervention framework/compiled from the case study projects (*Source* Authors)

an overlap of these various entities, with some projects specifying one entity (PN2, PN11) and others (e.g., PN1, PN6, PN10) referring to the resilience of multiple entities.

Climate-induced 'variability' and 'extremes' were common vulnerability factors and differed from country to country. The most common concerns were erratic rainfall, unpredictable monsoon season, high temperature, dry spells, droughts, and floods. Some other climate-induced biotic and abiotic stresses were also mentioned including water shortages, soil fertility loss, agricultural drainage issues, landslides, soil erosion, pest infestations, deforestation, and bushfires. These shocks and stresses were said to lead to yield/productivity decline, loss of income, loss of livelihood, and reduced food security. References to specific properties of related ecosystems were provided in some cases. Examples include: desert or arid ecosystems, special protected areas with valuable biodiversity, and fragile mountain ecosystems and grasslands (pasturelands). Vulnerability framing however was not limited to climatic reasons. Low income or poverty, discriminatory socio-cultural norms of exclusion (e.g., women, youth, and ethnic minority groups), lack of institutional capacities and services as well as inadequate awareness and knowledge of various actors regarding climatic change and appropriate adaptation options were argued to be key reasons for vulnerability.

The proposed interventions varied widely, with a common one being the promotion of new and improved agricultural technologies and practices. This included intensification and diversification of crop and animal species; crop rotations; Integrated Pest Management (IPM); better water management practices (e.g., polytunnels, microdams, irrigation schemes, rainwater harvesting, water mills, and drip irrigation systems), soil conservation (e.g., planting legumes and soil conservation practices); new planting time and techniques; agroforestry (e.g., planting native, fast-growing, and multipurpose perennials and/or tree species that provide food, fuel, timber, and other ecosystem services); conservation agriculture practices (e.g., minimum tillage and retention of crop residues); agricultural flood control techniques (e.g., semi-circular bunds, check dams, gully plugs, infiltration ditches, swales, and agroforestry plantations); drought tolerant field crops and livestock species; and integrated management of grasslands.

A few projects (e.g., PN8, PN9) mentioned risk management and market-related types of interventions such as market analysis and business plan development, contract farming, value chain development, agricultural insurance, microcredit, and new funding packages for farmers. Training of farmers about climatic changes, new agricultural practices, and new financial and market opportunities, was a common intervention.

All the projects showed a clear ambition to improve adaptive capacity at the institutional level. Enhancing the technical capacity of government and non-government institutions, especially meteorological and agricultural extension services, for monitoring local agroclimatic contexts and delivering location and weather-specific, timely, and climate-proof services (e.g., seasonal forecasts) featured prominently in all the projects. Staff training and development of new educational materials (e.g., manuals) appeared as common forms of intervention. Other interventions included: fostering learning and knowledge exchange through improved data analysis and management; communication and advocacy through short films, publications, printed materials, and press; forming new climate-smart farmer field schools or strengthening the capacity of existing ones; and demonstration plots. Engaging with and influencing policies through feedback and advocacy, and the mainstreaming of climate field schools were also among the institutional level interventions.

Interventions were targeted at the community level as well. These included: forming new resource user groups and/or strengthening the adaptive capacity of existing groups; training of local institutions (e.g., village *Panchayats* in India and *Village Development Committees* in Nepal) on climate change; developing community-based adaptation action plans; forming community grain banks, seed banks, and fodder banks; community-managed storage of agricultural produce, community forestry, and community-based drinking water and biogas facilities.

At both the household and community levels, there was a strong emphasis on creating new Income Generating Activities (IGAs) for target beneficiaries. Examples included: rabbit and duck rearing, medicinal and aromatic plants, oil extraction, organic farming for niche market, cultivation of high-value fruits and vegetables, milk processing, candle making, pickle making, growing herbs and mushrooms, and fuel-efficient cooking devices. An emphasis on women and youth was common in all these IGAs.

In all the projects, we found a strong emphasis on 'participatory learning and action' (PLA) involving project staff and beneficiaries. Almost all the projects abstained from top-down interventions and emphasized developing locally appropriate adaptation actions through participatory approaches. Consultation with stakeholders (including beneficiaries) during project preparation was a prerequisite for funding application, as evident in the Adaptation Fund project proposal template. To fulfill this requirement, the projects used a range of methods—including stakeholder meetings, workshops, interviews, focus groups, surveys, visits, and participant observations—at the proposal development stage. Some projects also conducted vulnerability assessments, or utilized past vulnerability assessment reports, and included them in the project proposals (e.g., PN2, PN4, PN9, PN10). Almost all projects emphasized learning and knowledge-sharing using a range of interventions, e.g., community radio, farmer field schools, enhancing the capacities of agricultural extension services, and continuous sharing of project results and best practices.

While all the projects explicitly stated resilience enhancement as a goal, the indicators proposed for evaluating such outcome varied widely and were questionable in some instances. The common outcome indicators proposed included: percent changes in farmers' income, percent of farmers adopting new agricultural practices, percent of the target population aware of climatic changes, decreasing livestock mortality on farms, percent of communities with better access to water, improvement in agricultural outputs, reduced incidents of downstream flooding and soil erosion; number of farmers with increased access to irrigation schemes, rainwater harvesting, and drip irrigation; number of farmers benefitting from soil conservation and land husbandry infrastructure, availability of local planning tools, percent of farmers reporting income loss and increase in supplementary income, and increased community awareness and knowledge. A number of rather vague indicators, such as 'percent of farmers with climate-resilient livelihoods', were proposed, for which no measurement methods were described. At the ecosystem level, some listed indicators were increased availability of forage and water for animals, increased native grassland biodiversity, and improvement in Vegetation Index. All these outcomes were expected to be achieved within each project's lifetime, ranging from 3 to 5 years (Table 7.1).

The outcome indicators were proposed to be evaluated based on reviews of project documents, focus groups, discussions, baseline and endline surveys, and government statistics. Of these, baseline and endline comparison through surveys was common. However, no project provided details of such evaluation methods, e.g., randomization, use of counterfactual and the associated statistical methods. Moreover, although adverse impacts of climatic changes on food security were commonly framed as a rationale for the projects, none provided any indicator or methods for assessing food security outcomes over time, which could have potentially manifested the resilience of the entities of interest in the projects.

## Discussion

In this chapter, we aimed to assess how the concept of 'resilience' has been integrated into discussions about climate change and food security by both academics and practitioners. We first performed a targeted review of the academic literature on climate change, food security, and resilience. We found an overwhelming increase in the use of this concept in the literature over the last two decades, while the concept of resilience itself was going through a transformation, from describing the stability of ecological systems to embodying the adaptive and reorganizational capacity of social-ecological systems. As such, resilience has become most of all an intuitive way for researchers to describe the ability of a system to bounce back to normal functioning (or even improve) after a shock. Our review of the literature suggests that the combination of a more advanced understanding of food security coupled with the concept of resilience has given support to researchers to investigate the many aspects of food systems that are vulnerable to climate change and to find and propose solutions. This is reflected in the myriad insights that the literature offers into how to address climate threats. This wealth of knowledge gets diluted, however, in a cacophony of methods, metrics and indexes and logical leaps when it is used to connect resilience to food security more formally or empirically. It appears that the problem is not in identifying what actions can potentially improve food security but rather in determining the pathways through which such actions translate into resilience, and then into food security. The result is that the academic community is still far from having a robust method for identifying an efficient and rational sequence of interventions that improves people's food security in response to one or multiple climate threats.

At the practitioner level, our case study of twelve Adaptation Fund projects revealed that, unlike the more recent trend in the academic literature, resilience in those projects is still commonly framed as an end goal, rather than a means to achieve other outcomes. Similar to that in the academic literature, the projects refer to the resilience of a number of entities, including farmers, farm households, agricultural or natural resource systems, and agricultural sectors. Such a diverse, and often overlapping focus makes it challenging to identify appropriate resilience-building interventions and to evaluate project outcomes, since the requisite evaluation indicators are likely to vary across interventions and must be able to account for synergies and trade-offs. Such multilevel indicators were clearly missing in the Adaptation Fund projects we reviewed. The interventions for achieving resilience appear to be multifaceted and systemic in nature, combining technical interventions with institutional interventions at multiple levels, and are coupled with a strong focus on participatory learning and action. These interventions were generally in line with the theories of resilience found in the academic literature, even though their breadth falls short of what would be expected from a food systems approach—increasingly identified as crucial for food security interventions in a range of contexts. More importantly, the key limitation in the Adaptation Fund projects is in the evaluation of resilience outcomes, both in terms of appropriate indicators and methods. There is a lack of standardization of the definition and metrics of resilience that coincides with the problems described in the academic literature. It is possible that the absence of a common language and metrics for measurement we found in the literature hampers the ability of the development community to prioritize interventions and to monitor and evaluate resilience-building programmes. Although the lack of agreement on definitions and metrics may allow flexibility and strategic manoeuvring for development projects, it poses the risk of making resilience yet another buzzword in the lexicon of development.

## Conclusion

Resilience appears to have contributed to what, in our opinion, is a positive shift in the paradigm underlying climate change and food security work in ways that favour integrated approaches and interventions. However, despite its appealing qualities, the operationalization of resilience in the context of climate change and food security remains problematic. It seems clear that further support is required to develop a new generation of operational definitions that harmonizes frameworks and metrics in order to create a common language to advance core ideas. This is necessary to translate high-level concepts of resilience into actionable information and to develop resilience-enhancing actions at multiple levels. Beyond the efforts of the academic community to offer practical and systematic guidance to decision-makers on how to use resilience in development work, it is important that partnerships between academic communities and practitioners are developed and strengthened to give more coherence to the formulation, implementation, and evaluation of resilience-building activities. At present, such collaborative designs and implementations are the exception rather than the norm in climate change and food security policies, projects, and programmes.

## REFERENCES

Adaptation Fund. (2021a). *Adaptation Fund—Governance*. https://www.adaptation-fund.org/about/governance/

Adaptation Fund. (2021b). *Adaptation Fund—Project sectors*. https://www.adaptation-fund.org/projects-programmes/project-sectors/

Adaptation Fund. (2021c). *Adaptation Fund—Projects & programmes*. https://www.adaptation-fund.org/projects-programmes/project-information/projects-table-view/

Aker, J. C., & Mbiti, I. M. (2010). Mobile phones and economic development in Africa. *Journal of Economic Perspectives, 24*(3), 207–232.

Alderman, H., Hoddinott, J., & Kinsey, B. (2006). Long term consequences of early childhood malnutrition. *Oxford Economic Papers, 58*(3), 450–474.

Alinovi, L., Mane, E., & Romano, D. (2008). Towards the measurement of household resilience to food insecurity: Applying a model to Palestinian household data. In R. Sibrian (Ed.), *Deriving food security information from national household budget surveys: Experiences, achievement, challenges* (pp. 137–152). United Nations Food and Agriculture Organization.

Altieri, M. A., Nicholls, C. L., Henao, A., & Lana, M. A. (2015). Agroecology and the design of climate change-resilient farming systems. *Agriculture and Sustainable Development, 35*, 869–890.

Ansah, I. G. K., Gardebroek, C., & Ihle, R. (2019). Resilience and household food security: A review of concepts, methodological approaches and empirical evidence. *Food Security, 11*(6), 1187–1203.

Arefi, H. I., Saffari, M., & Moradi, R. (2017). Evaluating planting date and variety management strategies for adapting winter wheat to climate change impacts in arid regions. *International Journal of Climate Change Strategies and Management, 9*(6), 846–863.

Arndt, C. (2019). Renewable energy: Bringing electricity to revitalize Africa's rural areas. In *Global food policy report 2019* (pp. 60–67). International Food Policy Research Institute.

Asfaw, M., Geta, E., & Mitiku, F. (2019). Economic efficiency of smallholder farmers in wheat production: The case of abuna Gindeberet district, Oromia national regional state, Ethiopia. *International Journal of Environmental Sciences and Natural Resource, 16*(2), 41–51.

Babu, S. C., De Pinto, A., & Paul, N. (2019). Strengthening institutional capacity for disaster management and risk reduction through climate-resilient agriculture. In V. Venkatachalam Anbumozhi, M. Breiling, & V. Reddy (Eds.), *Disasters, climate change and food security: Assessment, analysis and action for ASEAN* (pp. 269–289). Economic Research Institute for ASEAN and East Asia.

Battilani, P., Toscano, P., Van der Fels-Klerx, H. J., Moretti, A., Camardo Leggieri, M., Brera, C., Rortais, A., Goumperis, T., & Robinson, T. (2016). Aflatoxin B1 contamination in maize in Europe increases due to climate change. *Scientific Reports, 6*, 24328.

Beach, R. H., Sulser, T. B., Crimmins, A., Cenacchi, N., Cole, J., Fukagawa, N. K., Mason-D'Croz, D., Myers, M., Sarofim, S., Smith, M., & Ziska, L. H. (2019). A modeling approach combining elevated atmospheric $CO_2$ effects on protein, iron and zinc availability with projected climate change impacts on global diets. *The Lancet Planetary Health, 3*(7), 307–317.

Béné, C. (2020). Resilience of local food systems and links to food security—A review of some important concepts in the context of COVID-19 and other shocks. *Food Security, 12*, 805–822.

Béné, C., Arthur, R., Norbury, H., Allison, E. H., Beveridge, M. C., Bush, S., Campling, L., Leschend, W., Little, D., Squires, D., Thilsted, S. H., Troelli, M., & Williams, M. (2016). Contribution of fisheries and aquaculture to food security and poverty reduction: Assessing the current evidence. *World Development, 79*, 177–196.

Béné, C., Chowdhury, F. S., Rashid, M., Dhali, S. A., & Jahan, F. (2017). Squaring the circle: Reconciling the need for rigor with the reality on the ground in resilience impact assessment. *World Development, 97*, 212–231.

Béné, C., & Haque, M. A. B. (2021). Strengthening the resilience of vulnerable communities—Results from a quasi-experimental impact evaluation in coastal Bangladesh. *European Journal of Development Research, 34*(8). https://doi.org/10.1057/s41287-021-00399-9

Berlin, J., Sonesson, U., & Tillman, A. (2008). Product chain actors' potential for greening the product life cycle. *Journal of Industrial Ecology, 12*, 95–110.

Brenton, P., Chemutai, V., & Pangestu, M. (2022). Trade and food security in a climate change-impacted world. *Agricultural Economics, 53*, 580–591.

Brown, M. E., Carr, E. R., Grace, K. L., Wiebe, K., Funk, C. C., Attavanich, W., Backlund, P., & Buja, L. (2017). Do markets and trade help or hurt the global food system adapt to climate change? *Food Policy, 68*, 154–159.

Brown, M. E., & Kshirsagar, V. (2015). Weather and international price shocks on food prices in the developing world. *Global Environmental Change, 34*, 31–40.

Bryan, E., Theis, S., Choufani, J., De Pinto, A., Meinzen-Dick, R., & Ringler, C. (2017). Gender-sensitive, climate-smart agriculture for improved nutrition in Africa south of the Sahara. In A. De Pinto & J. M. Ulimwengu (Eds.), *A thriving agricultural sector in a changing climate: The contribution of climate-smart agriculture to Malabo and sustainable development goals* (pp. 114–135). International Food Policy Research Institute.

Chakrabarti, S., Kishore, A., Raghunathan, K., & Scott, S. P. (2018). Impact of subsidized fortified wheat on anaemia in pregnant Indian women. *Maternal & Child Nutrition, 15*(1), 12669.

Cochrane, L., & Cafer, A. (2018). Does diversification enhance community resilience? A critical perspective. *Resilience International Policies, Practices and Discourses, 6*(2), 129–143.

Cole, S., Giné, X., & Vickery, J. (2013). *How does risk management influence production decisions? Evidence from a field experiment* (Policy Research Working Paper, 6546). World Bank.

Constas, M., Frankenberger, T. R., Hoddinott, J., Mock, N., Romano, D., Béné, C., & Maxwell, D. (2014). *A common analytical model for resilience measurement—Causal framework and methodological options*. Resilience Measurement Technical Working Group (FSiN Technical Series Paper No. 2). World Food Program and Food and Agriculture Organization of the United Nations.

da Silva, G. J., & Fan, S. (2017). Smallholders and urbanization: Strengthening rural-urban linkages to end hunger and malnutrition. In *Global food policy report 2017* (pp. 14–23). International Food Policy Research Institute.

De Pinto, A., Cenacchi, N., Kwon, H.-Y., Koo, J., & Dunston, S. (2020). Climate smart agriculture and global food-crop production. *PLOS One, 15*(4), 0231764.

De Pinto, A., Seymour, G., Bryan, E., & Bhandari, P. (2020). Women's empowerment and farmland allocations in Bangladesh: Evidence of a possible pathway to crop diversification. *Climatic Change, 163*, 1025–1043.

Deb, P., Babel, M. S., & Denis, A. F. (2018). Multi-GCMs approach for assessing climate change impact on water resources in Thailand model. *Earth Systems and Environment, 4*, 825–839.

Dong, J., Gruda, N., Lam, S. K., Li, X., & Duan, Z. (2018). Effects of elevated $CO_2$ on nutritional quality of vegetables: A review. *Frontiers in Plant Sciences, 9*, 924.

Duensing, N., Sprink, T., Parrott, W. A., Fedorova, M., Lema, M. A., Wolt, J. D., & Bartsch, D. (2018). Novel features and considerations for ERA and regulation of crops produced by genome editing. *Frontiers in Bioengineering and Biotechnology, 6*, 79.

Elbehri, A., Elliot, J., & Wheeler, T. (2015). Climate change, food security and trade: An overview of global assessments and policy insights climate change and food systems. *Global assessments and implications for food security and trade* (pp. 1–19). Food and Agriculture Organization of the United Nations.

England, M. I., Dougill, A. J., Stringer, L. C., Vincent, K. E., Pardoe, J., Kalaba, F. K., Mkwambisi, D. D., Namaganda, E., & Afionis, S. (2018). Climate change adaptation and cross-sectoral policy coherence in southern Africa. *Regional Environmental Change, 18*, 2059–2071.

Fanzo, J., Davis, C., McLaren, R., & Choufani, J. (2018). The effect of climate change across food systems: Implications for nutrition outcomes. *Global Food Security, 18*, 12–19.

FAO. (1946). *World food survey*. Food and Agriculture Organization of the United Nations. https://www.fao.org/publications/card/en/c/cb6105en/

FAO. (2011). *Synthetic account of the second global plan of action for plant genetic resources for food and agriculture*. Food and Agriculture Organization of the United Nations. https://www.fao.org/3/i2650e/i2650e.pdf

FAO. (2016). *The state of food and agriculture: Climate change, agriculture and food security*. Food and Agriculture Organization of the United Nations. https://www.fao.org/3/i6030e/I6030E.pdf

FAO-WHO. (2014). *Conference outcome document: Rome declaration on nutrition*. Food and Agriculture Organization of the United Nations. https://www.fao.org/3/ml542e/ml542e.pdf

Garnett, T., Appleby, M. C., Balmford, A., Bateman, I. J., Benton, T. G., Bloomer, P., Burlingame, B., Dawkins, M., Doland, L., Fraser, D., Herrero, M., Hoffmann, I., Smith, P., Thornton, P. K., Toulmins, C., Vermeulen, S. J., & Godfray, H. C. J. (2013). Sustainable intensification in agriculture: Premises and policies. *Science, 341*(6141), 33–34.

Gil, J. D. B., Cohn, A. S., Duncan, J., Newton, P., & Vermeulen, S. (2017). The resilience of integrated agricultural systems to climate change. *Wires Climate Change, 8*, 461.

Giné, X., Townsend, R., & Vickery, J. (2008). Patterns of rainfall insurance participation in rural india. *World Bank Economic Review, 22*(3), 539–566.

Giné, X., & Yang, D. (2009). Insurance, credit, and technology adoption: Field experimental evidence from Malawi. *Journal of Development Economics, 89*(1), 1–11.

GLOPAN. (2016). *Food systems and diets: Facing the challenges of the 21st century*. Global Panel on Agriculture and Food Systems for Nutrition. http://glopan.org/sites/default/files/ForesightReport.pdf

Goddard, L. (2016). From science to service. *Science, 353*(6306), 1366–1367.

Grafton, R. Q., Doyen, L., Béné, C., Borgomeo, E., Brooks, K., Chu, L., Cumming, G. S., Dixon, J., Dovers, S., Garrick, D., & Helfgott, A. (2019). Realizing resilience for decision-making. *Nature Sustainability, 2*(10), 907–913.

Hallegatte, S., Vogt-Schilb, A., Bangalore, M., & Rozenberg, J. (2017). Unbreakable: Building the resilience of the poor in the face of natural disasters. *Climate change and development*. World Bank.

Hawkes, C., Harris, J., & Gillespie, S. (2017). Changing diets: Urbanization and the nutrition transition. *Global food policy report, 2017* (pp. 34–41). International Food Policy Research Institute.

Headey, D., Hirvonen, K., & Hoddinott, J. (2018). Animal sourced foods and child stunting. *American Journal of Agricultural Economics, 100*(5), 1302–1319.

Hill, R. V., Kumar, N., Magnan, N., Makhija, S., de Nicola, F., Spielman, D. J., & Ward, P. S. (2019). Ex ante and ex post effects of hybrid index insurance in Bangladesh. *Journal of Development Economics, 136*, 1–17.

Hill, R. V., Robles, M., & Ceballos, F. (2016). Demand for a simple weather insurance product in India: Theory and evidence. *American Journal of Agricultural Economics, 98*(4), 1250–1270.

Hogeboom, R. J., Borsje, B. W., Deribe, M. M., Van Der Meer, F. D., Mehvar, S., Meyer, M. A., Özerol, G., Hoekstra, A. Y., & Nelson, A. D. (2021). Resilience meets the water–energy–food nexus: Mapping the research landscape. *Frontiers in Environmental Science, 9*, 630395.

Högy, P., & Fangmeier, A. (2009). Atmospheric $CO_2$ enrichment affects potatoes: 2 tuber quality traits. *European Journal of Agronomy, 30*, 85–94.

Holling, C. S. (1973). Resilience and stability of ecological systems. *Annual Review of Ecology and Systematics, 4*, 1–23.

IFPRI. (2017). Food policy indicators: Tracking change. *Global food policy report* (pp. 84–110). International Food Policy Research Institute.

IPCC. (2018). Summary for policymakers. In V. Masson-Delmotte, P. Zhai, H. O. Portner, D. Roberts, J. Skea, P. R. Shukla, A. Pirani, W. Moufouma-Okia, C. Péan, R. Pidcock, S. Connors, J. B. R. Matthews, Y. Chen, X. Zhou, M. I. Gomis, E. Lonnoy, T. Maycock, M. Tignor, & T. Waterfield (Eds.), *Global warming of 1.5 °C. An IPCC special report on the impacts of global warming of 1.5 °C above pre-industrial levels and related global greenhouse gas emission pathways, in the context of strengthening the global response to the threat of climate change, sustainable development, and efforts to eradicate poverty*. World Meteorological Organization.

Jacxsens, L., Luning, P. A., der Vorst, J., Devlieghere, F., Leemans, R., & Uyttendaele, M. (2010). Simulation modelling and risk assessment as tools to identify the impact of climate change on microbiological food safety—The case study of fresh produce supply chain. *Food Research International, 43*, 1925–1935.

James, S. J., & James, C. (2010). The food cold-chain and climate change. *Food Research International, 43*, 1944–1956.

Kristjanson, P., Bryan, E., Bernier, Q., Twyman, J., Meinzen-Dick, R., Kieran, C., Ringler, C., Jost, C., & Doss, C. (2017). Addressing gender in agricultural research for development in the face of a changing climate: Where are we and where should we be going? *International Journal of Agricultural Sustainability, 15*(5), 482–500.

Leroy, J. L., & Frongillo, E. A. (2007). Can interventions to promote animal production ameliorate undernutrition? *The Journal of Nutrition, 137*(10), 2311–2316.

Levine, S. (2014). *Assessing resilience: Why quantification misses the point* (HPG working paper). https://www.odi.org/sites/odi.org.uk/files/odi-assets/publications-opinion-files/9049.pdf

Lin, B. B. (2011). Resilience in agriculture through crop diversification: Adaptive management for environmental change. *BioScience, 61*, 183–193.

Lipper, L., McCarthy, B., Zilberman, D., Asfaw, S., & Branca, G. (Eds.). (2018). *Climate smart agriculture: Building resilience to climate change.* Food and Agriculture Organization of the United Nation.

Lybbert, T. J., & Sumner, D. A. (2012). Agricultural technologies for climate change in developing countries: Policy options for innovation and technology diffusion. *Food Policy, 37*, 114–123.

Martins, V. J., Toledo Florêncio, T. M., Grillo, L. P., do Carmo, M., Franco, P., Martins, P. A., Clemente, A. P. G., Santos, C. D. L., de Fatima, M., Vieira, A., & Sawaya, A. L. (2011). Long-lasting effects of undernutrition. *International Journal of Environmental Research and Public Health, 8*(6), 1817–1846.

Martorell, R., Ascencio, M., Tacsan, L., Alfaro, T., Young, M. F., Yaw Addo, O., Dary, O., & Flores-Ayala, R. (2015). Effectiveness evaluation of the food fortification program of Costa Rica: Impact on anemia prevalence and hemoglobin concentrations in women and children. *American Journal of Clinical Nutrition, 101*(1), 210–217.

Mbow, C., Rosenzweig, C., Barioni, L. G., Benton, T. G., Herrero, M., Krishnapillai, M., Liwenga, E., Pradhan, P., Rivera-Ferre, M. G., Sapkota, T., Tubiello, F. N., & Xu, Y. (2019). Food security. In P. R. Shukla, J. Skea, E. Calvo Buendia, V. Masson-Delmotte, H. O. Pörtner, D. C. Roberts, P. Zhai, R. Slade, S. Connors, R. van Diemen, M. Ferrat, E. Haughey, S. Luz, S. Neogi, M. Pathak, J. Petzold, J. Portugal Pereira, P. Vyas, E. Huntley, K. Kissick, M. Belkacemi, & J. Malley (Eds.), *Climate change and land: An IPCC special report on climate change, desertification, land degradation, sustainable land management, food security, and greenhouse gas fluxes in terrestrial ecosystems.* Intergovernmental Panel on Climate Change, Geneva.

McGregor, D., Whitaker, S., & Sritharan, M. (2020). Indigenous environmental justice and sustainability. *Current Opinion in Environmental Sustainability, 43*, 35–40.

Meinzen-Dick, R., Bernier, Q., & Haglund, E. (2013). *The six 'ins' of climate-smart agriculture: Inclusive institutions for information, innovation, investment, and insurance* (CAPRi working paper 114). International Food Policy Research Institute.

Midega, C. A. O., Pittchar, J. O., Pickett, J. A., Hailu, G. W., & Khan, Z. R. (2018). A climate-adapted push-pull system effectively controls fall armyworm Spodoptera frugiperda (J. E. Smith), in maize in East Africa. *Crop Protection, 105*, 10–15.

Mojica, F. J., Díez-Villaseñor, C., García-Martínez, J., & Almendros, C. (2009). Short motif sequences determine the targets of the prokaryotic CRISPR defence system. *Microbiology, 155*, 733.

Molden, D., Oweis, T., Steduto, P., Bindraban, P., Hanjra, M. A., & Kijne, J. (2010). Improving agricultural water productivity: Between optimism and caution. *Agricultural Water Management, 97*, 528–535.

Moretti, C. L., Mattos, L. M., Calbo, A. G., & Sargent, S. A. (2010). Climate changes and potential impacts on postharvest quality of fruit and vegetable crops—A review. *Food Research International, 43*, 1824–1832.

Myers, S. S., Smith, M. R., Guth, S., Golden, C. D., Vaitla, B., Mueller, N. D., Dangour, A. D., & Huybers, P. (2017). Climate change and global food systems: Potential impacts on food security and undernutrition. *Annual Review of Public Health, 20*(38), 259–277.

Myers, S. S., Zanobetti, A., Kloog, I., Huybers, P., Leakey, A. D. B., Bloom, A. J., Carlisle, E., Dietterich, L. H., Fitzgerald, G., Hasegawa, T., Holbrook, N. M., Nelson, R. L., Ottman, M. J., Raboy, V., Sakai, H., Sartor, K. A., Schwartz, J., Seneweera, S., Tausz, M., & Usui, Y. (2014). Increasing $CO_2$ threatens human nutrition. *Nature, 510*, 139–142.

Nelson, G., Bogard, J., Lividini, K., Arsenault, J., Riley, M., Sulser, T. B., Mason-D'Croz, D., Power, B., Gustafson, D., Herrero, M., Wiebe, K., Cooper, K., Remans, R., & Rosegrant, M. (2018). Income growth and climate change effects on global nutrition security to mid-century. *Nature Sustainability, 1*, 773–781.

Nelson, G., Palazzo, A., Ringler, C., Sulser, T., & Batka, M. (2009). *The role of international trade in climate change adaptation* (ICTSD and Food & Agricultural Trade Issue Brief No. 4). International Centre for Trade and Sustainable Development (ICTSD) and International Food & Agricultural Trade Policy Council.

Nelson, G. C., Rosegrant, M. W., Palazzo, A., Gray, I., Ingersoll, C., Robertson, R., Tokgoz, S., Zhu, T., Sulser, T. B., Ringler, C., Msangi, S., & You, L. (2010). *Food security, farming, and climate change to 2050: Scenarios, results, policy options*. International Food Policy Research Institute.

Nicholls, R. J., & Cazenave, A. (2010). Sea-level rise and its impact on coastal zones. *Science, 328*(5985), 1517–1520.

Oliver, T., Boyd, E., Balcombe, K., Benton, T., Bullock, J., Donovan, D., Feola, G., Heard, M., Mace, G. M., Mortimer, S. R., Nunes, R. J., Pywell, R. F., & Zaum, D. (2018). Overcoming undesirable resilience in the global food system. *Global Sustainability, 1*(e9), 1–9.

Parry, M. L., Rosenzweig, C., Iglesias, A., Livermore, M., & Fischer, G. (2004). Effects of climate change on global food production under SRES emissions and socio-economic scenarios. *Global Environmental Change, 14*, 53–67.

Peterman, A., Behrman, J. A., & Quisumbing, A. R. (2014). A review of empirical evidence on gender differences in nonland agricultural inputs, technology, and services in developing countries. In A. R. Quisumbing, R. Meinzen-Dick, T. L. Raney, A. Croppenstedt, J. A. Ehrman, & A. Peteran (Eds.), *Gender in agriculture: Closing the knowledge gap* (pp. 145–186). Springer.

Phalkey, R. K., Aranda-Jan, C., Marx, S., Höfle, B., & Sauerborn, R. (2015). Systematic review of current efforts to quantify the impacts of climate change on undernutrition. *Proceedings of the National Academy of Science USA, 112*(33), 4522–4529.

Pingali, P., Alinovi, L., & Sutton, J. (2005). Food security in complex emergencies: Enhancing food system resilience. *Disasters, 29*, 5–24.

Portier, C. J., Thigpen, T. K., Carter, S. R., Dilworth, C. H., & Grambsch, A. E. (2013). *A human health perspective on climate change: A report outlining the research needs on the human health effects of climate change.* Environmental Health Perspectives/National Institute of Environmental Health Sciences, Research Triangle Park. https://digital.library.unt.edu/ark:/67531/metadc950189/

Prosperi, P., Allen, T., Cogill, B., Padilla, M., & Peri, I. (2016). Towards metrics of sustainable food systems: A review of the resilience and vulnerability literature. *Environment Systems and Decisions, 36*(1), 3–19.

Quandt, A., Neufeldt, H., & McCabe, J. T. (2017). The role of agroforestry in building livelihood resilience to floods and drought in semiarid Kenya. *Ecology and Society, 22*(3), 10.

Quandt, A., Neufeldt, H., & McCabe, J. T. (2019). Building livelihood resilience: What role does agroforestry play? *Climate and Development, 11*(6), 485–500.

Robinson, L. W., Ericksen, P. J., Chesterman, S., & Worden, J. S. (2015). Sustainable intensification in drylands: What resilience and vulnerability can tell us. *Agricultural Systems, 135*, 133–140.

Rockström, J. (2003). Resilience building and water demand management for drought mitigation. *Physics and Chemistry of the Earth, Parts A/B/C, 28*(20–27), 869–877.

Rockström, J., & Barron, J. (2007). Water productivity in rainfed systems: Overview of challenges and analysis of opportunities in water scarcity prone savannahs. *Irrigation Science, 25*, 299–311.

Rosegrant, M. W., Koo, J., Cenacchi, N., Ringler, C., Robertson, R., Fisher, M., Cox, C., Garrett, K., Perez, N. D., & Sabbagh, P. (2014). *Food security in a world of natural resource scarcity: The role of agricultural technologies.* International Food Policy Research Institute.

Rosenstock, T. S., Lamanna, C., Namoi, N., Arslan, A., & Richards, M. (2019). What is the evidence base for climate-smart agriculture in East and Southern Africa? A systematic map. In T. Rosenstock, A. Nowak, & E. Girvetz (Eds.), *The climate-smart agriculture papers* (pp. 141–151). Springer.

Rosenzweig, C., Mbow, C., Barioni, L. G., Benton, T. G., Herrero, M., Krishnapillai, M., Liwenga, E. T., Pradhan, P., Rivera-Ferre, M. G., Sapkota, T., Tubiello, F. N., Xu, Y., Mencos Contreras, E., & Portugal-Pereira, J. (2020). Climate change responses benefit from a global food system approach. *Nature Food, 1,* 94–97.

Ruel, M. T. (2001). *Can food-based strategies help reduce vitamin A and iron deficiencies? A review of recent evidence.* International Food Policy Research Institute.

Ruel, M. T., Garrett, J. L., & Yosef, S. (2017). Food security and nutrition: Growing cities, new challenges. *Global food policy report* (pp. 24–32). International Food Policy Research Institute.

Saito, N. (2013). Mainstreaming climate change adaptation in least developed countries in South and Southeast Asia. *Mitigation and Adaptation Strategies for Global Change, 18,* 825–849.

Schimmelpfennig, D. (2016). *Precision agriculture technologies and factors affecting their adoption.* United States Department of Agriculture. https://www.ers.usda.gov/amber-waves/2016/december/precision-agriculture-technologies-and-factors-affecting-their-adoption/

Schwan, S., & Yu, X. (2018). Social protection as a strategy to address climate-induced migration. *International Journal of Climate Change Strategies and Management, 10,* 43–64.

Serfilippi, E., & Ramnath, G. (2018). Resilience measurement and conceptual frameworks: A review of the literature. *Annals of Public and Cooperative Economics, 89*(4), 645–664.

Shi, L., Chu, E., & Debats, J. (2015). Explaining progress in climate adaptation planning across 156 U.S. municipalities. *Journal of the American Planning Association, 81*(3), 191–202.

Signorelli, S., Azzarri, C., & Roberts, C. (2016). *Malnutrition and climate patterns in the ASALs of Kenya: A resilience analysis based on a pseudopanel dataset.* Report prepared by the Technical Consortium, a project of the CGIAR (Technical Report Series No. 2: Strengthening the evidence base for resilience in the Horn of Africa). International Livestock Research Institute, Nairobi.

Silici, L., Ndabe, P., Friedrich, T., & Kassam, A. (2011). Harnessing sustainability, resilience and productivity through conservation agriculture: The case of Likoti in Lesotho. *International Journal of Agricultural Sustainability, 9*(1), 137–144.

Springmann, M., Mason-D'Croz, D., Robinson, S., Garnett, T., Charles, H., Godfray, J., Gollin, D., Rayner, M., Ballon, P., & Scarborough, P. (2016). Global and regional health effects of future food production under climate change: A modelling study. *The Lancet, 387*(10031), 1937–1946.

Stevanovic, M., Popp, A., Lotze-Campen, H., Dietrich, J. P., Müller, C., Bonsch, M., Schmitz, C., Bodirsky, B., Humpenöder, F., & Weindl, I. (2016). The impact of high-end climate change on agricultural welfare. *Science Advances, 2*(8), e1501452.

Stone, J., & Rahimifard, S. (2018). Resilience in agri-food supply chains: A critical analysis of the literature and synthesis of a novel framework. *Supply Chain Management, 23*(3), 207–238.

Thacker, S., Adshead, D., Fay, M., Hallegatte, S., Harvey, M., Meller, H., O'Regan, N., Rozenberg, J., Watkins, G., & Hall, J. W. (2019). Infrastructure for sustainable development. *Nature Sustainability, 2*, 324–331.

Tirado, M. C., Clarke, R., Jaykus, L. A., McQuatters-Gollop, A., & Frank, J. M. (2010). Climate change and food safety: A review. *Food Research International, 43*, 1745–1765.

Tirado, M. C., Hunnes, D., Cohen, M. J., & Lartey, A. (2015). Climate change and nutrition in Africa. *Journal of Hunger & Environmental Nutrition, 10*(1), 22–46.

Verchot, L., van Noordwijk, M., Kandji, S., Tomich, T., Ong, C., Albrecht, A., Mackensen, J., Bantilan, C., Anupama, K., & Palm, C. (2007). Climate change: Linking adaptation and mitigation through agroforestry. *Mitigation and Adaptation Strategies for Global Change, 12*, 901–918.

Vonthron, S., Dury, S., Fallot, A., & Bousquet, F. (2016). L'intégration des concepts de résilience dans le domaine de la sécurité alimentaire: Regards croisés. *Cahiers Agriculture, 25*, 64001.

Weindl, I., Lotze-Campen, H., Popp, A., Müller, C., Havlik, P., Herrero, M., Schmitz, C., & Rolinski, S. (2015). Livestock in a changing climate: Production system transitions as an adaptation strategy for agriculture. *Environmental Research Letters, 10*, 094021.

Wheeler, T., & von Braun, J. (2013). Climate change impacts on global food security. *Science, 341*, 508–513.

Wiebe, K. D., Lotze-Campen, H., Sands, R. D., Tabeau, A., van der Mensbrugghe, D., Biewald, A., Bodirsky, B., Islam, S., Kavallari, A., Mason-D'Croz, D., Müller, C., Popp, A., Robertson, R., Robinson, S., van Meijl, H., & Willenbocke, D. (2015). Climate change impacts on agriculture in 2050 under a range of plausible socioeconomic and emissions scenarios. *Environmental Research Letters, 10*, 85010.

Wilson, G. (2010). Multifunctional 'quality' and rural community resilience. *Transactions of the Institute of British Geographers, 35*(3), 364–381. http://www.jstor.org/stable/40890993

Woodfine, A. (2009). *The potential of sustainable land management practices for climate change mitigation and adaptation in Sub-saharan Africa*. Food and Agriculture Organization of the United Nations.

Xu, L., Marinova, D., & Guo, X. (2015). Resilience thinking: A renewed system approach for sustainability science. *Sustainability Science, 10,* 123–138.

Yin, R. K. (2009). *Case study research: Design and methods* (4th ed.). Sage.

Zilberman, D., Zhao, J., & Heiman, A. (2012). Adoption versus adaptation, with emphasis on climate change. *Annual Review of Resource Economics, 4,* 27–53.

**Open Access** This chapter is licensed under the terms of the Creative Commons Attribution 4.0 International License (http://creativecommons.org/licenses/by/4.0/), which permits use, sharing, adaptation, distribution and reproduction in any medium or format, as long as you give appropriate credit to the original author(s) and the source, provide a link to the Creative Commons license and indicate if changes were made.

The images or other third party material in this chapter are included in the chapter's Creative Commons license, unless indicated otherwise in a credit line to the material. If material is not included in the chapter's Creative Commons license and your intended use is not permitted by statutory regulation or exceeds the permitted use, you will need to obtain permission directly from the copyright holder.

CHAPTER 8

# Gender, Resilience, and Food Systems

*Elizabeth Bryan, Claudia Ringler, and Ruth Meinzen-Dick*

## INTRODUCTION

Resource-poor people face multiple, intersecting social, economic, health, political, and environmental risks and disturbances (Demetriades & Esplen, 2010; Hallegatte & Rozenberg, 2017; Leichenko & Silva, 2014). The concept of resilience has allowed researchers, policymakers, and practitioners to think more broadly about potential solutions to the confluence of challenges facing vulnerable communities, particularly where structural problems and inequalities, such as chronic poverty and gender inequalities, underlie persistent and recurrent shocks and stressors (Béné et al., 2014; Smyth & Sweetman, 2015; USAID, 2012). By lengthening the time frame for considering risks, the resilience lens has also helped

---

E. Bryan (✉) · C. Ringler · R. Meinzen-Dick
International Food Policy Research Institute, Washington, DC, USA
e-mail: E.Bryan@cgiar.org

C. Ringler
e-mail: C.Ringler@cgiar.org

R. Meinzen-Dick
e-mail: R.Meinzen-Dick@cgiar.org

to focus attention on the implications of humanitarian interventions on longer-term development and on safeguarding development gains against shocks, thereby bridging humanitarian and development efforts (Béné et al., 2016b; Frankenberger et al., 2014). Key elements for measuring the process of resilience include information on initial and subsequent states (well-being outcomes), disturbances (shocks and stressors), and capacities (Constas et al., 2014; Frankenberger et al., 2014).

While resilience is a complex concept that is understood and utilized in different ways by different disciplines, most definitions describe human resilience as the ability to draw upon a set of capacities to deal with shocks and stressors before, during, and after a disturbance, in a way that maintains or improves well-being outcomes (such as food security or adequate nutrition) (Frankenberger et al., 2014; Mercy Corps, 2016; USAID, 2012, 2017). In this paper, we use the definition of resilience as "the ability of people, households, communities, countries, and systems to mitigate, adapt to, and recover from shocks and stresses in a manner that reduces chronic vulnerability and facilitates inclusive growth" (USAID, 2012, p. 5). This approach prioritizes investments that build adaptive capacities, such as expanding economic opportunities, education, environmental sustainability, diverse livelihoods, and nutrition and health services, while also identifying and reducing risks.

Resilience approaches require studying and understanding the local context to ensure that interventions build on existing capacities and risk management institutions, and support people and institutions in pursuing their preferred strategies (Agrawal et al., 2010; Tschakert, 2007; Vaughan & Henly-Shepard, 2018). Attention to the specific context refers not only to a particular time and place, but also to the many social differences of people living in a specific geography at a given time. In designing and evaluating resilience-oriented programs and policies, development actors need to consider questions such as which kinds of capacities are important for building resilience in a particular context for specific groups of people, and how best to support people in developing these capacities and responding to shocks and stressors in a way that protects or improves well-being outcomes.

Growing evidence shows that men and women are differently exposed to and have different preferences and capacities to respond to shocks and stressors, which include some of the largest threats facing the global community like climate change, conflict, and COVID-19 (Jordan, 2019;

Smyth & Sweetman, 2015; Theis et al., 2019). Because shocks and stressors occur in local contexts with different power structures, institutions, infrastructure, and socio-cultural norms, it is difficult to generalize the ways in which men and women of different backgrounds are differently affected by and respond to these disturbances. Moreover, the ways in which men and women experience and react depends on the types of overlapping shocks and stressors that they experience and the individual capacities that they possess.

Underlying structural inequalities in society, including between men and women, shape the ways in which shocks and stressors are distributed and impact people, and the capacities and options men and women have at their disposal to respond to these disturbances (Njuki et al., 2022). Even members of the same household do not necessarily face the same set of risks or share the same capacities, vulnerabilities, preferences, and decision-making power. Similarly, men and women in rural farming communities face different challenges and have different resilience capacities and response options than poor urban consumers, as they are embedded in different food environments.[1] These differences can result in differential well-being outcomes, including food security and nutrition outcomes, and can exacerbate existing gender gaps in food systems as people experience and respond to multiple shocks and stresses. Thus, understanding the interactions between gender and resilience as they relate to food systems and food security is an important topic for the research for development agenda.

This chapter highlights the key gendered dimensions of resilience in the food system, drawing on evidence from the literature, including systematic reviews and global indicators, where available, as well as case study examples that highlight important linkages between gender, resilience, and food security. The chapter builds on a conceptual framework for gender and resilience (Theis et al., 2019) by incorporating a food systems lens that expounds dimensions related to gender and resilience at different stages of the food value chain from production to consumption and in different food environments (rural and urban). After presenting

---

[1] The food environment "refers to the physical, economic, political and socio-cultural context in which consumers engage with the food system to acquire, prepare and consume food" (HLPE, 2017, p. 11). Key elements of the environment are physical and economic access to food (proximity and affordability); food promotion, advertising and information; and food quality and safety.

the conceptual framework, this chapter applies it to case studies from the literature, which are used to highlight the components of the framework and how these link to key elements of food systems. It concludes by discussing how to strengthen attention to gender in resilience policies, programs, and investments and by highlighting remaining research and evidence gaps.

## Conceptual Framework

This chapter uses a conceptual framework developed by the authors for the Gender, Climate Change, and Nutrition Integration Initiative (GCAN), which characterizes the relationships between resilience, gender, and nutrition (Bryan et al., 2017; Theis et al., 2019). This chapter expands on this framework by integrating a food systems lens (Fig. 8.1). Food systems include all elements and activities related to the production, processing, distribution, preparation, and consumption of food, and the drivers and outputs of these activities, including food security and nutrition outcomes, as well as socioeconomic and environmental outcomes (HLPE, 2017). Specifically, this chapter integrates the three main elements of food systems—food value chains,[2] the food environment, and consumer behavior—which influence the ways in which shocks and stressors affect well-being outcomes—especially food security and nutrition outcomes.

The framework can be applied to different scales of analysis, including intrahousehold, household, community, and national levels. In this chapter, we mainly focus on individuals within households to illustrate that members of the same household do not necessarily share the same capacities, vulnerabilities, preferences, and decision-making power. While introduced here briefly, the interactions between gender and resilience are discussed in more detail in relation to each component of the framework in the following sections, drawing on evidence from the literature. The literature selected to illustrate the main components of the framework focuses more on gender dimensions at the household level; however, the enabling environment, including policies, investments, and interventions

---

[2] Food value (or supply) chains consist of all actors and activities involved in moving food from production to consumption including activities related to producing, processing, distributing, and preparing food (HLPE, 2017).

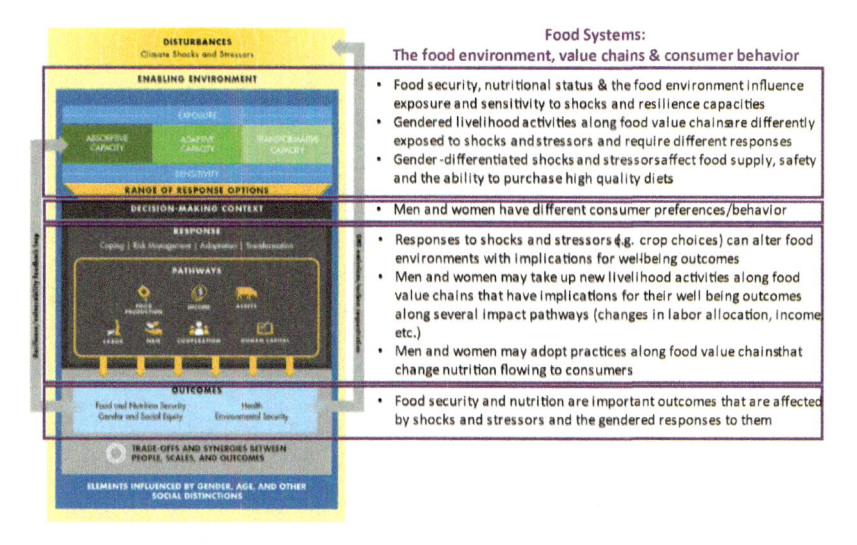

**Fig. 8.1** The Gender, Climate Change, and Nutrition Integration Initiative (GCAN) framework (*Source* Adapted by the authors from Bryan et al. [2017] and Theis et al. [2019])

also influence all the other elements of the framework from exposure and sensitivity to resilience outcomes.

We then discuss how a food system lens relates to each component of the framework. The added food system elements are summarized in purple boxes in Fig. 8.1 and linked to the components of the framework. Shocks and stressors affect people depending on their level of exposure and sensitivity to the disturbance, their resilience capacities to cope with, adapt to, and transform their livelihoods, and the enabling environment. Well-being outcomes also depend on intrahousehold decision-making processes which determine which responses are chosen and who benefits from these choices. Depending on these response choices, well-being outcomes follow along several potential impact pathways, including changes in food production practices, sources and allocation of income, asset dynamics, labor allocation, natural resource management, collective action, and human capital investments. The well-being outcomes resulting from current shocks and stressors then influence exposure, sensitivity, and resilience capacities to future disturbances through a feedback loop. Similarly, the greenhouse gas implications of different response options, such as consumption changes, changes in farming practices and land use changes, also influence future climate change, one of the key challenges facing vulnerable communities in low-income countries. Viewing this iterative cycle through a gender lens illustrates how differences between men and women emerge along each of these steps.

First, individuals are exposed to different disturbances (shocks and stressors) and experience the same shocks and stressors differently. Food security and nutritional (and health) status and the food environment can influence men's and women's sensitivity to shocks and stressors. Moreover, men and women's livelihood activities along agricultural value chains may be differently exposed to shocks and stressors (Limuva & Sinnevag, 2018).

Second, people have different resilience capacities (absorptive, adaptive, transformative), subject to gender and other social distinctions as well as the intersection of these identities, including those related to age, class, caste, ethnicity, marital status, and sexual identity, among others (Béné et al., 2014; Djoudi et al., 2016). Narratives that depict women as perpetually vulnerable and men as inevitably antagonistic toward gender equality ignore the ways in which women are agents of change and neglect the constraints faced by men as well as the opportunities to mobilize men as allies for gender and social equity (Doss et al., 2018). Nutritional status, food security, and the food environment are important determinants of men's and women's resilience capacities. In addition, the range of response options for men and women may differ, based on their individual capacities and livelihood activities in which they are engaged, including along agricultural value chains.

Third, within households, institutions, and communities, each response to a disturbance—even if that response is to do nothing—is the result of choice and negotiation, albeit among restricted options (Demetriades & Esplen, 2010; Quisumbing & Maluccio, 2003). Individuals do not all have the same preferences, knowledge, priorities, or decision-making power. The decision-making context, or an actor's ability to negotiate for a preferred response option within a household or community, is a key element within the process of resilience that has strong differences by gender but is often overlooked (Behrman et al., 2014). Importantly, men and women have different consumer preferences and behavior, including different roles in procuring food—with women generally being in charge of food preparation and cooking, but in many cases eating last (Hathi et al., 2021)—and in their dietary preferences (Fig. 8.1).

The process of negotiation can lead to a set of observed response choices that can be characterized in different ways. Response choices are reflective of the resilience capacities and livelihood roles of the actors (Ngigi et al., 2017). This framework, therefore, characterizes the responses along similar lines as coping, risk management, adaptive, and transformative responses (Béné et al., 2014; Keck & Sakdapolrak, 2013). Coping responses are usually short-term, ex post responses to experienced shocks or stresses and include actions like selling assets or changing consumption patterns, and, at larger scales, humanitarian interventions (Corbett, 1988; Dercon, 2002; Nguyen et al., 2020). While coping responses may aim to maintain well-being at pre-shock levels, they are often associated with a deterioration in well-being, such as poorer diets and increased indebtedness. Risk management strategies, like diversifying production or livelihood activities, and adaptive responses, like adopting new agronomic practices, tend to be proactive and aimed at avoiding or minimizing harmful impacts of shocks and stresses over the medium to long term (Corcoran-Nantes & Roy, 2018; Jost et al., 2016; Lawson et al., 2020). Transformative responses aim to change the fundamental attributes of a system or context to improve well-being outcomes, such as actions that directly address underlying social inequalities (Carr, 2020; McOmber et al., 2019).

Finally, responses to shocks and stressors can have differential impacts on the well-being outcomes of men, women, boys, and girls through several pathways, such as changes in gendered asset dynamics or labor allocation, and there may be trade-offs across different outcomes or different groups of people (Lee et al., 2021; Quisumbing et al., 2018). Men and women may take up new livelihood activities or practices along food value chains that affect well-being outcomes, including by influencing the flow of nutrition along food value chains (Brownhill et al., 2016; Gnisci, 2016). Drawing attention to outcome pathways helps uncover some of the key mechanisms driving well-being outcomes and how these outcomes are distributed among different groups of people.

## Application of the Framework to the Literature
### Gender Differences in Exposure and Sensitivity to Disturbances

Resilience-informed policy and programming require active investigation of how risks, and exposure and sensitivity to shocks and stressors differ within a local population. The literature discussed in this section illustrates gender differences in exposure and sensitivity to shocks and stressors and the reasons for these differences.

*Gender and Exposure*
In the case of natural disasters, several global reviews have found that women tend to have higher morbidity and reduced life expectancy compared to men following droughts, storms, earthquakes, and fires, especially where women have lower socioeconomic status, less access to information, and limited agency to make strategic life choices (Doocy et al., 2013; Erman et al., 2021; Neumayer & Plümper, 2007). Frankenberg et al. (2011) found that in Indonesia, women were twice as likely as men to die because of the 2004 Tsunami. Yet, other case studies found that men die at higher rates than women due to higher exposure to natural hazards given that they are overrepresented in high-risk occupations, such as construction (Delaney & Shrader, 2000; Erman et al., 2021; Zagheni et al., 2015). In Honduras and Nicaragua, more men than women died following Hurricane Mitch in 1998 (Delaney & Shrader, 2000).

Men's and women's differential experience with shocks and stressors such as climate change is reflected in the different ways in which they perceive and report the impacts of these disturbances, though patterns are not generalizable across contexts (Oloukoi et al., 2014; Twyman et al., 2014). In Nigeria, for example, while men were concerned with climate change impacts on yields of tuber and legume crops, women perceived a reduction in the availability of fruits, seeds, and herbs from community woodlots (Oloukoi et al., 2014).

There are also gender differences in the level of exposure to other types of shocks and stressors, such as violent conflict. Buvinic et al. (2013) outline gender differences in the direct and indirect impacts of armed conflict, including women's experience with sexual and gender-based violence, higher mortality rates among men and widowhood among women, greater displacement among women and children, and health impacts, such as the rate of HIV and AIDS. Shocks and stressors are often compounding—for instance, evidence suggests that weather shocks, like droughts, storms, and floods, exacerbate conflict and contribute to displacement and migration (Abel et al., 2019; IDMC, 2020).

COVID-19 has exacerbated gender disparities in how men and women experience the combined health and economic shocks of the pandemic, with considerable variation across countries. While men are at a higher risk of severe illness due to both biological and lifestyle factors (e.g., smoking and alcohol consumption) (Koo et al., 2021), women are likely to shoulder a larger care burden (IFPRI, 2021). IFPRI data from several rounds of phone surveys in selected countries in sub-Saharan Africa and South Asia suggest that both men and women in rural areas are hit hard by income losses, savings and asset depletion, indebtedness, and food insecurity because of the pandemic (ibid.). The extent to which families rely on men's or women's savings and assets to cope with the pandemic will have lasting implications for the gender asset gap and the resilience capacities of men and women to address future shocks and stressors. The World Economic Forum (2021) estimates that the time needed to close global gender gaps in economic opportunity, education, health, and political participation increased from 99.5 years to 135.6 years as a result of COVID-19.

*Gender and Sensitivity*

Sensitivity to particular shocks and stressors is also gendered and can change throughout the life cycle. For example, boys are biologically more vulnerable to shocks and stressors in utero, while young girls are more affected in early childhood (e.g., nutrition, schooling outcomes) due to greater social vulnerability (Erman et al., 2021). Later in life, the social and economic consequences of separation from or death of a partner are almost always more serious for women than they are for men because women risk losing access to land and other assets (Deere & Doss, 2006).

Gender disparities in the impacts of shocks and stressors can exacerbate gender inequality over the medium to long term. Studies of the gendered impacts of the 2008 food price crisis found that both the short- and long-term effects were greater on women smallholder farmers than men, including short-term food insecurity and hunger, and a long-term widening of the gender asset gap due to the lack of a gender-sensitive response to the crisis that addressed underlying structural inequalities in the food system (Botreau & Cohen, 2020; Quisumbing et al., 2011; Kumar & Quisumbing, 2013). Female-headed households appear to be particularly vulnerable to food price shocks, as demonstrated by a case study from Ethiopia that found that female-headed households were more likely to reduce meals and eat less preferred foods (Kumar & Quisumbing, 2013). In the absence of gender-responsive policies, investments, and interventions, the current global food crisis exacerbated by war in Ukraine continues to have disproportionate impacts on women and girls with potential long term, negative impacts (Bryan, Ringler, & Lefore, 2022).

*What Does a Food System Lens Add?*
Integrating a food system lens into the conceptual framework highlights important differences in exposure and sensitivity of men and women based not only on their level of food security or nutritional status, but also their role along agricultural value chains and the food environment in which they live. Men's and women's livelihood activities along food value chains may be differently exposed to disturbances and this determines what response options are most appropriate. In some cases, women's roles may be less vulnerable to shocks and stressors. For instance, women are more likely to raise local livestock breeds and smaller animals, which tend to be more resilient to the negative impacts of climate change (Chanamuto & Hall, 2015; Köhler-Rollefson, 2012). In other cases, women's livelihood activities may be more exposed to the harmful impacts of climate change. A case study from a peri-urban area around Magdalena, Mexico, shows that women in that environment were more affected by the negative impacts of climate change and associated water scarcity given that they rely on fruit and vegetable processing for their livelihoods, food security, and to maintain social ties (Buechler, 2009).

The types of shocks that men and women are exposed to in different food environments will also vary. A review of the literature on gendered vulnerabilities and water security found that while droughts negatively affect the farming activities on which many rural households depend, vulnerable urban households may experience more harmful impacts of flooding and other health-related risks, like cholera, due to poor water infrastructure and crowded conditions (Grasham et al., 2019). The review found that women from poor urban communities were particularly vulnerable to flood shocks in the absence of adequate water infrastructure and experienced greater loss of income, food insecurity, and increased risk of infection and disease (ibid.).

Sensitivity to disturbances also varies across different food environments. In the USA, food environments characterized by a high density of vendors selling unhealthy foods, called "food swamps," are stronger predictors of obesity than "food deserts" where access to healthy foods is limited (Cooksey-Stowers et al., 2017). Furthermore, shocks and stressors can alter food environments including the availability and quality of foods on which vulnerable populations depend. As an example, in response to the COVID-19 crisis, home food procurement (gardening, foraging, fishing, hunting, etc.) increased in many places across the world. A study in Vermont, USA, showed that food insecure households were more likely to engage in these activities but did not manage to significantly increase intake of fruits and vegetables, whereas intake increased significantly for already food secure households that engaged in home food procurement (Niles et al., 2021). A longer-term stressor is climate change, which is decreasing the nutritional content of staple crops—especially crops like wheat, rice, potatoes, and soy that comprise a large share of the diets in many low-income countries (Fanzo et al., 2018).

### *Gendered Resilience Capacities in Food Systems*

Investing in resilience capacities can enable people to expand and improve their range of options for dealing with disturbances. Investments in basic services, infrastructure, asset building, human capital development, and livelihood diversification were found to increase food consumption and dietary diversity during drought and input cost shocks in Malawi (Murendo et al., 2020). People with weak or limited resilience capacities may be forced to choose coping mechanisms that negatively influence their well-being or future adaptive capacity, such as reducing food consumption, taking children out of school, or drawing down assets (Theis et al., 2019). Individuals with greater resilience capacities have more options to protect and improve their livelihoods and well-being over the long term (Béné et al., 2016a). Men and women have different capacities to cope with shocks and stressors in the short to medium term (coping capacity) and to adjust their livelihood activities and strategies to address risks, seize opportunities, and reduce the negative impacts of shocks and stressors over the medium to long term (adaptive capacity) (Béné et al., 2015; Theis et al., 2019).

Resilience capacities also depend on other intersectional identities, such as age, class, caste, ethnicity, marital status, and sexual identity (Anderson,

2018; Carr & Thompson, 2014; Djoudi et al., 2016; Ravera et al., 2016; Tabaj & Spangler, 2017). One aspect of intersectionality that differentiates women's resilience capacities is marital status. In many contexts, female heads of household face greater limitations in access to land, capital, social networks, and labor, which limit their households' resilience (Mersha & Van Laerhoven, 2018; Van Aelst & Holvoet, 2016), while women in dual-headed households can, in some cases, benefit from access to these resources through male household members but may have less decision-making authority.

Many of the factors that contribute to women's empowerment also enhance resilience capacities at different levels (individual, household, and community). These include access to and control over resources like natural resources, physical assets, human capital, technology, and financial capital like savings and credit. Aspects of women's agency, like control over income, employment and livelihood choices, collective agency (including participation in groups), mobility, and work burden, also influence resilience capacities. The discussion below is not an exhaustive presentation of all these factors. Rather, it focuses on those capacities that are important for maintaining or improving food security following a disturbance.

*Access to and Control Over Resources for Women's Empowerment and Resilience*
Access to and control over assets is a key resilience capacity and a measure of women's empowerment (USAID, 2017). Assets function as a store of value and can be used to generate food and income or facilitate investment in better livelihood strategies (Johnson et al., 2016; Meinzen-Dick et al., 2011). Assets also influence social status and bargaining power at home and in the community (Johnson et al., 2016; Meinzen-Dick et al., 2011). Yet gender disparities in access to and control over productive assets, like land and livestock, mean that different approaches are required to effectively support women in building and safeguarding these assets (Ortiz-Ospina & Roser, 2018). In some cases, women's assets, such as jewelry, may be drawn down in response to shocks if the asset is less important for generating household income, the owner has weaker bargaining power within the household, or the asset is easier to sell (Quisumbing et al., 2018). Increasing women's control over productive assets, especially strengthening women's land rights, may, in some contexts, lead to adoption of practices that increase resilience, such as soil

and water conservation practices and agroforestry on agricultural lands (Meinzen-Dick et al., 2019). But expanding women's land rights, without raising awareness of what these rights entail, does not necessarily lead to changes in agricultural practices. In Ethiopia, Quisumbing and Kumar (2014) showed that women's knowledge of their land rights, following the land reforms in the early 2000s, was essential for the adoption of erosion control measures, and tree planting on plots managed by women.

Human capital is an important resilience capacity: people with better education, knowledge, and skills have more options to tap into government assistance programs (absorptive capacity), adopt new technologies (adaptive capacity), or diversify their livelihoods (transformative capacity). Moreover, women's education is associated with a range of positive outcomes, including reduced risk of child malnutrition and mortality (Sandiford et al., 1995). Even though more women are becoming educated today compared to 50 years ago, the gender gap in education persists, especially in countries that also have low levels of male educational attainment (Evans et al., 2019). There is evidence that some shocks, like COVID-19, negatively affect girls' education more than boys (Akmal et al., 2020), that the gender gap in education is associated with vulnerability to shocks and stressors, such as climate change, and that promoting girls' reproductive rights, education, and life skills would lead to greater resilience. Countries where girls have higher levels of schooling also have lower climate change vulnerability scores (Kwauk & Braga, 2017). Case study evidence from several contexts further shows that the risk of early marriage—which is associated with a series of negative outcomes, including maternal and child nutrition, and is often exacerbated by shocks—declines as years of girls' schooling increases (ibid.).

*Livelihood Roles, Employment, and Income*
Men's and women's ability to earn and control income through farming or other activities along agricultural value chains also determines their capacity to respond to the shocks and stressors that they face. In farm households, women tend to control crops that generate lower revenue relative to men's crops (Njuki et al., 2011). Women also have fewer economic opportunities along agricultural value chains and tend to be overrepresented in lower value nodes of the value chain given disparities in assets, time burden, and patriarchal norms (Coles & Mitchell, 2011; Dolan, 2001). Women are also less likely than men to participate

in the formal labor force and more likely to work in informal employment and in vulnerable, low-paid, or undervalued jobs than men (ILO, 2021). These constraints, which are particularly severe in the agricultural and food chain sectors, have implications for women's ability to earn and allocate income in ways that they prefer, generally toward purchases that increase household and children's well-being, such as spending on food, education, and medical expenses (Duflo & Udry, 2004; Hoddinott & Haddad, 1995; Malapit et al., 2015; Quisumbing, 2003)—investments which are generally considered to have positive implications for resilience.

Conversely, women's more limited ability to generate and control income has negative implications for resilience. A case study from Côte d'Ivoire illustrates how the gendered distribution of farming roles and expenditure obligations of men and women results in women having more limited income streams, which reduces the household's ability to cover unexpected shocks, such as medical expenses or consumption needs during the lean season (Kiewisch, 2015).

*Access to Services: Extension, Information, and Financial Services*
Access to information and extension services can increase resilience capacity, especially for addressing climate shocks and stressors, where information is needed to make appropriate response choices. Resilience programs that provide information services should consider gender differences in needs and preferences for information content (McOmber et al., 2019; Tall et al., 2014), channel of delivery (Jost et al., 2016; Partey et al., 2018; Tall et al., 2014; Twyman et al., 2014), and the ability of all end-users to understand and use the information (Ragasa et al., 2017). Moreover, gender differences in access to information vary across contexts depending on women's access to information and communication technology (ICT), participation in groups, and social norms affecting who is targeted for information dissemination (Gumucio et al., 2020). As an example, according to GSMA (2020), South Asia still has an urban-rural gap in access to mobile internet of 30%, while the gender gap in mobile internet use remains largest at 51%, suggesting large challenges for rural women to access key information, including on selling and purchasing food. In other regions, women's access is slowly improving. Kapinga and Montero (2017), for instance, note that the use of mobile technology was an effective tool to increase women entrepreneurs' access to market information and enabled women to expand their business networks while maintaining their family responsibilities in a case study of Tanzania. While

most studies of ICT use have focused on accessing formal information services (like extension or market information), ICTs are also important for maintaining familial and social ties, which can be important sources of resilience, such as reducing the risk of violent conflict, motivating cash remittances from urban workers, or rural family sending food to urban members in times of crisis (Bernier & Meinzen-Dick, 2014; Endris et al., 2017; Mukoya, 2020; Frankenberger et al., 2013).

Building financial capital and having access to financial services, such as insurance, credit, and savings, are essential for people to manage risk and mitigate the negative impacts of shocks and stressors (Moore et al., 2019). However, studies suggest that women are more likely to be excluded from the formal financial sector, including having a bank account, due to patriarchal social norms, limited legal rights, lower financial literacy, the gender gap in education, the lack of access to information, and other factors (Fletschner & Kenney, 2014; Hasler & Lusardi, 2017; Morsy, 2020). Women's lack of access to credit relative to men affects their access to food, especially during times of crisis.

*Time Burden and Resilience*

The gendered burden of unpaid work within households falls heavily on girls and women and limits their ability to generate income and build their resilience capacities like investing in human capital and accumulating assets (Kes & Swaminathan, 2006). In addition, women's workloads can be further exacerbated by shocks and stressors as well as the responses to them. Climatic shocks, including droughts and floods, often decrease water available for domestic uses. In 2017, one in ten people lacked basic services[3] for domestic water, including the 144 million people who drank untreated surface water, mostly in sub-Saharan Africa (UNICEF & WHO, 2019). This reduces the learning opportunities of women and children, who are responsible for most water collection in rural areas and undermines their health and livelihood opportunities (Geere & Cortobius, 2017). Women may be more reluctant to adopt some adaptation options because they entail too heavy a workload (Jost et al., 2016) and take time away from food preparation and other care practices (Komatsu et al., 2018).

---

[3] This is defined as an improved source within 30 minutes round trip collection time (UNICEF & WHO, 2019, p. 27).

*Food Security and Nutritional Status*

Food security and nutritional status are not only affected by shocks and stressors like climate change, but also play a role in determining individuals' capacity to respond to disturbances. Better child nutrition builds human capital, given its association with future cognitive and educational performance and positive impact on productivity in adulthood (Victora et al., 2008). Adequate nutrition during pregnancy, infancy, and childhood has also been shown to improve physical capabilities, which increase productivity of manual labor (Haas et al., 1995; Kolctzko et al., 2019; Rivera et al., 1995; Victora et al., 2008). Based on Food Insecurity Experience Scale data collected during 2014 and 2018 for more than 140 countries, women have a 27% higher chance than men of being severely food insecure and the gender gap in food insecurity is more prevalent among the poor, the less educated, and the suburban dwellers of large cities (FAO et al., 2020). Moreover, the gap in food insecurity increased during the COVID-19 pandemic, illustrating the unequal impact of the pandemic on men and women (FAO, IFAD, UNICEF, WFP, & WHO, 2022). At the same time, women are at a higher risk of being obese, which can affect health and ability to work. In 2016, 390 million women and 281 million men were obese (NCD-RisC, 2017).

## *The Decision-Making Space and Responses to Disturbances*

Within households and communities, men and women often have different preferences regarding how to use resources, what risks to take, and how to respond to specific shocks and stressors, which are often tied to the different roles men and women play in securing livelihoods (Bernier et al., 2015; Ravera et al., 2016). Because interests are not homogeneous within households, institutions, and communities, people must negotiate for their desired response when faced with a disturbance. Although equitable decision-making can be classified as a transformative resilience capacity (Vaughan & Henly-Shepard, 2018), some degree of agency and decision-making power is needed to exercise any resilience capacity—whether absorptive, adaptive, or transformative—for any preferred response. Thus, women's agency—including their intrinsic motivations, and their ability to make strategic life choices and carry out these decisions—is essential to ensure that their needs and preferences are reflected in the observed responses to shocks and stressors (Béné et al., 2019).

*Decision-Making in Household and Community Spaces*
Women tend to have less bargaining power and limited influence over important household decisions, including how their own income is spent (Ortiz-Ospina & Roser, 2018). Decisions, such as a woman's choice to pursue an income-generating activity or wage employment or participate in a group or program activity, can be subject to their husband's or other family members' consent. In some farming systems, men control agricultural land and may choose whether to allocate land to women, and the quantity and quality of land that they will farm. Furthermore, even when women have access to credit or own assets, women's credit acquisition may require male approval, and the sale of a woman's own assets may not be her decision (Pradhan et al., 2019).

Joint decision-making within households can help families better plan, prepare, and respond to shocks and stressors in a way that benefits the entire family, by accounting for different household members' needs and drawing on women's knowledge and abilities. Studies have found that women's increased bargaining power is associated with increases in household expenditures on child health and education—human capital investments that are recognized to increase resilience in the long run (Quisumbing & Maluccio, 2003). In Somalia, Mercy Corps (2014) found that women's involvement in household decision-making was strongly linked with household dietary diversity and a reduction in negative coping mechanisms. In Bangladesh, women's role in production decisions increased the diversity of crops grown on the farm, away from rice to other more nutritious crops (De Pinto et al., 2020). Evidence suggests that such diversified cropping systems provide multiple benefits, including reducing climate risks to production and livelihoods and better environmental sustainability (Asfaw et al., 2019; Birthal & Hazrana, 2019; Makate et al., 2016). Nutrition-sensitive agricultural programs have found it important to not only train mothers of young children on healthy diets, but also include their husbands and mothers-in-law (Malapit et al., 2021).

At the community level, local institutions play multiple roles in building resilience to climate shocks and stressors, including mobilizing, pooling, or regulating the use of shared resources, distributing benefits from interventions, using indigenous knowledge, and facilitating social learning (Agrawal, 2010; Béné, 2020a; Choudhury et al., 2021). When institutions that establish rules around the use and management of community resources and social protection programs—such as village councils or water user committees—do not represent the needs and priorities of the most marginal, they can serve to reinforce intracommunity inequality and curtail the response options of the most vulnerable groups. Women are often excluded from decisions at the community level in local decision-making bodies, which may be the result of discriminatory norms about who can actively participate, lack of social capital, the inconvenient timing and location of meetings, and exclusive membership criteria, such as a requirement that members of a water user association own land or be literate (Pandolfelli et al., 2007). Women's greater involvement in community-based organizations could contribute to community resilience by tapping into their specific knowledge and ability to reach certain networks—for instance, in determining where to situate a well, identifying vulnerable households, or sharing information with other women (Demetriades & Esplen, 2010). In particular, collective action through women's self-help groups builds social capital and provides important resources, such as group savings or shared labor, that increase community resilience (Cabot Venton et al., 2021). In addition, evidence suggests that participation in self-help groups builds individual women's resilience capacities by improving psychosocial outcomes, increasing social capital, and providing economic opportunities (ibid.).

*Gendered Preferences for Response Choices and the Implications for Food Security*

Men's and women's preferences for alternative response choices may vary across different food environments and different livelihood activities, including along food value chains. The food environment determines which response strategies may be most effective at addressing the most pressing food security and nutritional challenges (Downs et al., 2020; Herforth & Ahmed, 2015; HLPE, 2017; Turner et al., 2018). Poor men's and women's preferences for cash transfers versus vouchers or in-kind food assistance depend on the degree of control women have over household resources and their role in the household, as well as whether markets and public distribution systems work effectively, which often varies between rural and urban areas (Alderman et al., 2017). Household and community garden programs promoting production of fruits, vegetables, livestock, and fish may be important where such micronutrient-rich foods are not widely available or affordable. Strengthening value chains, particularly for nutrient-rich foods, is increasingly recognized as a way to leverage agriculture to improve nutrition (Ruel et al., 2013). However, value chains need to be considered more broadly beyond farm-level production and include the way foods are produced, processed, distributed, and marketed; how shocks and stressors, like climate change, may affect these activities; and what response options are needed to minimize any negative effects on food security, nutritional status, and the environment (Fanzo et al., 2018). These interventions may include increased access to improved, stress-tolerant crop and livestock varieties, improved agronomic and livestock feeding practices at the production stage, and better food storage and processing practices that ensure food safety, preservation of nutritional content and reduced food waste during later value chain stages (ibid.).

Women have important roles to play throughout food systems—both along food value chains and as consumers making food purchases—and are, thus, well-positioned to adopt practices that improve food security and nutrition outcomes. In Bangladesh, women were more likely to adopt practices like improved livestock feeding practices and grain storage, given that these are livelihood activities for which they are responsible (Bryan et al., 2021). A recent study of four value chains in the Philippines showed that small-scale retail entailed a large time burden with little payoff for women processors and traders in the abaca, coconut, and seaweed value chains (Malapit et al., 2020). Grace et al. (2015) analyzed gender roles and food safety in 20 livestock and fish value chains in Africa and Asia and observed a clear gender division of labor in the individual stages of the value chain—men were more likely than women to work in production and slaughtering activities, while women were more likely to work in processing and retail activities. In West Africa, women were responsible for smoking or drying fish and producing traditional dairy products while both women and men were involved in the processing of offal in Nigeria and South Africa and the making of biltong (a form of dried, cured meat) in South Africa (ibid.).

Post-harvest management practices are critical for household food security and resilience-building. They affect the nutrient content, food availability (as a result of food loss) and stability, and food safety outcomes. Men's and women's roles in post-harvest management are usually determined by context-specific cultural norms, which may change over time in response to technological and demographic changes. An example is aflatoxin[4] risk, which is increasing with climate change (Thomas et al., 2019), with detrimental effects on health (Kensler et al., 2011) and child growth (Khlangwiset et al., 2011; PACA, 2014). Women have key roles to play in minimizing contamination during the post-harvest period through improved drying and storage practices. Moreover, women are also key decision-makers on foods that are prepared for home consumption and in that role can guide diversification of diets away from aflatoxin-contaminated foods toward more fruits, vegetables, legumes, and animal-source foods, if they receive information on aflatoxin risks (Brown, 2018). The above case studies demonstrate that women have important contributions to make along food value chains to secure the supply of healthy, nutritious foods and as food consumers to provide healthy diets for their families.

Women also play important roles in maintaining household food security and increasing resilience in urban food environments. The diversity of cultural values and gender norms often found in cities can enable greater flexibility in gender roles, responsibilities, and behavior, including greater income-earning and schooling opportunities and increased access to services (Hovorka et al., 2009). Women are often heavily involved in urban agriculture activities, which is increasingly important given ongoing trends toward urbanization and climate change that can cause income losses, supply chain disruptions, and food insecurity (Hovorka et al., 2009; Olivier & Heinecken, 2017). Women also dominate processing activities in peri-urban and urban areas and benefit from having greater access to a wider range of marketing channels (Buechler, 2009).

Women also comprise a larger share of food vendors in urban areas, playing an important function in the food system by distributing foods more widely throughout urban areas, including to poor consumers (Giroux et al., 2020). A case study from India found that even though women street vendors were less likely to be the managers or decision-makers in street vendor enterprises, those who led street vendor enterprises expressed a greater desire to invest in expanding their enterprise than their male counterparts, highlighting the entrepreneurial nature of

---

[4] Aflatoxins are toxins produced by the mold Aspergillus flavus that contaminate certain crops, like maize and groundnuts, in the field, especially under drought conditions, and can continue to grow during post-harvest if crops are not properly processed and stored. When consumed, aflatoxins can cause illness, namely liver damage and liver cancer, in humans and animals.

women in this context (Patel et al., 2014). In Vietnam, women rely on social relations to operate in the informal food system and these relations provide an important source of resilience, enabling women to cope more easily with and overcome economic shocks than men, who rely more on market interactions (Kawarazuka et al., 2018). On the other hand, women remain more vulnerable than men to policy efforts to formalize and regulate the informal food system (Kawarazuka et al., 2018).

### *Pathways from Response Choices and Interventions to Differential Well-Being Outcomes*

Applying a gender lens to research on resilience highlights the importance of considering gender differentials in well-being outcomes and the pathways contributing to these differentials. Responses to shocks and stressors can have differential impacts on men's and women's well-being outcomes, especially when the decision-making context is characterized by large power differentials or exclusion and lack of representation. The resilience lens highlights the dynamic nature of these well-being outcomes as men and women face shocks and stressors over time. Well-being outcomes measured at aggregated levels (e.g., households or communities) or in the short term may obscure the different ways in which responses to shocks and stressors affect individuals' well-being outcomes over time.

### *Impact Pathways*

There are several pathways through which responses to disturbances can have differential effects on well-being outcomes, including changes in food production, income allocation, consumption patterns, asset dynamics, labor allocation, and human capital investments (Theis et al., 2019). To illustrate how one of these pathways can lead to gender-differentiated outcomes, consider how shocks and stressors induce distress sales of assets. Whose assets are sold has implications for differential well-being outcomes. Women's personal assets, such as small livestock or jewelry, may be sold for household liquidity, which can reduce women's intrahousehold bargaining power and economic independence and widen the gender asset gap (Pradhan et al., 2019; Quisumbing et al., 2017). Gendered asset dynamics also depend on the type of shock (e.g., covariate or idiosyncratic) and whose livelihood activities are affected (Rakib & Matz, 2016).

Another key pathway relates to changes in agricultural production practices, adoption of new technologies, and labor allocation. Many agricultural technologies and practices that can assist in building resilience also redistribute family labor and control over income with different implications for men's and women's time burden. For example, while conservation agriculture is often touted as an important climate-smart practice, several studies have demonstrated that it increases women's labor burden (Beuchelt & Badstue, 2013; Nelson & Stathers, 2009). Similarly, in Tanzania, Lee et al. (2021) found that heat stress caused families to shift labor allocation in ways that increased disparities in men's and women's agricultural labor burden in dual-headed households and significantly increased the labor burden of female household heads.

The ways in which these pathways play out depend on the context, including the local food environment. The potential for livelihood diversification as a pathway to improve well-being and resilience varies in rural and urban food environments, as well as for men and women in both contexts. In rural areas of many low-income countries, livelihoods revolve around agricultural production and related activities and there are fewer economic opportunities that would enable livelihood diversification to spread risk and increase resilience (Mulwa & Visser, 2020). Even fewer opportunities exist for women, especially in rural contexts where their involvement in production decisions, access to productive resources, and mobility is limited (Jost et al., 2016; Perez et al., 2015). Some opportunities like seasonal migration for employment are available mainly to men but have both positive and negative implications for the women who remain behind, such as an increase in work burden on the negative side and greater control over decision-making on the positive side (Djoudi & Brockhaus, 2016; Erman et al., 2021; Mueller et al., 2014).

There is some evidence that men and women navigate a different set of challenges in the urban food environment to maintain food security in the face of multiple shocks. While urban areas may offer greater employment opportunities, especially in non-agricultural sectors, not everyone in urban areas may benefit from these opportunities (Pingali et al., 2019). A case study from Kenya shows that women living in urban slums with limited livelihood options and very little access to basic infrastructure and services (like water and health facilities) have much lower resilience capacities and tend to adopt coping strategies that meet short-term survival needs but compromise long-term well-being, like withdrawing children from school, engaging in transactional sex, and resorting to crime (Beyer et al., 2016).

*Well-Being Outcomes, Trade-Offs, and Feedback Loops*
Every response option carries some degree of trade-off among people and across outcomes and spatial and temporal scales. That is, responses to disturbances may improve economic outcomes for people in the short term at the expense of long-term well-being, others, or the environment. For example, taking girls out of school may save cash for the household's immediate subsistence needs but can hamper girls' long-term human capital development and lead to a widening gender gap in educational and economic outcomes (Duryea et al., 2007). Coping responses to address income losses related to the COVID-19 pandemic, such as reducing consumption or selling productive assets, further illustrate the temporal trade-off between meeting short-term health and survival needs, and long-term health and well-being (Béné, 2020b). In particular, reducing consumption in the short term can have long term, even intergenerational, implications—short maternal stature (a consequence of poor nutrition in childhood) is associated with low birth weight and child stunting, which in turn has implications for adolescent nutritional status, thus perpetuating the cycle of undernutrition (Martorell & Zongrone, 2012).

The degree of trade-offs that the most vulnerable people are compelled to accept depends on their resilience capacities. Resource poor households may lower expectations and aspirations for what they can achieve in acceptance of their deprivation (Béné et al., 2014). Coulthard (2012) emphasizes the difference between agency and necessity in describing fishers' selection of inadequate coping strategies to address increased livelihood vulnerability and the well-being trade-offs they are forced to accept as a result, including women having to take on additional livelihood activities like seaweed farming to supplement declining income from fishing.

There are also important trade-offs between livelihood and environmental sustainability. Jarvis et al. (2011) emphasize that risk management strategies are not always adaptive over the long term and may pose environmental costs. For instance, expanding agricultural lands, overapplication of fertilizer, and agricultural intensification to address yield and income declines can have detrimental environmental impacts, such as declining soil fertility, increased greenhouse gas emissions, and deterioration of water quality (Smith et al., 2016).

Potential synergies also exist between well-being outcomes. Households with both male and female economic activity can spread risk across different livelihood activities and reduce exposure to idiosyncratic risks (Eriksen et al., 2005). Several climate-smart practices and technologies that decrease drudgery and women's time burden are also associated with higher productivity, incomes, and in some cases positive environmental outcomes (Khatri-Chhetri et al., 2020).

## Gender-Responsive Policies and Programs for Greater Resilience

This chapter has focused on the resilience capacities, preferences, and responses of individuals and households. However, other factors, including policies, interventions, investments, infrastructure, and institutions, also have a role to play in creating an enabling environment for resilience and in reducing gender inequalities in food systems. Conversely, the lack of gender sensitivity of policies, investments (e.g., infrastructure), and interventions, can reinforce gender inequalities and miss opportunities for increasing resilience. Targeted actions are needed to reduce inequalities and support the most vulnerable members of society given their limited capacity to respond to disturbances, shocks, and stressors like climate change (Huyer & Partey, 2020).

While policies increasingly integrate gender, some important gaps remain. A policy review of climate change policies across countries in Latin America found a lack of gender inclusion in climate change planning despite a relatively higher level of gender integration in the agriculture and food security sector (Gumucio & Rueda, 2015). The absence of gender sensitivity in climate change policies and interventions may be attributed to a lack of capacity on gender within institutions (Bryan et al., 2018). Even when gender analyses are conducted at the start of a project, ensuring gender equality in outcomes requires integrating gender into the design, planning, and monitoring of the intervention. A case study from Senegal demonstrates that this is not always the case—despite assessing gendered roles and contributions to urban agricultural production, a project was designed around male-dominated activities in production rather than women's roles in transport, processing, and marketing of the products (Gaye & Ndong Touré, 2009).

While many limitations remain, more resources are being directed toward ensuring economic inclusion, in ways that have positive implications for resilience. A global assessment of economic inclusion programming—including social safety nets, livelihoods and jobs programs, and financial inclusion programs—found that women's economic empowerment is a key feature of nearly 90% of programs surveyed having a gender focus (Andrews et al., 2021). A review of the impact of these programs found that they strengthened women's economic opportunities and income, increased asset ownership, and led to subtle shifts in gender norms, such as increased mobility (ibid.). While social protection programs alone have shown limited impacts on agricultural production or nutrition, bundling cash or in-kind transfers with high-quality complementary programming, such as agricultural extension or nutrition behavioral change communications does show significant improvements in agriculture and nutrition outcomes (PIM, 2021).

In addition to social protection, there are other promising interventions to increase women's empowerment and resilience capacities. Tabaj and Spangler (2017) point to a set of potential activities to increase the gender-responsiveness of resilience programs including building women's social capital and collective action and increasing women's access to and control over productive resources and essential services. Interventions designed to promote greater joint decision-making within the household around labor allocation, production, budgeting, and resource management can contribute to both women's empowerment and resilience goals (Farnworth et al., 2018; Mercy Corps, 2018). An intervention in Niger aimed to increase women's involvement in decision-making through facilitated household dialogues on issues like gendered division of labor (Mercy Corps, 2018). Qualitative research conducted following the intervention found that it increased women's participation in decisions about food purchases, women's mobility, access to information, and financial services and reduced labor burden for women (ibid.).

Similarly, interventions that seek to build women's assets and reduce gendered resource gaps are also an effective approach to build resilience (Theis et al., 2019). An assessment of eight agricultural development projects that aimed to increase asset levels in Africa and South Asia found that all projects were successful at building assets and that projects that distributed assets directly to women were able to increase women's control over assets (Johnson et al., 2016). These projects provided other benefits in terms of increasing women's role in decision-making and household income and food security (ibid.). Natural resources like land and water are also essential assets to which women often lack access. Reviews on women's land rights (Meinzen-Dick et al., 2019) and decision-making on irrigation through participation in water user associations (Meinzen-Dick & Zwarteveen, 1998) suggest that not only women but households and the communities would benefit from closing resource access gaps.

Furthermore, programs that facilitate and support women's groups can strengthen an important source of resilience for women, as these groups provide a variety of benefits like shared savings and loans, shared labor, and collective agency (Cabot Venton et al., 2021). Women members of self-help groups in India were more politically engaged, more aware of public entitlements, and more likely to benefit from public entitlement schemes than non-members (Kumar et al., 2019). This suggests that women's groups can help women access resources and services when faced with shocks and stressors.

Other types of interventions may be adapted to increase women's empowerment and resilience. Certification processes, for example, have been widely used for agroforestry crops, such as cocoa and coffee, with a general focus on environmental objectives, such as shade-grown coffee. Social objectives typically include child-labor free coffee but could be expanded to reduce gendered gaps in well-being outcomes from cultivating these crops. Women-only coffee cooperatives like Café Femenino in Peru and Las Hermanas in Honduras are successfully empowering women and supplying large coffee retailers interested in fulfilling the pro-social objectives demanded by their customers (Rubin & Manfre, 2014).

The impacts of women's empowerment programs are highly context-specific: impacts are more muted in contexts where social norms are more restrictive of women (Kabeer et al., 2012) and, in some cases, programs can have negative unintended consequences, such as an increase in domestic violence (Buller et al., 2018; Holmes & Jones, 2013). Programs should be designed in ways that address areas of women's disempowerment in a given context, boost program effectiveness, and minimize unintentional consequences, such as exacerbated time poverty, reinforced traditional gender roles, and gender-based violence (Andrews et al., 2021).

## Conclusions

Structural inequalities in society influence the resilience capacities of men and women. These include inequalities in men's and women's participation in formal and informal governance institutions at multiple scales and access to critical services like information, extension, and financial services. Encouragingly, the global community increasingly recognizes the importance of tackling structural inequalities to end poverty and hunger and achieve inclusive and sustainable food systems (Freistein & Mahlert, 2016). While women's empowerment is not synonymous with increasing resilience, and there is little evidence (to date) of the direct linkages, the literature summarized here suggests that many synergies exist: women's empowerment can increase resilience of women themselves, as well as their households and communities; and well-designed resilience interventions, such as social protection programs, can also increase women's empowerment. Similarly, improving women's access to sustainable healthy diets can improve individual, household and community well-being indicators and improve overall resilience to future risks.

As investments and interventions to increase resilience of vulnerable populations to shocks and stressors intensify, it is important to identify and address capacity gaps of different groups within a target population. Identifying differing resilience capacities is critical to determine how programs and policies can strengthen or diversify the resilience capacities of all groups. In addition, understanding existing constraints to building and exercising capacities is important so that policymakers and practitioners can ensure that programs and services are accessible to and relevant for all groups.

Building capacity among institutions operating at larger scales to integrate thinking about gender into resilience policies and programs is also essential so that these policies and programs meet the needs of the most vulnerable groups. More effort is also needed to measure changes in women's empowerment to track gender-differentiated outcomes of resilience interventions and to build evidence on the extent to which women's empowerment and resilience are mutually reinforcing.

# REFERENCES

Abel, G. J., Brottrager, M., Crespo Cuaresma, J., & Muttarak, R. (2019). Climate, conflict and forced migration. *Global Environmental Change, 54*, 239–249.

Agrawal, A. (2010). Local institutions and adaptation to climate change. In R. Mearns & A. Norton (Eds.), *Social dimensions of climate change: Equity and vulnerability in a warming world* (pp. 173–197). World Bank.

Akmal, M., Hares, S., & O'Donnell, M. (2020). *Gendered impacts of COVID-19 school closures: Insights from Frontline Organizations* (CGD Policy Paper, 175). Center for Global Development.

Alderman, H., Gentilini, U., & Yemtsov, R. (2017). The evolution of food as social assistance: An overview. In H. Alderman, U. Gentilini, & R. Yemtsov (Eds.), *The 1.5 billion people question: Food, vouchers, or cash transfers?* The World Bank. https://doi.org/10.1596/978-1-4648-1087-9

Anderson, A. (2018). *Resilience in action technical brief: Gender equity and social inclusion*. Resilience Evaluation, Analysis and Learning (REAL) Associate Award series. Mercy Corps.

Andrews, C., de Montesquiou, A., Arévalo Sánchez, I., Vasudeva Dutta, P., Varghese Paul, B., Samaranayake, S., Heisey, J., Clay, T., & Chaudhary, S. (2021). *The state of economic inclusion report 2021: The Potential to scale*. World Bank.

Asfaw, S., Scognamillo, A., Di Caprera, G., Sitko, N., & Ignaciuk, A. (2019). Heterogenous impact of livelihood diversification on household welfare: Cross-Country evidence from Sub-Saharan Africa. *World Development, 117*, 278–295.

Behrman, J., Bryan, E., & Goh, A. (2014). Gender, climate change, and group-based approaches to adaptation. In C. Ringler, A. Quisumbing, E. Bryan, & R. Meinzen-Dick (Eds.), *Climate change, collective action, and women's assets: Enhancing women's assets to manage risk under climate change—Potential for group-based approaches* (pp. 3–6). International Food Policy Research Institute.

Béné, C. (2020a). Are we messing with people's resilience? Analysing the impact of external interventions on community intrinsic resilience. *International Journal of Disaster Risk Reduction, 44*, 101431.

Béné, C. (2020b). Resilience of local food systems and links to food security—A Review of some important concepts in the context of COVID-19 and other shocks. *Food Security, 12*, 805–822.

Béné, C., Frankenberger, T., Griffin, T., Langworthy, M., Mueller, M., & Martin, S. (2019). "Perception matters": New insights into the subjective dimension of resilience in the context of humanitarian and food security interventions. *Progress in Development Studies, 19*(3), 186–210.

Béné, C., Al-Hassan, R. M., Amarasinghe, O., Fong, P., Ocran, J., Onumah, E., Ratuniata, R., Van Tuyen, T., McGregor, J. A., & Mills, D. J. (2016a). Is resilience socially constructed? Empirical evidence from Fiji, Ghana, Sri Lanka, and Vietnam. *Global Environmental Change, 38*, 153–170.

Béné, C., Headey, D., Haddad, L., & von Grebmer, K. (2016b). Is resilience a useful concept in the context of food security and nutrition programmes? Some conceptual and practical considerations. *Food Security, 8*(1), 123–138.

Béné, C., Frankenberger, T., & Nelson, S. (2015). *Design, monitoring and evaluation of resilience interventions: Conceptual and empirical considerations* (IDS Working Paper, 459). Institute of Development Studies.

Béné, C., Newsham, A., Davies, M., Ulrichs, M., & Godfrey-Wood, R. (2014). Review article: Resilience, poverty and development. *Journal of International Development, 26*(5), 598–623.

Bernier, Q., & Meinzen-Dick, R. (2014). *Resilience and social capital* (2020 Conference Paper, 4). International Food Policy Research Institute (IFPRI).

Bernier, Q., Meinzen-Dick, R., Kristjanson, P., Haglund, E., Kovarik, C., Bryan, E., Ringler, C., & Silvestri, S. (2015). *Gender and institutional aspects of climate-smart agricultural practices: Evidence from Kenya* (CCAFS Working Paper 79). CGIAR Research Program on Climate Change, Agriculture and Food Security.

Beuchelt, T. D., & Badstue, L. (2013). Gender, Nutrition and climate-smart food production: Opportunities and trade-offs. *Food Security, 5*, 709–721.

Beyer, L. I., Chaudhuri, J., & Kagima, B. (2016). Kenya's focus on urban vulnerability and resilience in the midst of urban transitions in Nairobi. *Development Southern Africa, 33*(1), 3–22.

Birthal, P. S., & Hazrana, J. (2019). Crop diversification and resilience of agriculture to climatic shocks: Evidence from India. *Agricultural Systems, 173*, 345–354.

Botreau, H., & Cohen, M. (2020). Gender Inequality and food insecurity: A dozen years after the food price crisis, rural women still bear the brunt of poverty and hunger. In M. Cohen (Ed.), *Advances in food security and sustainability* (Chapter 2, Vol. 5, pp. 53–117). Elsevier.

Brown, L. R. (2018). *Aflatoxins in food and feed: Impacts, risks, and management strategies* (GCAN Policy Note, 9). International Food Policy Research Institute.

Brownhill, L., Njuguna, E. M., Mungube, E., Nzioka, M., & Kihoro, E. (2016). Nested economies: Gendered small-livestock enterprise for household food security. In L. Brownhill, E. M. Njuguna, K. L. Bothi, B. Pelletier, L. W. Muhammad, & G. M. Hickey (Eds.), *Food security, gender and resilience: Improving smallholder and subsistence farming*. Routledge.

Bryan, E., Kato, E., & Bernier, Q. (2021). Gender differences in awareness and adoption of climate-smart agriculture practices in Bangladesh. In J. Eastin &

K. Dupuy (Eds.), *Gender, Climate change and livelihoods: Vulnerabilities and adaptations*. CABI.

Bryan, E., Bernier, Q., Espinal, M., & Ringler, C. (2018). Making climate change adaptation programs in Sub-Saharan Africa More gender-responsive: Insights from implementing organizations on the barriers and opportunities. *Climate and Development, 10*(5), 417–431.

Bryan, E., Theis, S., Choufani, J., De Pinto, A., Meinzen-Dick, R., & Ringler, C. (2017). Gender-Sensitive, climate-smart agriculture for improved nutrition in Africa South of the Sahara. In A. De Pinto & J. M. Ulimwengu (Eds.), *ReSAKSS annual trends and outlook report: The contribution of climate-smart agriculture to Malabo and Sustainable development goals* (pp. 113–145). International Food Policy Research Institute.

Bryan, E., Ringler, C., & Lefore, N. (2022). To ease the world food crisis, focus resources on women and girls: The global effects of the Ukraine war hit girls and women the hardest, exacerbating inequalities. Aid programmes must adapt. *Nature comment*. https://www.nature.com/articles/d41586-022-02312-8

Buechler, S. (2009). Gender dynamics of fruit and vegetable production and processing in peri-urban Magdalena, Sonora, Mexico. In A. Hovorka, H. de Zeeuw, & M. Njenga (Eds.), *Women Feeding cities: Mainstreaming gender in urban agriculture and food security*. Practical Action Publishing.

Buller, A. M., Peterman, A., Ranganathan, M., Bleile, A., Hidrobo, M., & Heise, L. (2018). A Mixed-method review of cash transfers and intimate partner violence in low- and middle-income countries. *World Bank Research Observer, 33*(2), 218–258.

Buvinic, M., Das Gupta, M., Casabonne, U., & Verwimp, P. (2013). Violent conflict and gender inequality: An overview. *The World Bank Research Observer, 28*, 110–138.

Cabot Venton, C., Prillaman, S. A., & Kim, J. (2021). *Building resilience through self help groups: Evidence review*. The Resilience Evaluation, Analysis, and Learning (REAL) Award.

Carr, E. R. (2020). Resilient livelihoods in an era of global transformation. *Global Environmental Change, 64*, 102155.

Carr, E. R., & Thompson, M. C. (2014). Gender and climate change adaptation in Agrarian settings: Current thinking, new directions and research frontiers. *Geography Compass, 8*(3), 182–197.

Chanamuto, N. J. C., & Hall, S. J. G. (2015). Gender equality, resilience to climate change, and the design of livestock projects for rural livelihoods. *Gender and Development, 23*(3), 515–530.

Choudhury, M.-U.-I., Haque, C. E., Nishat, A., & Byrne, S. (2021). Social learning for building community resilience to cyclones: Role of indigenous and local knowledge, power, and institutions in coastal Bangladesh. *Ecology and Society, 26*(1), 5.

Coles, C., & Mitchell, J. (2011). *Gender and agricultural value chains. A review of current knowledge and practice and their policy implications* (ESA Working Paper No. 11-05).

Cooksey-Stowers, K., Schwartz, M. B., & Brownell, K. D. (2017). Food swamps predict obesity rates better than food deserts in the United States. *International Journal of Environmental Research and Public Health, 14*(11), 1366. https://doi.org/10.3390/ijerph14111366

Constas, M. A., Frankenberger, T. R., Hoddinott, J., Mock, N., Romano, D., Béné, C., & Maxwell, D. (2014). *A common analytical model for resilience measurement: Causal framework and methodological options* (Technical Series No. 2., n.p.). Food Security Information Network.

Corbett, J. (1988). Famine and household coping strategies. *World Development, 16*(9), 1099–1112.

Corcoran-Nantes, Y. & Roy, S. (2018). Gender, climate change, and sustainable development in Bangladesh. In J. McIntyre-Mills, N. Romm, & Y. Corcoran-Nantes (Eds.), *Balancing individualism and collectivism* (pp 163–179). Springer.

Coulthard, S. (2012). Can we be both resilient and well, and what choices do people have? Incorporating agency into the resilience debate from a fisheries perspective. *Ecology and Society, 17*(1), 4.

Deere, C. D., & Doss, C. (2006). The gender asset gap: what do we know and why does it matter? *Feminist Economics, 12*(1–2), 1–50.

Delaney, P. L., & Shrader, E. (2000). *Gender and post-disaster reconstruction: The case of Hurricane Mitch in Honduras and Nicaragua.* World Bank.

Demetriades, J., & Esplen, E. (2010). The gender dimensions of poverty and climate change adaptation. *IDS Bulletin, 39*(4), 24–31.

De Pinto, A., Seymour, G., Bryan, E. & Bhandary, P. (2020). Women's empowerment and farmland allocations in Bangladesh: Evidence of a possible pathway to crop diversification. *Climatic Change*, published online:https://doi.org/10.1007/s10584-020-02925-w

Dercon, S. (2002). Income risk, coping strategies, and safety nets. *The World Bank Research Observer, 17*(2), 141–166.

Djoudi, H., Locatelli, B., Vaast, C., Asher, K., Brockhaus, M., & Sijapati, B. (2016). Beyond dichotomies: Gender and intersecting inequalities in climate change studies. *Ambio, 45*(Suppl. 3), S248–S262.

Djoudi, H., & Brockhaus, M. (2016). Unveiling the complexity of gender and adaptation: The "feminization" of forests as a response to drought-induced men's migration in mali. In C. J. Pierce Colfer, B. Sijapati Basnett, & M. Elias (Eds.), *Gender and forests: Climate change, tenure, value chains and emerging issues.* Routledge.

Dolan, C. S. (2001). The 'good wife': Struggles over resources in the Kenyan horticulture sector. *Journal of Development Studies, 37*(3), 39–70.

Doocy, S., Daniels, A., Murray, S., & Kirsch, T. D. (2013). The human impact of floods: A historical review of events 1980–2009 and systematic literature review. *PLOS Currents Disasters, 1*. https://doi.org/10.1371/currents.dis.f4deb457904936b07c09daa98ee8171a

Doss, C., Meinzen-Dick, R., Quisumbing, A., & Theis, S. (2018). Women in agriculture: Four myths. *Global Food Security, 16*, 69–74.

Downs, S., Ahmed, S., Fanzo, J., & Herforth, A. (2020). Food environment typology: Advancing an expanded definition, framework, and methodological approach for improved characterization of wild, cultivated, and built food environments toward sustainable diets. *Foods, 9*, 532. https://doi.org/10.3390/foods9040532

Duflo, E., & Udry, C. (2004). *Intrahousehold resource allocation in Cote d'Ivoire: Social norms, separate accounts and consumption choices* (Working Paper No. w10498). National Bureau of Economic Research.

Duryea, S., Lam, D., & Levison, D. (2007). Effects of economic shocks on children's employment and schooling in Brazil. *Journal of Development Economics, 84*(1), 188–214.

Endris, G. S., Kibwika, P., Hassan, J. Y., & Obaa, B. B. (2017). Harnessing social capital for resilience to livelihood shocks: Ethnographic evidence of indigenous mutual support practices among rural households in Eastern Ethiopia. *International Journal of Population Research, 4513607*, 26.

Eriksen, S. H., Broan, K., & Kelly, P. M. (2005). The dynamics of vulnerability: Locating coping strategies in Kenya and Tanzania. *The Geographic Journal, 171*(4), 287–305.

Erman, A., De Vries Robbe, S. A., Thies, S. F., Kabir, K., & Maruo, M. (2021). *Gender dimensions of disaster risk and resilience: Existing evidence*. The World Bank and the Global Facility for Disaster Reduction and Recovery.

Evans, D. K., Akmal, M., & Jakiela, P. (2019). *Gender gaps in education: The long view* (CGD Working Paper 523). Center for Global Development. https://www.cgdev.org/publication/gender-gaps-education-long-view

FAO, IFAD, UNICEF, WFP, & WHO. (2020). *The state of food security and nutrition in the world 2020*. Transforming Food Systems for Affordable Healthy Diets. FAO.

FAO, IFAD, UNICEF, WFP, & WHO. (2022). *The state of food security and nutrition in the world 2022*. Repurposing food and agricultural policies to make healthy diets more affordable. FAO. https://doi.org/10.4060/cc0639en

Fanzo, J., Davis, C., McLaren, R., & Choufani, J. (2018). The effect of climate change across food systems: Implications for nutrition outcomes. *Global Food Security, 18*, 12–19.

Farnworth, C. R., Stirling, C. M., Chinyophiro, A., Namakhoma, A., & Morahan, R. (2018). Exploring the potential of household methodologies to strengthen gender equality and improve smallholder livelihoods: Research in Malawi in maize-based systems. *Journal of Arid Environments, 149,* 53–61.

Fletschner, D., & Kenney, L. (2014). Rural women's access to financial services: Credit, savings, and insurance. In A. Quisumbing, R. Meinzen-Dick, T. Raney, A. Croppenstedt, J. Behrman, & A. Peterman (Eds.), *Gender in agriculture.* Springer. https://doi.org/10.1007/978-94-017-8616-4_8

Frankenberg, E., Gillespie, T., Preston, S., Sikoki, B., & Thomas, D. (2011). Mortality, the family and the Indian Ocean Tsunami. *The Economic Journal, 121*(554), F162–F182.

Frankenberger, T., Mueller M., Spangler T., & Alexander S. (2013). *Community resilience: Conceptual framework and measurement feed the future learning agenda.* Westat.

Frankenberger, T., Constas, M., Nelson, S., & Starr, L. (2014). *Current approaches to resilience programming among nongovernmental organizations* (Building Resilience for Food and Nutrition Security 2020 Conference Paper 7). International Food Policy Research Institute.

Freistein, K., & Mahlert, B. (2016). The potential for tackling inequality in the sustainable development goals. *Third World Quarterly, 37*(12), 2139–2155.

Gaye, G., & Ndong Touré, M. (2009). Gender and urban agriculture in Pikine, Senegal. In A. Hovorka, H. de Zeeuw, & M. Njenga (Eds.), *Women feeding cities: Mainstreaming gender in urban agriculture and food security.* Practical Action Publishing.

Geere, J. A., & Cortobius, M. (2017). Who carries the weight of water? Fetching water in rural and urban areas and the implications for water security. *Water Alternatives, 10*(2), 513–540.

Giroux, S., Blekking, J., Waldman, K., Resnick, D., & Fobi, D. (2020). Informal vendors and food systems planning in an emerging African city. *Food Policy, 103,* 101997.

Gnisci, D. (2016). *Women's roles in the West African food system: Implications and prospects for food security and resilience* (West African Papers, No. 3). OECD Publishing. https://doi.org/10.1787/5jlpl4mh1hxn-en

Grace, D., Roesel, K., Kang'ethe, E., Bonfoh, B., & Theis, S. (2015). *Gender roles and food safety in 20 informal livestock and fish value chains* (IFPRI Discussion Paper 01489). IFPRI.

Grasham, C. F., Korzenevica, M., & Charles, K. J. (2019). On considering climate resilience in urban water security: A review of the vulnerability of the urban poor in sub-Saharan Africa. *Wires Water, 6*(3), e1344.

GSMA. (2020). *The state of mobile internet connectivity 2020.* GSMA. Accessed at https://www.gsma.com/r/wp-content/uploads/2020/09/GSMA-State-of-Mobile-Internet-Connectivity-Report-2020.pdf

Gumucio, T., & Rueda, M. T. (2015). influencing gender-inclusive climate change policies in Latin America. *Journal of Gender, Agriculture and Food Security, 1*(2), 42–61.

Gumucio, T., Hansen, J., Huyer, S., & van Huysen, T. (2020). Gender-responsive rural climate services: A review of the literature. *Climate and Development, 12*(3), 241–254.

Haas, J. D., Martinez, E. J., Murdoch, S., Conlisk, E., Rivera, J. A.,& Martorell, R. (1995). Nutritional supplementation during the preschool years and physical work capacity in adolescent and young adult guatemalans. *The Journal of Nutrition, 125*(4 Suppl.), 1078S–1089S. http://www.ncbi.nlm.nih.gov/pubmed/7722710. Accessed June 8, 2017.

Hallegatte, S., & Rozenberg, J. (2017). Climate change through a poverty lens. *Nature Climate Change, 7*, 250–256. https://doi.org/10.1038/nclimate3253

Hasler, A., & Lusardi, A. (2017). *The gender gap in financial literacy: A global perspective*. Global Financial Literacy Excellence Center, The George Washington University School of Business.

Hathi, P., Coffey, D., Thorat, A., & Khalid, N. (2021). When women eat last: Discrimination at home and women's mental health. *PLoS ONE, 16*(3), e0247065. https://doi.org/10.1371/journal.pone.0247065

Herforth, A., & Ahmed, S. (2015). The food environment, its effects on dietary consumption, and potential for measurement within agriculture-nutrition interventions. *Food Security, 7*, 505–520.

HLPE. (2017). *Nutrition and food systems*. A Report by the High-Level Panel of Experts on Food Security and Nutrition of the Committee on World Food Security.

Hoddinott, J., & Haddad, L. (1995). Does female income share influence household expenditures? Evidence from Côte d'Ivoire. *Oxford Bulletin of Economics and Statistics, 57*, 77–96. https://doi.org/10.1111/j.1468-0084.1995.tb00028.x

Holmes, R., & Jones, N. (2013). *Gender and social protection in the developing world: Beyond mothers and safety nets*. Zed.

Hovorka, A., de Zeeuw, H., & Njenga, M. (2009). Gender in urban agriculture: An introduction. In A. Hovorka, H. de Zeeuw, & M. Njenga (Eds.), *Women feeding cities: Mainstreaming gender in urban agriculture and food security*. Practical Action Publishing.

Huyer, S., & Partey, S. (2020). Weathering the storm or storming the norms? Moving gender equality forward in climate-resilient agriculture. *Climatic Change, 158*, 1–12. https://doi.org/10.1007/s10584-019-02612-5

IDMC. (2020). *Global report on internal displacement*. Internal Displacement Monitoring Centre (IDMC).

IFPRI (International Food Policy Research Institute). (2021). *COVID-19 impact on rural men and women (multiple rounds of data from Ghana, Kenya, Nepal, Niger, Nigeria, Rwanda, Senegal, Uganda)*. International Food Policy Research Institute.

ILO (International Labor Organization). (2021). *Gender and financial inclusion*. https://www.ilo.org/empent/areas/social-finance/WCMS_737729/lang--en/index.htm. Accessed April 9, 2021.

Jarvis, A., Lau, C., Cook, S., Wollenberg, E., Hansen, J., Bonilla, O., & Challinor, A. (2011). An integrated adaptation and mitigation framework for developing agricultural research: Synergies and trade-offs. *Experimental Agriculture, 47*, 185–203.

Johnson, N. L., Kovarik, C., Meinzen-Dick, R., Njuki, J., & Quisumbing, A. (2016). Gender, assets, and agricultural development: Lessons from eight projects. *World Development, 83*, 295–311.

Jordan, J. C. (2019). Deconstructing resilience: Why gender and power matter in responding to climate stress in Bangladesh. *Climate and Development, 11*(2), 167–179.

Jost, C., Kyazze, F., Naah, J., Neelormi, S., Kinyangi, J., Zougmore, R., Aggarwal, P., Bhatta, G., Chaudhury, M., Tapio-Bistrom, M.-L., Nelson, S., & Kristjanson, P. (2016). Understanding gender dimensions of agriculture and climate change in smallholder farming communities. *Climate and Development, 2*(8), 133–144.

Kabeer, N., Huda, K., Kaur, S., & Lamhauge, N. (2012). *Productive safety nets for women in extreme poverty: Lessons from pilot projects in India and Pakistan* (Discussion Paper 28/12). Centre for Development Policy and Research, School of Oriental and African Studies.

Kapinga, A. F., & Montero, C. S. (2017). Exploring the Socio-Cultural challenges of food processing women entrepreneurs in IRINGA, TANZANIA and strategies used to tackle them. *Journal of Global Entrepreneurship Research, 7*(1), 17.

Kawarazuka, N., Béné, C., & Prain, G. (2018). Adapting to a new urbanizing environment: Gendered strategies of Hanoi's street food vendors. *Urbanization and Environment, 30*(1), 233–248.

Keck, M., & Sakdapolrak, P. (2013). What is social resilience? Lessons learned and ways forward. *Erdkunde, 67*(1), 5–19.

Kensler, T. W., Roebuck, B. D., Wogan, G. N., Groopman, J. D. (2011). Aflatoxin: A 50-year Odyssey of mechanistic and translational toxicology. *Toxicological Sciences, 120*(Suppl. 1), S28–S48.

Kes, A., & Swaminathan, H. (2006). Gender and time poverty in Sub-Saharan Africa. In C. M. Blackden & Q. Wodon (Eds.), *Gender, time use, and poverty in Sub-Saharan Africa* (Chapter 2, World Bank Working Paper No. 73). The World Bank.

Khatri-Chhetri, A., Regmi, P. P., Chanana, N., & Aggarwal, P. K. (2020). Potential of climate-smart agriculture in reducing women farmers' drudgery in high climatic risk areas. *Climatic Change, 158*, 29–42.

Khlangwiset, P., Shephard, G. S., & Wu, F. (2011). Aflatoxins and growth impairment: A review. *Critical Reviews in Toxicology, 41*(9), 740–755.

Kiewisch, E. (2015). Looking within the household: A study on gender, food security, and resilience in cocoa-growing communities. *Gender and Development, 23*(3), 497–513.

Köhler-Rollefson, I. (2012). *Invisible guardians—Women manage livestock diversity* (FAO Animal Production and Health Paper 174). FAO.

Koletzko, B., Godfrey, K. M., Poston, L., Szajewska, H., van Goudoever, J. B., de Waard, M., Brands, B., Grivell, R. M., Deussen, A. R., Dodd, J. M., Patro-Golab, B., & Zalewski, B. M. (2019). Nutrition during pregnancy, lactation and early childhood and its implications for maternal and long-term child health: The early nutrition project recommendations. *Annals of Nutrition and Metabolism, 74*, 93–106. https://doi.org/10.1159/000496471

Komatsu, H., Malapit, H. J. L., & Theis, S. (2018). Does women's time in domestic work and agriculture affect women's and children's dietary diversity? Evidence from Bangladesh, Nepal, Cambodia, Ghana, and Mozambique. *Food Policy, 79*, 256–270.

Koo, J., Azzarri, C., Ghosh, A., & Quabili, W. (2021). *Assessing the risk of COVID-19 in feed the future countries*. Gender, climate change, and nutrition integration initiative (GCAN) Policy Note 13.

Kumar, N., Raghunathan, K., Arrieta, A., Jilani, A., Chakrabarti, S., Menon, P., & Quisumbing, A. R. (2019). Social networks, mobility, and political participation: The potential for women's self-help groups to improve access and use of public entitlement schemes in India. *World Development, 114*, 28–41. https://doi.org/10.1016/j.worlddev.2018.09.023

Kumar, N., & Quisumbing, A. R. (2013). Gendered impacts of the 2007–2008 food price crisis: Evidence using panel data from rural Ethiopia. *Food Policy, 38*, 11–22.

Kwauk, C., & Braga, A. (2017). *Three platforms for girls' education in climate strategies* (Brooke Shearer Series, 6). The Brookings Institution.

Lawson, E. T., Alare, R. S., Salifu, A. R. Z., & Thompson-Hall, M. (2020). Dealing with climate change in Semi-Arid Ghana: Understanding intersectional perceptions and adaptation strategies of women farmers. *GeoJournal, 85*, 439–452.

Lee, Y., Haile, B., Seymour, G., & Azzarri, C. (2021). The heat never bothered me anyway: Gender-specific response of agricultural labor to climatic shocks in Tanzania. *Applied Economic Perspectives and Policy, 43*(2), 732–749.

Leichenko, R., & Silva, J. A. (2014). Climate change and poverty: Vulnerability, impacts, and alleviation strategies. *Wires Climate Change, 5*(4), 539–556.

Limuwa, M. M., & Synnevag, G. (2018). Gendered perspective on the fish value chain, livelihood patterns and coping strategies under climate change—insights from Malawi's small-scale fisheries. *The African Journal of Food, Agriculture, Nutrition and Development, 18*(2), 13521–13540.

Makate, C., Wang, R., Makate, M., & Mango, N. (2016). Crop diversification and livelihoods of smallholder farmers in Zimbabwe: Adaptive management for environmental change. *SpringerPlus, 5*, 1135.

Malapit, H., Heckert, J., Scott, J., Padmaja, R., & Quisumbing, A. (2021). Nutrition-sensitive agriculture for gender equality. In R. Pyburn & A. van Eerdewijk (Eds.), *Advancing gender equality through agricultural and environmental research: Past, present, and future*. International Food Policy Research Institute.

Malapit, H. J., Ragasa, C., Martinez, E. M., Rubin, D., Seymour, G., & Quisumbing, A. (2020, May). Empowerment in agricultural value chains: Mixed methods evidence from the philippines. *Journal of Rural Studies, 76*, 240–253. https://doi.org/10.1016/j.jrurstud.2020.04.003

Malapit, H. J., Kadiyala, S., Quisumbing, A. R., Cunningham, K., & Tyagi, P. (2015). Women's empowerment mitigates the negative effects of low production diversity on maternal and child nutrition in Nepal. *The Journal of Development Studies, 51*(8), 1097–1123.

Martorell, R., & Zongrone, A. (2012). Intergenerational influences on child growth and undernutrition. *Pediatric and Perinatal Epidemiology, 26*(Suppl. 1), 302–314.

McOmber, C., Audia, C., & Crowley, F. (2019). Building resilience by challenging social norms: Integrating a transformative approach within the BRACED consortia. *Disasters, 43*(S3), S271–S294.

Meinzen-Dick, R., Quisumbing, A., Doss, C., & Theis, S. (2019). Women's land rights as a pathway to poverty reduction: Framework and review of available evidence. *Agricultural Systems, 172*, 72–82.

Meinzen-Dick, R., Johnson, N., Quisumbing, A., Njuki, J., Behrman, J., Rubin, D., Peterman, A. & Waithanji, E. (2011). *Gender, assets, and agricultural development programs: A conceptual framework* (CAPRi Working Paper 99). CAPRI.

Meinzen-Dick, R., & Zwarteveen, M. (1998). Gendered participation in water management: Issues and illustrations from water users' associations in South Asia. *Agriculture and Human Values, 15*, 337–345.

Mercy Corps. (2014). *Rethinking resilience: Prioritizing gender integration to enhance household and community resilience to food insecurity in the Sahel.*

Mercy Corps. (2018). *Priming resilience with a women's "empowerment package".*

Mercy Corps. (2016). *Our resilience approach to relief, recovery and development.*

Mersha, A., & van Laerhoven, F. (2018). Interplay between planned and autonomous adaptation in response to climate change: Insights from rural

Ethiopia. *World Development, 107*, 87–97. https://doi.org/10.1016/j.wor lddev.2018.03.001

Moore, D., Niazi, A., Rouse, R., & Kramer, B. (2019). *Building resilience through financial inclusion: A review of existing evidence and knowledge gaps.* Innovations for Poverty Action.

Morsy, H. (2020). Access to finance—Mind the gender gap. *The Quarterly Review of Economics and Finance, 78*, 12–21.

Mueller, V., Gray, C., & Kosec, K. (2014). Heat stress increases long-term human migration in rural Pakistan. *Nature Climate Change, 4*, 182–185.

Mukoya, F. (2020). ICTs as enablers of resilient social capital for ethnic peace. *COMPASS '20: Proceedings of the 3rd ACM SIGCAS Conference on Computing and Sustainable Societies*, pp. 116–126.

Mulwa, C. K., & Visser, M. (2020). Farm diversification as an adaptation strategy to climatic shocks and implications for food security in northern Namibia. *World Development, 129*, 104906.

Murendo, C., Kairezi, G., & Mazvimavi, K. (2020). Resilience capacities and household nutrition in the presence of shocks. Evidence from Malawi. *World Development Perspectives*, p. 100241.

NCD Risk Factor Collaboration (NCD-RisC). (2017). Worldwide trends in body-mass index, underweight, overweight, and obesity from 1975 to 2016: A pooled analysis of 2416 population-based measurement studies in 128·9 million children, adolescents, and adults. *Lancet, 390*, 2627–2642.

Nelson, V., & Stathers, T. (2009). Resilience, power, culture and climate and new research directions: A case study from Semi-Arid Tanzania. *Gender and Development Journal, 17*, 81–94.

Neumayer, E., & Plümper, T. (2007). The gendered nature of natural disasters: The impact of catastrophic events on the gender gap in life expectancy, 1981–2002. *Annals of the Association of American Geographers, 97*(3), 551–566.

Ngigi, M. W., Mueller, U., & Birner, R. (2017). Gender differences in climate change adaptation strategies and participation in group-based approaches: An intra-household analysis from rural Kenya. *Ecological Economics, 138*, 99–108.

Nguyen, T. T., Nguyen, T. T., & Grote, U. (2020). Multiple shocks and households' choice of coping strategies in rural Cambodia. *Ecological Economics, 167*, 106442.

Niles, M. T., Wirkkala, K. B., Belarmino, E. H., & Bertmann, F. (2021). Home food procurement impacts food security and diet quality during COVID-19. *BMC Public Health, 21*, 945. https://doi.org/10.1186/s12889-021-109 60-0

Njuki, J., Eissler, S., Malapit, H., Meinzen-Dick, R., Bryan, E., & Quisumbing, A. (2022). A review of evidence on gender equality, women's empowerment, and food systems. *Global Food Security, 33*, 100622. https://doi.org/10.1016/j.gfs.2022.100622

Njuki, J., Kaaria, S., Chamunorwa, A., & Chiuri, W. (2011). Linking smallholder farmers to markets, gender and intra-household dynamics: Does the choice of commodity matter? *European Journal of Development Research, 23,* 426–443. https://doi.org/10.1057/ejdr.2011.8

Olivier, D. W., & Heinecken, L. (2017). Beyond food security: Women's experiences of urban agriculture in Cape Town. *Agriculture and Human Values, 34,* 743–755.

Oloukoi, G., Fasona, M., Olorunfemi, F., Adedayo, V., & Elias, P. (2014). A gender analysis of perceived climate change trends and ecosystems-based adaptation in the Nigerian wooded savannah. *Agenda: Empowering Women for Gender Equity, 28*(3), 16–33.

Ortiz-Ospina, E., & Roser, M. (2018). *Economic inequality by gender.* Published online at OurWorldInData.org. https://ourworldindata.org/economic-inequality-by-gender. Accessed April 9, 2021.

PACA. (2014). *A potential relationship between aflatoxin exposure and childhood stunting, Addis Ababa, Ethiopia.* http://aflatoxinpartnership.org/uploads/PACAStuntingPolicyBrief.pdf

Pandolfelli, L., Meinzen-Dick, R., & Dohrn, S. (2007). *Gender and collective action: A framework for analysis* (CAPRi Working Paper 64). International Food Policy Research Institute.

Partey, S. T., Dakorah, A. D., Zougmore, R. B., Ouedraogo, M., Nyasimi, M., Nikoi, G. K., & Huyer, S. (2018). Gender and climate risk management: Evidence of climate information use in Ghana. *Climatic Change, 158,* 61–75.

Patel, K., Guenther, D., Wiebe, K., & Seburn, R. A. (2014). Promoting food security and livelihoods for urban poor through the informal sector: A case study of street food vendors in Madurai, Tamil Nadu, India. *Food Security, 6,* 861–878.

Perez, C., Jones, E. M., Kristjanson, P., Cramer, L., Thornton, P. K., Förch, W., & Barahona, C. (2015). How resilient are farming households and communities to a changing climate in Africa? A gender-based perspective, *Global Environmental Change, 34,* 95–107. https://doi.org/10.1016/j.gloenvcha.2015.06.003

PIM (CGIAR Research Program on Policies, Institutions, and Markets). (2021). *Social protection for agriculture and resilience: Highlights, lessons learned, and priorities for One CGIAR* (PIM Flagship Insights 4). International Food Policy Research Institute (IFPRI). https://doi.org/10.2499/p15738coll2.134376

Pingali P., Aiyar, A., Abraham, M., & Rahman, A. (2019). Rural livelihood challenges: Moving out of agriculture. In *Transforming food systems for a rising India*. Palgrave Studies in Agricultural Economics and Food Policy. Palgrave Macmillan. https://doi.org/10.1007/978-3-030-14409-8_3

Pradhan, R., Meinzen-Dick, R., & Theis, S. (2019). Property rights, intersectionality, and women's empowerment in Nepal. *Journal of Rural Studies, 70*, 26–35. https://doi.org/10.1016/j.jrurstud.2019.05.003

Quisumbing, A. R. (2003). *Household decisions, gender, and development: A synthesis of recent research.* International Food Policy Research Institute (IFPRI).

Quisumbing, A. R., & Maluccio, J. (2003). Resources at marriage and intrahousehold allocation: Evidence from Bangladesh, Ethiopia, Indonesia, and South Africa. *Oxford Bulletin of Economics and Statistics, 65*(3), 283–327.

Quisumbing, A., Meinzen-Dick, R., Behrman, J., & Basset, L. (2011). Gender and the global food-price crisis. *Development in Practice, 21*(4–5), 488–492. https://doi.org/10.1080/09614524.2011.561283

Quisumbing, A., & Kumar, N. (2014). *Land rights knowledge and conservation in Rural Ethiopia mind the gender gap* (IFPRI Discussion Paper, 1386). International Food Policy Research Institute.

Quisumbing, A., Kumar, N., & Behrman, J. (2018). Do shocks affect men's and women's assets differently? Evidence from Bangladesh and Uganda. *Development Policy Review, 36*(1), 3–34. https://doi.org/10.1111/dpr.12235

Ragasa, C., Aberman, N. L., & Alvarez Mingote, C. (2017). *Does providing agricultural and nutrition information to both men and women improve household food security? Evidence from Malawi* (IFPRI Discussion Paper 1653). IFPRI.

Rakib, M., & Matz, J. A. (2016). The impact of shocks on gender-differentiated asset dynamics in Bangladesh. *The Journal of Development Studies, 52*(3), 377–395.

Ravera, F., Martín-López, B., Pascual, U., & Drucker, A. (2016). The diversity of gendered adaptation strategies to climate change of Indian farmers: A feminist intersectional approach. *Ambio, 45*(Suppl. 3), S335–S351.

Rivera, J. A., Martorell, R., Ruel, M. T., Habicht, J. P., & Haas, J. D. (1995). Nutritional supplementation during the preschool years influences body size and composition of Guatemalan adolescents. *The Journal of Nutrition, 125*(4 Suppl.), 1068S–1077S.

Rubin, D., & Manfre, C., & Nichols Barrett, K. (2014). Promoting gender-equitable agricultural value chains: Issues, opportunities, and next steps. In A. Quisumbing (Ed.), *Gender in agriculture: Closing the knowledge gap* (pp. 287–314). Springer.

Ruel, M. T., Alderman, H., & The Maternal and Child Nutrition Study Group. (2013). Nutrition-sensitive interventions and programmes: How can they help to accelerate progress in improving maternal and child nutrition? *The Lancet, 382*(9891), 536–551.

Sandiford, P., Cassel, J., Montenegro, M., & Sanchez, G. (1995). The impact of women's literacy on child health and its interaction with access to health services. *Population Studies, 49*(1), 5–17.

Smith, P., House, J. I., Bustamante, M., Sobocká, J., Harper, R., Pan, G., West, P. C., Clark, J. M., Adhya, T., Rumpel, C., Paustian, K., Kuikman, P., Cotrufo, M. F., Elliott, J. A., McDowell, R., Griffiths, R. I., Asakawa, S., Bondeau, A., Jain, A. K., ... Pugh, T. A. M. (2016). Global change pressures on soils from land use and management. *Global Change Biology, 22*(3), 1008–1028.

Smyth, I., & Sweetman, C. (2015). Introduction: gender and resilience. *Gender and Development, 23*(3), 405–414.

Tabaj, K., & Spangler, T. (2017). *Integrating gender into resilience analysis: A conceptual overview*. Resilience Evaluation, Analysis and Learning (REAL) Associate Award series. Save the Children.

Tall, A., Hansen, J., Jay, A., Campbell, B., Kinyangi, J., Aggarwal, P. K., & Zougmoré, R. (2014). *Scaling up climate services for farmers: Mission possible—Learning from good practice in Africa and South Asia* (CCAFS Report No. 13). CGIAR Research Program on Climate Change, Agriculture and Food Security.

Theis, S., Bryan, E., & Ringler, C. (2019). Addressing gender and social dynamics to strengthen resilience for all. In A. Quisumbing, R. Meinzen-Dick, & J. Njuki (Eds.), *2019 Annual Trends and Outlook Report (ATOR): Gender equality in rural Africa. From commitments to outcomes*. International Food Policy Research Institute (IFPRI).

Thomas, T. S., Robertson, R. D., & Boote, K. J. (2019). *Evaluating the risk of climate change-induced aflatoxin contamination in groundnuts and maize: Result of modeling analyses in six countries* (GCAN Policy Note 12). International Food Policy Research Institute (IFPRI).

Tschakert, P. (2007). Views from the vulnerable: Understanding climatic and other stressors in the Sahel. *Global Environmental Change, 17*(3/4), 381–396.

Turner, C., Aggarwal, A., Walls, H., Herforth, A., Drewnowski, A., Coates, J., Kalamatianou, S., & Kadiyala, S. (2018). Concepts and critical perspectives for food environment research: A global framework with implications for action in low- and middle-income countries. *Global Food Security, 18*, 93–101. https://doi.org/10.1016/j.gfs.2018.08.003

Twyman, J., Green, M., Bernier, Q., et al. (2014). *Gender and climate change perceptions, adaptation strategies, and information needs: Preliminary results from four sites in Africa* (CCAFS Working Paper no. 83). CGIAR Research Program on Climate Change, Agriculture and Food Security.

UNICEF, & WHO. (2019). *Progress on household drinking water, sanitation and hygiene, 2000–2017. Special focus on inequalities*. UNICEF, WHO. https://www.who.int/water_sanitation_health/publications/jmp-2019-full-report.pdf

USAID (United States Agency for International Development). (2012). *Building resilience to recurrent crisis: USAID policy and program guidance*.

USAID (United States Agency for International Development). (2017). *Objective 2: Strengthened resilience among people and systems* (Global Food Security Strategy Technical Guidance Brief).

Vaughan, E., & Henly-Shepard, S. (2018). *Risk and resilience assessments. resilience measurement practical guidance note series 1.* Mercy Corps.

Van Aelst, K., & Holvoet, N. (2016). Intersections of gender and marital status in accessing climate change adaptation: Evidence from rural Tanzania. *World Development, 79*(C), 40–50.

Victora, C.G., Adair, L., Fall, C., Hallal, P. C., Martorell, R., Richter, L., & Singh Sachdev, H. (2008). Maternal and child undernutrition: Consequences for adult health and human capital. *The Lancet, 371*(9609), 340–357.

World Economic Forum. (2021). *Global gender gap report 2021.* World Economic Forum.

Zagheni, E., Muttarak, R., & Striessnig, E. (2015). Differential mortality patterns from hydro-meteorological disasters: Evidence from cause-of-death data by age and sex. *Vienna Yearbook of Population Research, 13*(1), 47–70.

**Open Access** This chapter is licensed under the terms of the Creative Commons Attribution 4.0 International License (http://creativecommons.org/licenses/by/4.0/), which permits use, sharing, adaptation, distribution and reproduction in any medium or format, as long as you give appropriate credit to the original author(s) and the source, provide a link to the Creative Commons license and indicate if changes were made.

The images or other third party material in this chapter are included in the chapter's Creative Commons license, unless indicated otherwise in a credit line to the material. If material is not included in the chapter's Creative Commons license and your intended use is not permitted by statutory regulation or exceeds the permitted use, you will need to obtain permission directly from the copyright holder.

CHAPTER 9

# COVID-19, Household Resilience, and Rural Food Systems: Evidence from Southern and Eastern Africa

*Joanna Upton, Elizabeth Tennant, Kathryn J. Fiorella, and Christopher B. Barrett*

## INTRODUCTION

Few people living today have experienced anything like the COVID-19 pandemic. Mass global transmission of a highly infectious disease has caused many millions of confirmed deaths as of the time of writing and is still rising rapidly (Dong et al., 2020). This toll far exceeds that of other recent international viral outbreaks, such as severe acute respiratory syndrome (SARS) in 2003, Middle East respiratory syndrome (MERS) in 2013, and Ebola in 2014. Yet, the story of the food security impacts of the

---

J. Upton (✉)
Charles H. Dyson School of Applied Economics and Management, Cornell University, Ithaca, NY, USA
e-mail: jbu3@cornell.edu

E. Tennant
Department of Economics, Cornell University, Ithaca, NY, USA
e-mail: ejt58@cornell.edu

© The Author(s) 2023
C. Béné and S. Devereux (eds.), *Resilience and Food Security in a Food Systems Context*, Palgrave Studies in Agricultural Economics and Food Policy, https://doi.org/10.1007/978-3-031-23535-1_9

COVID-19 pandemic in rural areas of LMICs—on which we focus—is strikingly familiar.

Response to the COVID-19 pandemic, and to the stresses caused by the growing frequency and severity of other adverse shocks—due to conflicts, macroeconomic disruptions, natural disasters, pandemics, etc. and combinations of them—has become a defining feature of contemporary development and humanitarian policy. When facing a multi-stressor, multi-shock environment (Béné et al., 2016), what can governments, non-governmental organizations (NGOs), donors, private firms, communities, households, and individuals do—independently and collectively—to reduce exposure to and damage from catastrophic shocks and to accelerate recovery after shocks occur? This general line of inquiry increasingly falls under the amorphous label of "resilience", often qualified as "development resilience" (Barrett & Constas, 2014; Barrett, Ghezzi-Kopel, et al., 2021).

Parallel to the rise of resilience as an organizing theme of much development and humanitarian discourse, the food security community has increasingly embraced systems-based understandings of the complex web of relations that influence individual nutritional and food security status. The history of this trajectory is covered in detail within this volume, explicating the origins of food systems approaches (Fanzo, Chapter 2 in this volume) and the nexus of food security, food systems, and resilience (Constas, Chapter 5), among other key dimensions. From pre-production phenomena like climate change and soil nutrient depletion (Gashu et al., 2021; Vermeulen et al., 2012; Willett et al., 2019), to post-harvest value chain intermediary behaviors (Barrett, Benton, et al., 2021; Gómez et al.,

---

K. J. Fiorella
Department of Public and Ecosystem Health, College of Veterinary Medicine, Cornell University, Ithaca, NY, USA
e-mail: kfiorella@cornell.edu

C. B. Barrett
Charles H. Dyson School of Applied Economics and Management, Department of Economics, Department of Global Development, Jeb E. Brooks School of Public Policy, Cornell Atkinson Center for Sustainability, Cornell University, Ithaca, NY, USA
e-mail: cbb2@cornell.edu

2013), and everything in between, a complex web of actors and actions mediate food security outcomes.

In this chapter, we emphasize four central ideas of the food systems' framing of food security issues. First, interconnected ecosystems, institutions, and markets create multiple entry points for both shocks and interventions. Second, multiple iterations of endogenous responses by the many actors throughout the system have feedbacks locally and further away. Third, the feedback between individual and household behaviors at the most micro-level, meso-level collective and market behaviors, and macro-level policy responses underscores the need for multi-scalar analysis. Fourth, and integrating the prior points, shocks often have indirect impacts that far exceed their direct impacts in the aggregate, precisely because of the multiple mechanisms that link individuals to features of the social-ecological systems and the endogenous, multi-scalar behavioral responses by many people and organizations.

Those four ideas have proved highly salient during the COVID-19 pandemic, which has underscored the deep interconnectedness of communities around the globe. Such "telecoupling"—meaning the ties between distant communities' socioeconomic and environmental interactions (Hull & Liu, 2018; Liu et al., 2007)—has become apparent as policies in one nation have acutely impacted other places (e.g., far-reaching economic and epidemiological consequences of the Wuhan lockdown). The pandemic has also illuminated the vast response capacity a major shock can elicit from governments, NGOs, multilateral agencies, firms, and private individuals. The Coronavirus Government Response Tracker, a tool managed and regularly updated by Oxford researchers, reported almost 67,000 distinct policy responses by 183 different governments over the course of 2020 (Hale et al., 2021). These range from border or school closures, stay-at-home orders, and other policies such as mask mandates or capacity limits intended to slow the spread of the disease, to public health interventions, to new social protection measures intended to support those who lost jobs or businesses. Indeed, a separate study identified that at least 3,333 different social protection measures were planned or implemented in 222 countries or territories as of mid-May 2021, most of them since January 1, 2021, as the depth and duration of the pandemic's impact necessitated additional response (Gentilini et al., 2021). Social protection coverage has been lowest and most variable in those countries with the highest poverty rates, however, and especially in harder-to-reach rural areas, like the sites we studied in Malawi, Madagascar, and Kenya.

Despite social protection measures, food security generally deteriorated globally during the pandemic, albeit with significant variation over time and across and within countries (FAO et al., 2021). Clean longitudinal comparisons are nonetheless difficult because among the few panel studies that track the same measures pre- and post-pandemic among the same households, virtually all had to switch survey methods, from in-person to telephone or online elicitation, and often to shorter survey instruments as a result, rarely with more than one observation after the pandemicbegan (Adjognon et al., 2021; Ahmed et al., 2021; Amare et al., 2021; Egger et al., 2021).[1] In the limited cases where large-scale, high-frequency data are consistently available, there is evidence of considerable fluctuations in the prevalence of food security over the course of the pandemic. In the US, for example, the share of adults who reported that their household sometimes or often did not have enough food to eat peaked at about 12% in July 2020, subsided sharply as the daily infection and hospitalization rates ebbed, and then returned to that peak again from late December through January 2021 as infection rates hit their highest levels to date.[2]

In this chapter, we harness unusual, high-frequency household panel survey data from rural areas in three countries—Malawi, Madagascar, and Kenya—that afford an in-depth look into the resilience of household food security status over the course of the pandemic, and the mechanisms seemingly most strongly associated with those patterns. We emphasize the descriptive nature of the evidence we present. We do not assess the causal impacts of the pandemic—which are easily confounded by seasonality and other shocks, such as drought in southern Madagascar, a bumper maize harvest in southern Malawi (Upton et al., 2021), and flooding in Kenya—much less of various policy responses to the covariate shock. We show how a seemingly unique, catastrophic shock like the COVID-19 pandemic has played out in distressingly familiar ways that reinforce

---

[1] Aggarwal et al. (2021) is an important exception. They analyze household phone survey data collected from purposively selected locations in Liberia and Malawi starting before COVID-19 began and continuing well into the pandemic period.

[2] US Census Bureau Household Pulse survey data available for April 23, 2020–March 29, 2021, at https://www.census.gov/programs-surveys/household-pulse-survey/data.html.

the need to improve our understanding of resilience to a wide variety of shocks that rural populations routinely (and increasingly) suffer.

We begin by laying out a conceptual framework for analogous shocks and the ways in which COVID-19 has played out through familiar and inter-locking mechanisms. We then provide background on our contexts, data, and COVID-19-related policies in these study areas. The subsequent section lays out the results, first observing food security outcomes and then illustrating the impact mechanisms as put forward in the framework. We conclude with a brief discussion and conclusion.

## Conceptual Framework

Resilience represents the capacity to "persist in the face of change, to continue to develop in ever changing environments" (Folke, 2016, p. 45) or, applied more specifically to the international development context, as the "capacity that ensures stressors and shocks do not have long-lasting adverse development consequences" (Constas et al., 2014, p. 6). The COVID-19 pandemic was an indisputably major shock piled on top of a host of other stressors faced by rural communities in low- or middle-income countries (LMICs). Yet the emerging empirical literature on the pandemic's impacts suggest that in these settings the pandemic has perhaps impacted food security—and other well-being indicators—more through familiar, systemic pathways, for example by negatively impacting physical and financial access to food (see Béné et al., 2021 for a review), than by directly impacting households' health or causing excess mortality within communities. We therefore look to resilience theory as well as the antecedent literature on rural development and resilience to other large, covariate shocks to develop a conceptual framework for understanding resilience in the context of the COVID-19 pandemic.

Following Barrett and Constas (2014), we view resilience as the sufficient probability of attaining and maintaining an acceptable level of well-being, here reflected in food security measured using several different indicators. Importantly, this implies that households that achieve stability but remain below a normative threshold are not considered resilient; a chronically hungry household is not resilient even if hunger levels are steady in the face of shocks. This outcome is enabled by diverse features of households and communities often termed "resilience capacities" (Ansah et al., 2019; Barrett, Ghezzi-Kopel, et al., 2021). As per Barrett and Constas (2014, p. 14625), the concept of resilience adds value primarily in that "it compels a coherent, multi-disciplinary, and rigorous explanation of the interrelated dynamics of risk exposure, multi-scalar human standards of living, and broader ecological processes". In developing our framework and empirical analysis, we conceptualize resilience specifically in terms of food security. Food security is a critical and universal aspect of well-being, and the four widely recognized axioms of food security—the availability, access, utilization of food, and its stability over time—share key properties with definitions of resilience (FAO, 1996; Upton et al., 2016).

Empirical studies of development resilience frequently focus on a specific shock or class of shocks (Barrett, Ghezzi-Kopel, et al., 2021). The most common shocks studied have been natural hazards, such as meteorological (extreme weather events) or climatological (droughts) phenomena (Barrett, Ghezzi-Kopel, et al., 2021). The next most common are economic shocks, such as global financial crises, and sociopolitical shocks, such as conflict. Studying resilience in reference to a specific class of shock has benefits for answering certain types of research questions, for example to inform policies designed to mitigate the direct effects of a particular type of shock to which a place or population is predisposed (Manyena, 2006). This can help us to understand the unique and (sometimes) severe direct effects of certain types of shocks that may emerge, for example when lives are lost and communities razed during a tropical cyclone, when civilians are killed or forced to flee civil conflict, or the morbidity and mortality from HIV/AIDS, Ebola, or COVID-19.

Our framework instead highlights meso-level mechanisms—e.g., local institutions and local and regional markets—that shape the impact of macro-scale, covariate shocks on households. Framing around a specific event or class of shocks can be limiting if households and communities are often most affected by meso-level, systemic mechanisms, at least some of which can be linked to any number of causes. We focus instead on the common intermediary pathways with relevance across multiple classes of large covariate shocks—including new and emerging threats, such as the COVID-19 pandemic.

While in principle the pandemic is a shock to human health and health systems, beginning in March/April 2020 and for nearly all of the subsequent two years, we do not observe detectable increases in morbidity attributable to COVID-19 in the communities we study. The main shocks arose rather from the economic impacts of global, national, and local policy responses. This is consistent with emerging empirical literature on food security in the COVID-19 pandemic in many LMICs (see Béné et al., 2021). The predominant mechanisms that have proved salient in the COVID-19 pandemic context in our settings are mobility restrictions, livelihood disruptions, and price shocks.

Figure 9.1[3] lays out the basic conceptual model we employ. At the top (in capital letters), one finds a range of major, covariate shocks. Some may be distal to a rural community—like global financial shocks or an epidemic/pandemic experienced mainly in distant cities and countries—while others may be more proximate, like conflict or natural hazards. A defining feature of a covariate shock is that exposure is correlated across many households. The scale of these shocks commonly induces policy responses at local, national, regional, or even global level(s). Combined with the initial covariate shock itself, induced policy responses impact a range of individual and collective behavioral responses. These manifest in the meso-level mechanisms that govern livelihoods, food production, distribution (including markets), and consumption, and the provision of public services, including health care (indicated in red, with the dominant categories of mechanisms in boxes aligned with the shock(s) with which it is most commonly associated; and arrows below indicating where there are posited causal relationships). Together, those meso-level mechanisms impact the four pillars of food security (in blue): availability, access, utilization, and stability, resulting in observable changes in food security indicators. A pandemic like COVID-19 is simply a special case that has a more direct, visible effect on health but quickly propagates through other meso-level mechanisms, including employment, mobility, and price effects.

In what follows, we trace out the impacts of the COVID-19 pandemic in the three different low-income rural settings we study. We ask the question, what does pre-COVID-19 evidence under each of these shocks/stressors suggest about the resilience of households and food systems to the pandemic? We outline likely direct and indirect impact pathways in each area and glean insights by examining these using our data and current evidence in the broader literature.

---

[3] All figures sourced from authors. With the exception of Fig. 9.2 (as cited), data used are those described in the Contexts section.

288   J. UPTON ET AL.

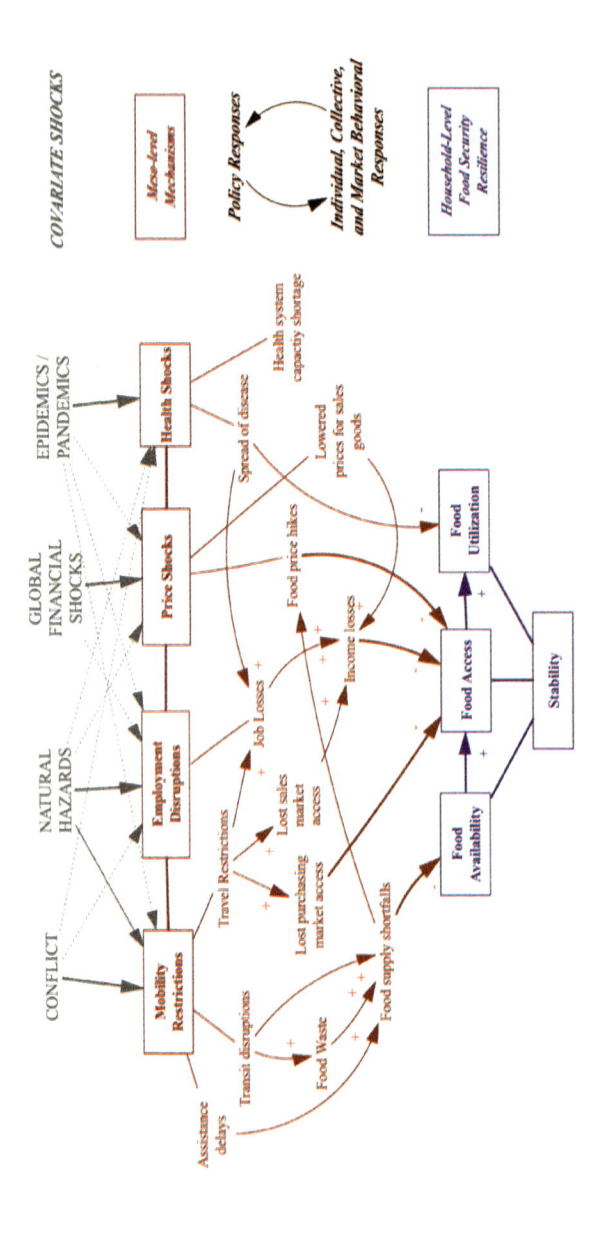

**Fig. 9.1** Conceptual framework of covariate shocks, mechanisms, and food security resilience (*Source* Authors)

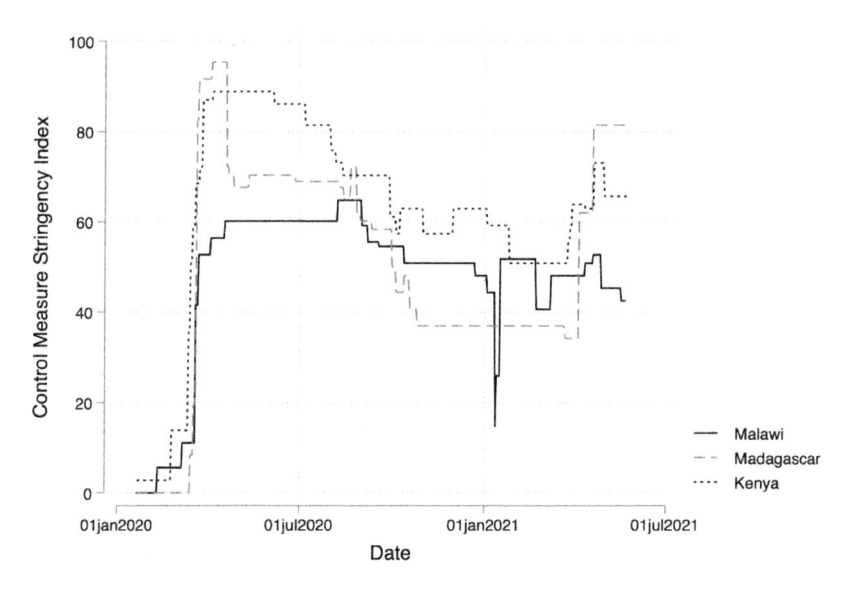

**Fig. 9.2** COVID-19 control measures' stringency varied over time in Malawi, Madagascar and Kenya—original figured designed from Oxford COVID-19 Government Response Tracker data (Johns Hopkins University, 2021; Hale et al., 2021)

## Contexts and the COVID-19 Pandemic in Rural Malawi, Madagascar, and Kenya

We tap high-frequency survey and market monitoring data from three distinct study areas in Malawi, Madagascar, and Kenya (Table 9.1) to illustrate the application of our conceptual framework. These data include five food security indicators (Table 9.2). Within each country, COVID-19 policy measures varied considerably over space and time (Fig. 9.2). In this section, we provide brief background on the context within each country, including an overview of dynamics of COVID-19 and related control measures, and on the data collection platforms on which we draw.

### *Malawi*

Rural Malawi is emblematic of the challenges inherent in building resilience. It includes diverse and dynamic socio-ecological systems and

**Table 9.1** Data collection start dates, locations, and sample sizes

|  | Start date | Region | District | Households sampled |
|---|---|---|---|---|
| Malawi | September 2017 | Southern | Blantyre Rural<br>Chikwawa<br>Nsanje | 750<br>750<br>750 |
| Madagascar | July 2018 | Androy | Tsihombe<br>Beloha | 364<br>237 |
| Kenya | March 2020 | Western | Homa Bay and Kisumu Counties | 90 |

*Source* Authors

**Table 9.2** Food security indicators collected, by location

| Indicator | Description | Citation | Collected monthly in… |
|---|---|---|---|
| Household Dietary Diversity Score (HDDS) | Sum across 11 food groups (unweighted, 24-hour recall) | Swindale and Bilinsky (2006), USAID/FANTA | Madagascar, Kenya |
| Food Consumption Score (FCS) | Sum of 8 food groups, weighted for quality and frequency (7 day recall) | Weismann et al. (2009), WFP/VAM | Madagascar |
| Household Hunger Scale (HHS) | Weighted sum of three extreme strategies, over last month (no food at all available in household, going to sleep without eating, and going a full day and night without eating) | Ballard et al. (2011), USAID/FANTA—III | Malawi, Madagascar |
| Reduced Coping Strategies Index (rCSI) | Weighted sum of 5 less severe strategies, over the last week (loan, reduction, less-preferred foods …) | Maxwell et al. (2008), USAID | Malawi, Madagascar |
| Household Food Insecurity Access Scale (HFIAS) | Summed score or weighted categorization of food insecurity severity | Coates et al. (2007) | Kenya |

is characterized by multiple shocks and stressors that trap households in food insecurity and poverty. The predominant livelihood in the study area is rainfed agriculture. Roughly, 75% of households own land for agriculture, with home-grown crops, especially maize, accounting for the primary income source for 40%, and agricultural labor for another 40%. Rates of food insecurity in Malawi are typically high, often as high as 70%, including "moderate" food insecurity, with nearly half of these severely food insecure (Fisher & Lewin, 2013; IPC, 2021). The high-frequency measures we analyze reveal that food insecurity is dynamic, with respect to both averages and *who* is food insecure. In the three southern districts, we have studied over the past three years, 64% of households changed their monthly food security status over time, with 56% classified as "crisis" *or worse* in over half of all periods.

In response to the increasing severity of weather-related shocks threatening food security, Catholic Relief Services (CRS) partnered with Cornell University to develop the Measurement Indicators for Resilience Analysis (MIRA) platform, a system for high-frequency monitoring to understand resilience following a recommended sentinel site survey design (Headey & Barrett, 2015).[4] MIRA's use of decentralized enumerators and a Computer-Assisted Personal Interviews (CAPI) platform allowed data collection to continue uninterrupted throughout the pandemic lockdown. See sample information in Table 9.1, and food security indicators collected in Table 9.2.

### *Madagascar*

The Great South of Madagascar is relatively unique within the island's patchwork of ecological systems in its susceptibility to damaging dry spells, drought, and water scarcity, and suffers high chronic food insecurity. The region is often rated highest risk by monitoring reports, with the highest rates in the country of households suffering crisis levels and worse

---

[4] See sample information in Table 9.1, and food security indicators collected in Table 9.2. Upton and Knippenberg (2019) for a description of MIRA, and work on MIRA data to date include new insights into resilience measurement (Knippenberg et al., 2019), early warning (Knippenberg et al., 2020), and identification of the impacts of the Covid-19 pandemic and policy response (Upton et al., 2021).

(Government of Madagascar and IPC, 2020).[5] In our study sample, nearly 100% of households changed monthly food security status over time, with 64% classified as "crisis" *or worse* in over half of all periods. As in Malawi, the predominant livelihood is rainfed agriculture; 95% of households in the study area report owning land, and 78% report crop agriculture as a primary livelihood. Given the coastal geography, relatively more rely also on livestock and fishing (13%), though fishing employs rudimentary technologies in open ocean, so presents its own challenges (ibid.).

Following initial success with the MIRA approach in Malawi, CRS and Cornell rolled out a data collection platform in southern Madagascar very similar to that in Malawi, starting in July 2018. The sample is described in Table 9.1, and food security indicators (an expansion on those collected in Malawi) in Table 9.2.[6]

*Kenya*

Lake Victoria fisheries have undergone nearly a century of ecological and social change following the introduction of non-native Nile perch (Pringle, 2005), rapid commercialization of an export fishery (Abila et al., 2003), and high incidence of HIV within regional fishing communities (Seeley & Allison, 2005). Despite longstanding concerns of overharvest, today Lake Victoria's fisheries supply the largest lake harvest of fish in the world (FAO, 2016) amid ongoing vulnerabilities to a suite of contemporary ecological challenges, including climate and ecological change (Aura et al., 2020), and persistent rates of food insecurity within fishing regions (KNBS, 2018).

Prior to the COVID-19 pandemic, eight fishing communities in Kisumu and Homa Bay counties were selected for an in-person survey that aimed to examine perceptions of harmful algal blooms. Communities around Kisumu Bay were purposively selected based on exposure to high intensity of algal blooms as detected in satellite remote sensing data. We

---

[5] These Integrated Food Insecurity Phase Classification reports are periodically updated, Government of Madagascar and IPC (2020), cited as an example, issued in December 2020.

[6] See Upton and Knippenberg (2019) for a report describing MIRA and the Madagascar expansion process, with further detail on the protocol and questionnaires.

randomly selected 30 households within each community. All questionnaires were administered to the "food manager" within each household, defined as the person mainly responsible for food preparation.

The survey was halted on March 21, 2020, after Kenya experienced its first case of COVID-19, and after we had completed data collection for three of the eight communities: Dunga, Lela and Kananga.

In June 2020, we began a follow-up phone survey with the 90 previously surveyed households to understand the changes in food security and food consumption patterns of households during the COVID-19 pandemic and the coping strategies adopted by households.[7] We collected information on household food security and household food consumption using the same questionnaires as the March 2020 survey. An additional module was administered to measure the perceived impact of COVID-19 and associated movement restrictions on food access, and also to explore coping strategies used by households during the pandemic.

### COVID-19 and Policy Responses

We tracked the implications of COVID-19 and the associated global and local policy responses in the study areas through both the surveys and remote key informant interviews, starting in April 2020 and continuing through (and beyond) the time of writing.

#### Malawi

Our data and informants show that, through the end of 2020, the disease itself seems to have made very few people ill in southern Malawi; reports of experienced illness reflect a seasonal pattern (due to periods of cholera and malaria) that is nearly identical to the prior years surveyed. There have, however, been diverse impacts on mobility and markets that have had significant impacts on households' lives and livelihoods. The first and most dramatic change was the closure of international borders on April 17, 2020. Borders did not reopen until just over a year later, with traffic limited to essential goods and returning locals screened and

---

[7] We reached out to all 90 participants by phone and explained the follow-up study to them. Those who gave their verbal consent were re-enrolled into the study and interviewed by phone. Out of the 90 baseline participants, we re-enrolled and collected data on 88 (one participant could not be located and one participant had died prior to the follow-up study). Additional methodological details available in Fiorella et al. (2021).

quarantined on entry. The maize harvest in the area had been exceptionally good, meaning that maize was readily available internally. After the border closure, however, the prices of other, imported goods immediately rose, and there was concern that farmers would deplete maize stocks by bartering at very low prices for other goods. The national marketing board (ADMARC) stepped in to declare a sales price for maize in an attempt to curtail the bartering. But prices for other goods continued to be higher than usual. Internal movement was not restricted per se, but social distancing requirements and forced closure of certain types of businesses, predominantly in the service sector but also some small retail operations, led to reduced economic activity and job losses. Schools also closed—first from 23 March to 13 July 2020, and then again from 17 January to 22 February 2021—with limitations on classroom capacity during periods of instruction and job losses for private school teachers and other support staff.

*Madagascar*

In Madagascar, as in Malawi, household reports offer no evidence of significant spread of the virus throughout 2020; survey reports of respiratory and other health problems are under 5% over the period, and track at or below levels reported since 2018. But global and national policy and behavioral responses have had considerable impacts, as we detail below. Inter-regional transit was briefly officially restricted in late June. Although official local restrictions were lifted relatively soon thereafter, the international borders remained closed for a much longer period, and local travel continued to be affected by quarantines, high fuel prices, and a state of uncertainty. Schools were closed for only brief periods, but the school calendar was disrupted. Some households reported receiving aid, but we see no evidence that aid increased in our Madagascar (or Malawi) study area during the pandemic period.

*Kenya*

The Kenyan government responded to the COVID-19 pandemic with a number of policies to restrict movement among the population, and particularly within urban centers. Measures ranged from closing of schools to restrictions on maximum seating capacity in public transportation to a mandatory curfew, which was initially strictly enforced.[8] Throughout the pandemic, Kenya's government largely imposed more stringent policy responses than Madagascar or Malawi (Fig. 9.2). In May 2020, the Kenyan government announced a substantial aid stimulus package to help its population to cope with the economic disruption stringent pandemic response measures had caused.

By early July, the Kenyan government changed course to reopen the country, despite the implications for COVID-19 risk and consequent waves of infections. The first wave of COVID-19 cases and deaths in Kenya peaked at the beginning of August 2020 (Dong et al., 2020). In October 2020, schools were abruptly permitted to reopen and movement restrictions were further eased; a second, larger wave of COVID-19 infections and deaths soon followed, peaking in mid-November 2020. The third, and largest, wave of COVID-19 infections peaked in March 2021.

## Food Security Outcomes and Identified Mechanisms

In this section, we describe the net effect of these complex and varied policy, market, and individual responses on household-level food security indicators. We ask the question: have the households in the study areas been resilient to the health impacts, mobility restrictions, livelihood disruptions, and changes in food prices driven by the COVID-19 pandemic and other coincident shocks and institutional responses? We find that in none of our study settings do all households achieve and maintain food security amid the pandemic; these populations suffer from chronic food insecurity, which pre-dates the pandemic. This finding echoes concerns in the pre-COVID-19 resilience literature that poverty is both a driver and predictor of low levels of resilience in household food security (for a review of evidence, see Barrett, Ghezzi-Kopel, et al., 2021). In sum, the duration, magnitudes, and mechanisms through which the COVID-19 pandemic impacted resilience seem to vary among and within our study areas.

---

[8] Policies were put forward in a series of press releases and executive orders, described on the Kenya Ministry of Health website (available online here: https://www.health.go.ke).

## Food Security Outcomes

Assessing the true, strictly causal effects of the pandemic on food security—much less unpacking the mechanisms and estimating each's contribution to the net effect—is a daunting task because the meso-level mechanisms necessarily integrate other, contemporaneous shocks (e.g., from global commodity markets, local weather, etc.), which can mask or exaggerate the true pandemic effects in simple before-and-after comparisons (Upton et al., 2021). The richness of the data across the three sites, however, enables an informed discussion of the associations we observe and plausible (and implausible) mechanisms behind the observed net food security changes over the course of the pandemic crisis.

In Malawi, where we measure coping strategies indices—summarized in the Household Hunger Scale—food security status in 2020 was comparable to, or even slightly better than, prior years (Fig. 9.3). That outcome is almost surely due to a strong harvest that supported farm and farmworker incomes and kept staple food prices relatively low. Further in-depth analysis leaning on the full duration of the panel data and developing a counterfactual predictive model, however, suggests that food security was not as good as it should have been, and further that the pandemic-specific effects were regressive, affecting the worst off more severely (Upton et al., 2021). This finding is consistent qualitatively with the diverse likely impact mechanisms we identify (discussed below).

In Madagascar, in contrast, household food security deteriorated during the COVID-19 pandemic period—which coincided with a severe drought in southern Madagascar (Fig. 9.4). Descriptively, coping strategies indices were not worse on average than in prior years, although local experts attribute this largely to households already being at the limit of what they could do to improve their situation, for example, having already reduced meal sizes and frequencies to a minimum. As regards dietary diversity, the situation clearly worsened relative to prior years, with many attributing this primarily to COVID-related market restrictions and price hikes. This is consistent with the broader, cross-national storyline one commonly hears (e.g., Egger et al., 2021). Due to the compound nature of the pandemic and the drought shock in Madagascar, and overlaps in the meso-level mechanisms through which they impact food systems (discussed in the following sections), we are unable to parse their relative contribution to the deepening humanitarian crisis currently impacting this region.

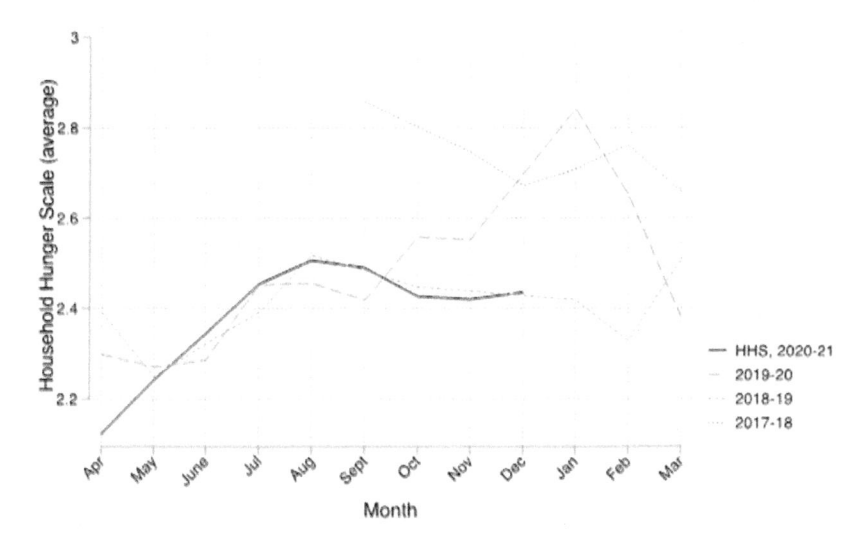

**Fig. 9.3** Mean Household Hunger Scale, over time, Malawi (*Note* Range is 0–6 with higher worse; >=1 is considered 'stressed,' and >=3 'crisis')

In Kenya, the situation has been less grave in our sites, even if food insecurity increased initially, as reflected in HFIAS measures (Fig. 9.5) from March (pre-COVID-19) to June 2020 (during the COVID-19 lockdown), but returned roughly to pre-pandemic levels over the next several months. The relatively quick return to pre-pandemic levels of food insecurity is noteworthy given that communities within the region experienced not only the COVID-19 pandemic but acute flooding amid rising lake levels, which also threatened food security (Aura et al., 2020). Food insecurity increases in these Lake Victoria coastal communities were in part a function of large upticks in worry, and in people not being able to eat preferred foods. Market closures by government restrictions also affected households, with over half reporting in June 2020 that they had to go more places than usual to find food. Finally, food aid during this period was initially accessed by 17% of households in June 2020 and vacillated (range: 1–12% of households) from July 2020 to May 2021, and may also have provided a safety net.

Households activated coping strategies, including borrowing from friends and relatives, purchasing food on credit, and reducing portion sizes and meals in the early months of the pandemic. However, use of

**Fig. 9.4** Mean Food Consumption Score, over time, Madagascar. Dashed horizontal lines represent standard thresholds for "Food Secure" (35) and "Borderline" (21). Years run April–March. A higher FCS is better

these coping strategies reduced and then plateaued around September 2020. Overall, dietary diversity did not significantly change, although it was low at baseline, but diets shifted from nutrient-rich perishables such as eggs and fresh fish to more shelf-stable products (e.g., beans).

In this setting, the easing of movement restrictions in mid-2020 and reopening of schools in October and November was associated with a second wave of COVID-19 cases. This increase in case rates, in turn, increased worry about COVID-19 infections. At the same time, these policy shifts appear to have stabilized rates of food insecurity and use of coping strategies by households in lakeside communities, with fewer households reporting reductions of work and incomes over the same period. In other words, easing of movement restrictions and reopening of markets—both far-reaching, meso-level impacts of the pandemic— allowed some households to get back to work, earn a fishing income, and participate in fish trading with consistent demand for fish within the region supporting this rebound. This highlights a key trade-off between

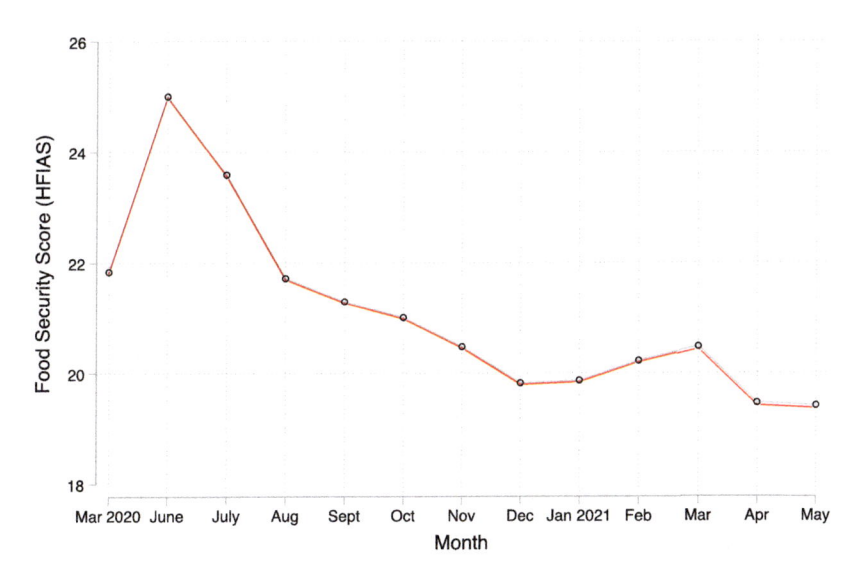

Fig. 9.5 Food Security Score (HFIAS), March 2020–March 2021, Kenya

meso-level and direct effects of shocks: increased exposure to health shocks occurred in conjunction with reduced livelihood disruptions.

The general story is thus one of marked variation among and within sites in the evolution of food security over the course of the pandemic's first year. These mixed outcomes—with unchanged, or even slightly improved food security outcomes in southern Malawi, much worse ones in southern Madagascar, and initially worse and then largely unchanged outcomes in western Kenya—reflect the mixing of the pandemic shock, and its associated policy and market responses, with other shocks. Where markets, governments and communities function relatively well—as in the Lake Victoria basin of western Kenya—or where nature blesses farmers with good weather—as in southern Malawi—the likely net-negative impacts of the pandemic can be partly or fully offset, limiting the damage except to the relatively few who directly experience the shock through morbidity or mortality. Conversely, where infrastructure is poor and a climate shock compounds the damages wrought by the pandemic, as in southern Madagascar, an already-bad situation gets made much worse. In this way, the COVID-19 pandemic shock's effects on food security in these rural African communities bear striking similarity to the broader profile of vulnerability and resilience to other shocks, and underscores the central importance of the meso-level institutions that mediate the changes households and individuals experience (Barrett, 2005; Barrett & Swallow, 2006; Béné, 2020).

## Meso-Level Mechanism 1: Health Shocks

Many large covariate shocks have both direct and indirect impacts on human health. For example, the Ebola virus infected an estimated 28,000 people and killed over 11,000 during the 2014–2015 West Africa outbreak, but the cost in lives of the associated breakdown in healthcare access and utilization, although more difficult to measure, was likely orders of magnitude more deadly (see Elston et al., 2017 for a review). The emerging literature on COVID-19 indicates that—in addition to the morbidity and mortality associated with infections—health system disruptions, including decreases in childbirth interventions, treatment of (non-COVID) infectious diseases, as well as care for chronic conditions have likely resulted in large-scale increases in mortality (Jain & Dupas, 2020; Roberton et al., 2020).

Here, we assess how shocks to human health and health systems may have impacted food systems in our study areas during the COVID-19 pandemic. We consider both direct and indirect mechanisms. For example, increased morbidity and mortality resulting from infection and/or disruptions in health systems may adversely affect household labor supply. This could result in income losses and reduced food access, as well as caregiving and decreased nutrient absorption that compromise food utilization, leading to negative food security impacts of the pandemic. At a systems level, the strain on the healthcare system may divert societal resources away from more general social protection toward support for an overburdened health system, further compromising food access, availability, and utilization.

### *Findings from Our Sites*

While early in the pandemic there was no documented spread of the disease to our study areas, there was also a complete lack of testing. We therefore considered it prudent to track both reports of health incidents (asked as "shocks" in the questionnaires) and more specific questions regarding incidence in the households of symptoms attributable to COVID-19. Even under normal conditions, disease prevalence is relatively high and varies seasonally and spatially. So one would like to know whether the pandemic increased household reports of illness or death relative to non-pandemic periods. In both Malawi and Madagascar, high percentages of households reported health-related problems in prior years, in Malawi with typically high incidence (~15–20% of households) in February–April, often due to seasonal spikes in cholera or malaria. Reports of health shocks were no higher than usual in 2020–2021, and in fact declined over the early pandemic period. In Madagascar, a cholera outbreak in 2018–2019 affected many households, relative to which the experience of non-respiratory illnesses was comparatively very low in 2020–2021 after the pandemic began. The trends we see are suggestive that there has been little direct morbidity from COVID-19, while also highlighting that properly evaluating any eventual direct health impact could be confounded by the already high and heterogeneous burden of disease in typical years.

Particularly within the context of a pandemic, fear of contracting an illness may also play a substantial role in shaping households' behavior. Reports from key informants in Madagascar confirm that fear of contracting the illness in transit and in larger cities contributed to reduced rates of migration for work. Especially within communities with scant savings, disease control measures that limit households' ability to work may also be an interrelated source of worry. In Kenya, worry about contracting COVID-19 was initially ubiquitous and fluctuated with periods of COVID-19 waves and high case counts (left panel of Fig. 9.6). Initially (June to September 2020) nearly 80% of households reported some degree of worry about the impact of COVID-19 control measures. This share fell in October 2020 as movement restrictions were lifted and the country reopened (right panel of Fig. 9.6). Fear and lack of trust, as well as movement restrictions and disrupted health service availability, may also hamper access to routine care. COVID-19-related health system disruptions likely affected prenatal and antenatal care and treatment for non-COVID infectious and chronic diseases, with potentially large health implications.

## Meso-Level Mechanism 2: Mobility Restrictions

Natural disasters such as cyclones, earthquakes, floods, and tsunamis can damage transport infrastructure, directly disrupting mobility among and

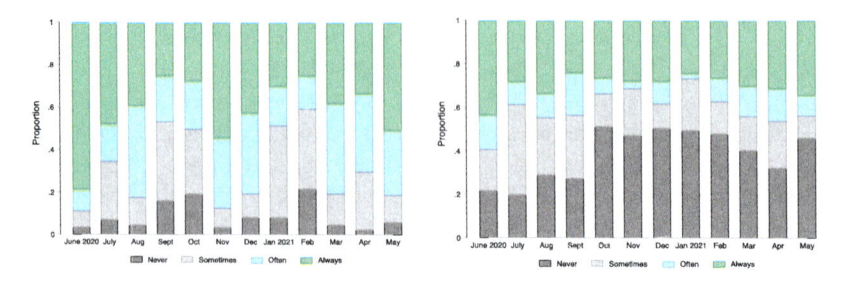

**Fig. 9.6** Reported worry about impacts of (a) COVID-19 and (b) COVID-19 control measures, Kenya, June 2020–March 2021

within countries, often for extended periods (Hallegatte et al., 2019). Infectious human or animal diseases and some types of civil unrest can elicit government movement restrictions. Mobility restrictions imposed by governments or resulting from hazards are then complemented by voluntarily reduced mobility as fear and uncertainty induce caution among people who might otherwise travel (Elston et al., 2017; Gates et al., 2012). Mobility restrictions can impact food access directly, for example limiting access to markets, but also cross over into healthcare access, livelihood disruptions, food supply chain issues (and associated price shocks), and other disruptions to the longer-term accumulation of human capital (Béné et al., 2021).

In the case of COVID-19 in LMICs, mobility restrictions would result directly from government policy responses to the pandemic, such as inter-regional travel restrictions and border closures. This would disrupt migration for employment, causing income losses for both workers who typically migrate and for employers who typically rely on in-migrant workers. Transport restrictions can also disrupt food shipments, causing prices to rise or fall for products imported into and exported from the locale, respectively.

Less direct pathways for mobility restrictions could result from household behavioral change akin to what occurs in the case of conflict. Households could choose to limit their movement and/or change economic behaviors out of fear and uncertainty, both fear of infection due to the disease itself and in response to uncertainty around government policies and their enforcement.

### Findings from Our Sites

In Malawi, according to key informants, international and domestic restrictions (as described above) affected goods and markets, as well as migrants' ability to travel (predominantly to higher-paying jobs in South Africa). Even where there were no official travel restrictions within and between regions, there were requirements for appropriate distancing in transit vehicles and mask wearing. The sanctions for violations were unevenly applied, and many people hesitated to circulate for fear of both the disease and of uncertainty in policies and crack-downs. Many of these more intangible impacts lessened starting around September, as concerns about the disease abated, only to pick up again by January 2021 (see Fig. 9.7). Note that there was little variation across months, almost all households reported some impacts, and 15–20% reported being unable to travel as needed. The more commonly reported impacts, however, are more relevant to employment and market access issues discussed below.

In Madagascar, the government did impose restrictions on travel between regions. These official restrictions were short lived, however, lasting for only roughly two weeks, from late June through early July 2020. However, household reports of inability to travel where needed

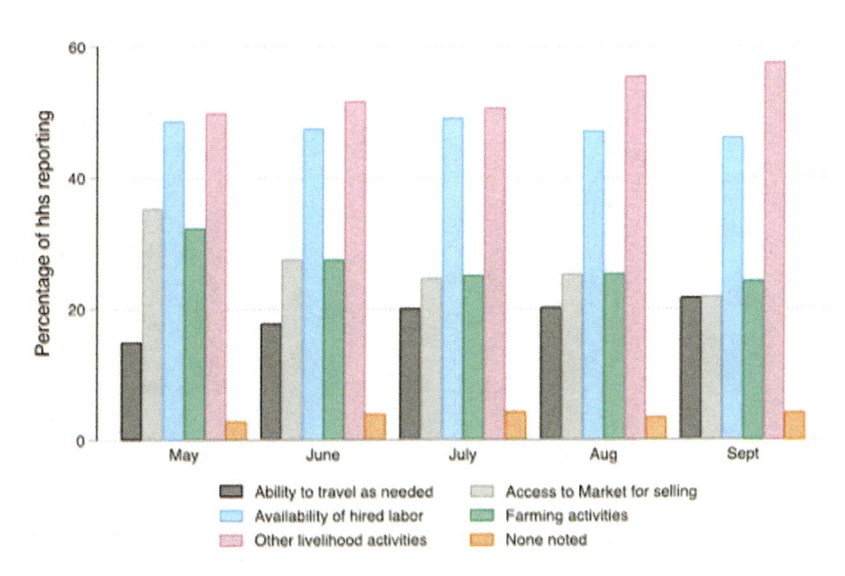

**Fig. 9.7** Reported impacts of the pandemic, May–September 2020, Malawi

*increased* throughout the period, peaking at fully 90% of all households in January 2021 (Fig. 9.8). According to key informants, the de facto restrictions on travel were several-fold. First, policies such as inter-regional screening and quarantines, and uncertainty around these policies, delayed or discouraged even major commercial trucks (and all the more, individual travelers). Second, the towns themselves were economically depressed relative to typical years, making work less available. This reduced travel and further reinforced the unavailability of transit options. And, finally, people had the impression that towns were at much greater risk of infection with the disease, which further reduced transit demand and supply, making it less available even to people who did wish to travel. The remoteness and limited, rudimentary infrastructure of the Great South of Madagascar gave far greater prominence to mobility restrictions in this site, as compared to the more accessible ones in Kenya or Malawi.

In Kenya, while the strictest initial enforcement of movement restrictions within major metro areas did not extend to our study site, school closures and national curfews affected all communities and fomented high levels of concern about contracting COVID-19. In June 2020, 75% of

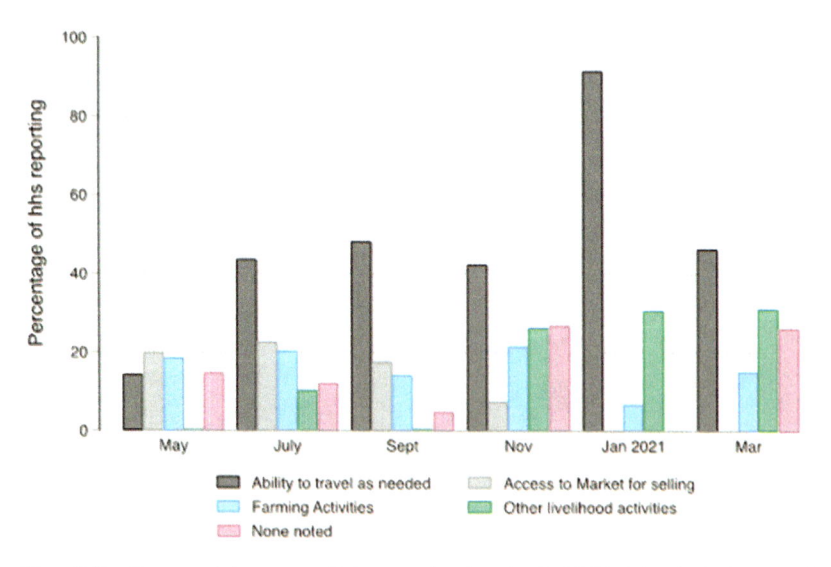

**Fig. 9.8** Reported impacts of the pandemic, April 2020–March 2021, Madagascar

households reported they had reduced trips to get food; in May 2021, 38% were still reporting reduced trips to access food. Curfews also had a disproportionate effect—especially on fishers operating in the *dagaa* fishery. *Dagaa* are fished at nighttime, using lights to attract fish, and curfews forced fishers to either curtail activities or break the rules. A subset of participants reported that their regular activities were highly disrupted in the early months of the pandemic, and disruptions endured throughout the year. In the months of June, July, and August 2020, more than 85% of fishers reported their fish catch was often or always reduced compared to the preceding year, a figure that fell to 56% in September 2020 as restrictions on movement began to ease and markets reopened.

## Meso-Level Mechanism 3: Livelihood Disruptions

Large covariate shocks may directly impact individuals' capacity to earn income, for example through the destruction of a productive asset or due to death or illness in the household (de Waal & Whiteside, 2003). But as these shocks propagate through the economy the indirect livelihood disruptions are often much more widespread, realized especially through general equilibrium effects in labor markets (Devereux, 2007; Gates et al., 2012; Hallegatte et al., 2017; Kodish et al., 2019). In the context of COVID-19, initial concern frequently focused on income shocks to wage laborers as strict lockdowns took effect (e.g., Adjognon et al., 2021). However, emerging evidence indicates that the COVID-19 pandemic has also resulted in large-scale employment disruptions in rural areas of LMICs, with impacts falling disproportionately on the poorest subpopulations who depend most heavily on wage earnings for income (Egger et al., 2021), although not all studies find such effects (Aggarwal et al., 2021).

Here, we explore the multiple pathways through which the COVID-19 pandemic may have resulted in employment disruptions and associated loss of income in our study areas. We posit that market-mediated occupations might be disrupted both by official policies as well as employers' behavioral responses to those policies. Employment disruptions may also be closely linked to mobility disruptions, for example as transport disruptions and border closures prevent migrant labor or disrupt agricultural markets. In turn, and similar to the case of mobility restrictions, less direct changes could result from households' choices due to fear and uncertainty, e.g., unwillingness to hire workers due to fear of infection, and/or not reopening a business when one could due to fear that government policy would change again and shut it down, resulting in greater losses.

## Findings from Our Sites

Initial reports from Malawi posited that the pandemic impacts would be much more important in urban areas, in large part due to reliance on market-mediated incomes such as services and wage labor (e.g., Furbush et al., 2021). Given the strong maize harvest in 2020, some were less concerned about the rural areas. Nearly 90% of the rural households in our sample, however, reported impacts of the pandemic on incomes (Fig. 9.9). The source of this income loss is heterogeneous; for many, it was linked to closure of places of business, whereas for others it directly affected farming activities through labor unavailability and the high cost of inputs.

In Madagascar, the pandemic and response coincided with unanticipated drought conditions in late 2019 (a long dry spell following

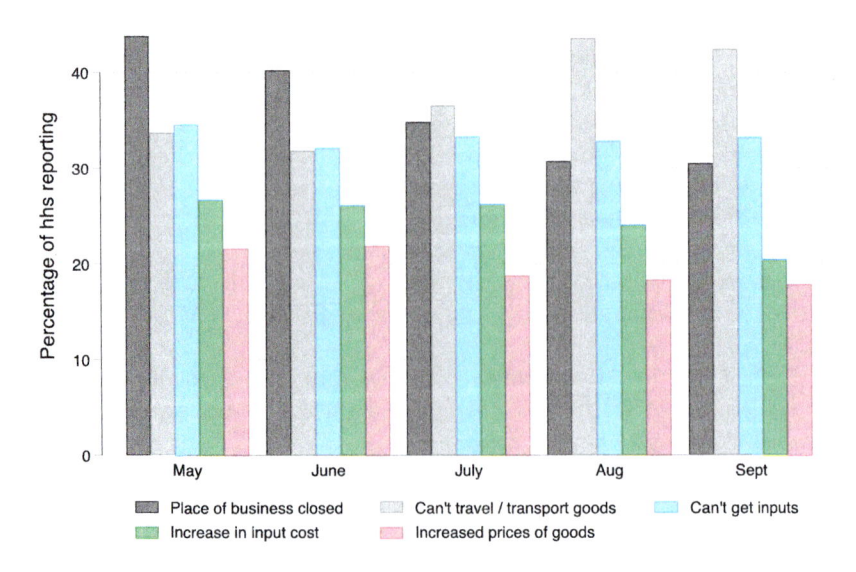

**Fig. 9.9** Reports of reasons for lost income among those reporting losses, May 2020–September 2020, Malawi. Of responding households, 87–89% reported lost income in every period, the vast majority (77–80%) a male income earner

initial rain), that nearly completely destroyed the anticipated March–April harvest. Agencies scrambled to recalibrate aid decisions and make food available, even as the pandemic unfolded and restrictions were put in place. The poor harvest had a clear and direct effect on households' food availability, as well as on income for those selling post-harvest, that would have been at that stage unaffected by the pandemic. All the same, households all reported a combination of negative impacts of the pandemic and policy response on their income, in parallel with upward impacts on expenditures. Some of these were due to closure of businesses, particularly early on, and some to unavailability of inputs, later and into the 2020 planting season. The greatest singular impact, however, was on prices of basic necessities (due essentially to border closures and other availability shocks).

In Kenya, initial movement restrictions receded in mid-2020 to allow for reopening of markets, greater movement, and reenergizing of the economy. While households (84%) in our study initially reported experiencing income reductions due to fear of COVID-19, the proportion of households experiencing income reductions steadily declined (Fig. 9.10) with a brief, but relatively small, increase again in November (36% of households) in response to the second wave of COVID-19 cases and deaths, during which the pandemic was believed to be spreading to a larger degree within rural areas. Notably, the third wave of infections to hit Kenya in 2021 saw these rates increase again to 25% in May 2021, but had an even more diminished impact. Through August 2020, over 85% of fishing households also reported reduced fish catch compared to the previous year. As markets reopened, however, fish catch improved starting in September 2020. Catch translated to improved incomes because demand for fish in Kenya's domestic markets remained strong, bolstered in part by declines in availability of imported tilapia from China. Underscoring this demand, fish farms within Kenya's young aquaculture sector even reported growth in 2020 (Benhamo, 2021).

## Meso-Level Mechanism 4: Price Shocks

Sudden and sharp food price changes can be triggered by processes that are local to global in nature—including regional growing conditions, disasters, epidemics or pandemics, conflict, and other geopolitical events—with institutions playing an important mediating role (Dillon & Barrett, 2016; Headey & Fan, 2008). The poor are particularly vulnerable

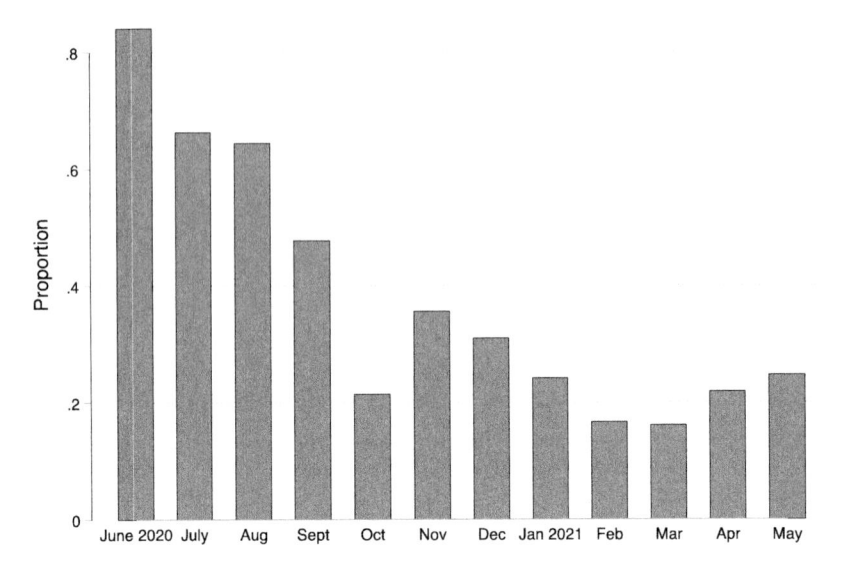

**Fig. 9.10** Household reports of working or earning less to avoid COVID-19 infection, June 2020–March 2021, Kenya

to food price shocks because they spend a large share of their income on food; most smallholder farmers and farm workers in eastern and southern Africa buy more food than they sell (Barrett, 2008). Consistent with previous global food price shocks and the importance of local policy and institutions in mediating their local effects, the COVID-19 pandemic's effects on food prices have been heterogeneous; prices fell in some places and rose in others (Béné et al., 2021).[9]

In rural areas of LMICs, we hypothesize that the nature of price shocks and their impacts on food security during the COVID-19 pandemic will vary according to local context. The direction and magnitude of price shocks will depend upon the extent to which transport was disrupted as well as the resilience of local markets to these disruptions. Unlike in urban

---

[9] The strong recovery in global agricultural commodity prices since May 2020 has little to do with the pandemic. Rather, it seems attributable more to depreciation in the US dollar, weather-related harvest shortfalls in several major export countries, relaxation in US-China trade tensions, and economic recovery in China, not least of which from the African swine fever outbreak that decimated its hog population in 2019–2020, leading to a collapse in global feed grains and oilseeds demand.

areas that uniformly import food, the impacts of price shocks will depend on whether a community is a net exporter or importer of staple foods.

### *Findings from Our Sites*

Our sites reflect contrasts among rural African food markets and underscore the need to unpack the meso-level mechanisms behind observed food price shocks. While the harvest in Malawi was good and maize price followed fairly normal seasonal trends, the overland border closures led to increases in prices of other products of primary necessity, such as cooking oil, sugar, and soap. Roughly 20% of households reported that these goods increased in price, particularly from May through June, and these price increases were correspondingly reported as a major source of the (real) income shock for roughly 20% of households. For this net food exporting region, transport disruptions during a good harvest year resulted in a drop in crop prices, hurting the incomes of many net seller farmers. In the COVID-19 period, however, the greater number of net buyers most likely could not take advantage of maize prices during a favorable harvest because of rising prices of other, imported necessities.

By contrast, in southern Madagascar, transit restrictions combined with the poor harvest led to abnormally high food prices as the year wore on. These price increases affected all food items, but especially imported rice, the most commonly consumed food (Fig. 9.11). Meanwhile, livestock prices fell in response to distress sales, pinching households' income on both sides, as sales of small livestock are a common coping strategy when households need to purchase food. Prices began to normalize with the second harvest and greater transit opening in August–September 2020, but then began to increase again into October, earlier than in typical years (as is evident in the right-hand panel of the figure). While one cannot discern the cause of these increases, they were widely understood to be exacerbated by the transit disruption.

## DISCUSSION AND CONCLUSION

We view COVID-19 through the lens of the mechanisms that mediate the impacts of large, covariate shocks on individual households and communities. This highlights how the pandemic was a major new shock, but impacted households in our study areas through recurring processes of structural deprivation activated by the myriad shocks and stressors faced

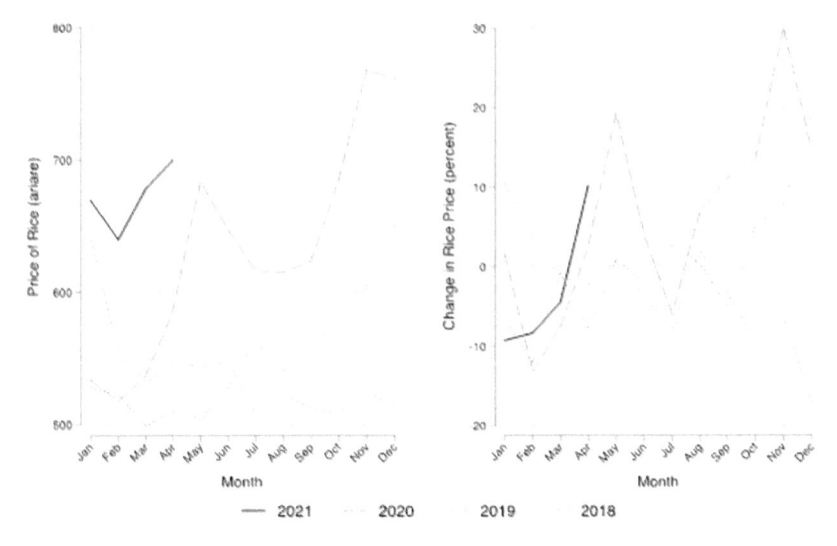

**Fig. 9.11** Month-to-month changes in imported rice price, comparing years, Madagascar

in low-income, rural communities (Béné et al., 2016). We build on pre-pandemic resilience theory and evidence to outline a conceptual framework for examining the linkages between households in low-income, rural regions and local, regional, and international food systems during the COVID-19 pandemic. We then apply that framework to empirical data from high-frequency monthly surveys of rural households in Malawi, Madagascar, and Kenya. We see the COVID-19 pandemic as the latest major covariate shock that lays bare systemic vulnerabilities that burden many rural populations situated within complex food systems in the low-income world. Rural households are exposed to food system shocks not only in their roles as food producers, but equally as food consumers or as workers within the broader agri-food value chain. Individual, collective, and behavioral responses to both shocks and responding policies too often aggravate underlying structural problems that compromise food security among a large population not directly affected by the shock.

Our findings underscore how the systematic mechanisms that mediate a large covariate shock can dramatically amplify the reach and magnitude of its negative impacts. This is consistent with important lessons from the pre-COVID-19 resilience literature. The detrimental development consequences of armed conflict on poverty, food security, and human health are well established (Gates et al., 2012; Rockmore, 2017, 2020). Although the impacts of the direct experience of violence (e.g., abduction, injury, or death of an immediate family member) are more severe at the individual level, relatively few people are directly impacted by violence while conflict disrupts movement and behavior broadly, also with adverse consequences, albeit of lower magnitude. The end result is that the aggregate impacts of violence on development indicators such as per capita expenditures or food security status come more through the indirect, systemic effects than due to direct experience of violence (Rockmore, 2017, 2020). Natural hazards and disease outbreaks have likewise been shown to impact food security and other aspects of well-being through multiple channels (Adger, Hughes, et al., 2005; Devereux, 2007; Hallegatte et al., 2017; Stoop et al., 2021).

The framework and analysis in this chapter focus on how macro-level policies and meso-level mechanisms matter for the resilience of households' food security to a large covariate shock. In the case of the COVID-19 pandemic, our analysis illustrates how for many households these intermediate pathways have dominated their experience of the shock. We are not the first to observe the multi-scalar nature of the mechanisms that reinforce vulnerability, poverty, and food insecurity, a theme that emerges in several strands of literature including, for example, work on fractal poverty traps (Barrett & Swallow, 2006; Radosavljevic et al., 2021), system dynamics (e.g., Reyers et al., 2018), adaptation to climate change (e.g., Adger, Arnell, et al., 2005), and food security resilience (Smith & Frankenberger, 2018; Upton et al., 2016; Vaitla et al., 2020). Yet much of the empirical development resilience literature to date focuses on the importance of individual or household level characteristics or "capacities" for coping with adverse shocks and stressors (for a review, see Barrett, Ghezzi-Kopel, et al., 2021). This chapter highlights why this evidence base is insufficient. A household's ability to draw down savings, liquidate assets, or borrow to remain food secure during a price shock is desirable. But better yet, timely policies and well-functioning markets would ensure that such price shocks do not arise and provide a robust safety net for vulnerable households.

In developing timely policies in response to covariate shocks, governments are tasked with the formidable challenge of redressing and rebalancing exposure to myriad shocks in real time. Such efforts may involve complex trade-offs in not only navigating among different shock exposures (e.g., health shocks, livelihood shocks), but shaping who is exposed to which shocks (e.g., health shocks were initially most intense in urban areas). Unforeseen consequences, or, conversely, a bit of good luck can also play a big role. For example, nations with similar policies have faced or avoided, thus far, COVID-19 virus variants that alter the burden of covariate shocks. In the early phase of the pandemic, our data indicate health shocks have not been the predominant risk faced by poor, rural communities in Malawi, Madagascar, and Kenya.

Resilience is a forward-looking concept about the stochastic dynamics of well-being in the presence of shocks or stressors. Food security is one of the most basic manifestations of well-being and is absent or at risk for far too many households in the rural communities we study, and many like them throughout the Global South. In the case of a major covariate shock like the COVID-19 pandemic, the food systems within which they operate suffer a range of shocks that affect them in their multiple roles as food consumers, producers, and workers.

Food availability shocks due directly to the disease have been negligible. Rather, to date, food availability—i.e., supply—disruptions arose mainly due to mobility restrictions and associated transport bottlenecks. Those supply shocks can lead to sharp and sudden relative price swings that impact food access, which has been more variable over the course of the pandemic. Downward price adjustments have hurt net sellers, e.g., of maize in Malawi or chickens in southern Madagascar. Meanwhile, upward price adjustments have made net buyers worse off, e.g., of most processed foods for these rural households. And relative price changes have induced consumers to substitute foods in their diets. Meanwhile, the same meso-level drivers that have disrupted the flow of food supplies have caused employment and income losses that have too often necessitated a rise in adverse household coping strategies. While social protection programs have scaled up dramatically in these and other low- and lower middle-income countries over the course of the pandemic (Gentilini et al., 2021), to date there has been scant evidence of expanded food or livelihood assistance in the rural communities we study.

Ultimately, we see food insecurity indicators fluctuating significantly over time in these communities. Things have not gotten uniformly worse. Indeed, in some places and for some people, things seem to be somewhat better, at least temporarily, for example thanks to bumper maize harvests in southern Malawi or reduced fish imports that increased domestic demand in Kenya. This merely underscores how the pandemic shock is not the only—or even always the most salient—covariate shock impacting poor rural populations. The many different shocks and stressors they face all get mixed together in the complex food systems within which they operate. The meso-level mechanisms of those systems ultimately regulate food availability, access, utilization, and stability for these rural households. Multiple adverse shocks hitting simultaneously—as in southern Madagascar—can cause mass hardship.

The upshot is that governments, NGOs, and donor agencies need to monitor both households and system-level indicators, watching for any of a wide range of shocks or new stressors that might disrupt the delicate equilibria that commonly regulate the food security status of these populations. Shoring up labor markets, transport systems, safety nets, and other meso-level mechanisms that mediate between covariate shocks and households' food security status is crucial not just to pandemic response but to building household- and systems-level resilience more broadly in the varied systems of rural Africa. The issue is not so much resilience to a specific shock or stressor—pandemic, drought, floods, or something else—as it is resilience to the many perils faced in everyday existence. And there are no one-size-fits-all solutions. Context matters enormously, as manifest in the markedly different experiences of the rural communities we study in Malawi, Madagascar, and Kenya. They may all be poor rural African communities facing the same global pandemic at the same time. But, as we have shown, the different ways in which meso-level mechanisms have mediated the common, global disease and macroeconomic shock have manifested in different, time-varying food security profiles in these communities.

## REFERENCES

Abila, R. (2003). *Fish trade and food security: Are they reconcilable on Lake Victoria?* (FAO Fisheries Report No. 708). Expert Consultation on International Fish Trade and Food Security.

Adger, W. N., Arnell, N. W., & Tompkins, E. L. (2005). Successful adaptation to climate change across scales. *Global Environmental Change, 15*(2), 77–86. https://doi.org/10.1016/j.gloenvcha.2004.12.005

Adger, W. N., Hughes, T. P., Folke, C., Carpenter, S. R., & Rockström, J. (2005). Social-ecological resilience to coastal disasters. *Science, 309*(5737), 1036–1039. https://doi.org/10.1126/science.1112122

Adjognon, G. S., Bloem, J. R., & Sanoh, A. (2021). The coronavirus pandemic and food security: Evidence from Mali. *Food Policy, 101*, 102050. https://doi.org/10.1016/j.foodpol.2021.102050

Aggarwal, S., Jeong, D., Kumar, N., Park, D. S., Robinson, J., & Spearot, A. (2021). *Did Covid-19 market disruptions disrupt food security? Evidence from households in rural Liberia and Malawi* (Unpublished Working Paper).

Ahmed, F., Islam, A., Pakrashi, D., Rahman, T., & Siddique, A. (2021). Determinants and dynamics of food insecurity during COVID-19 in rural Bangladesh. *Food Policy, 101*, 102066. https://doi.org/10.1016/j.foodpol.2021.102066

Amare, M., Abay, K. A., Tiberti, L., & Chamberlin, J. (2021). COVID-19 and food security: Panel data evidence from Nigeria. *Food Policy, 101*, 102099. https://doi.org/10.1016/j.foodpol.2021.102099

Ansah, I. G. K., Gardebroek, C., & Ihle, R. (2019). Resilience and household food security: A review of concepts, methodological approaches and empirical evidence. *Food Security, 11*(6), 1187–1203. https://doi.org/10.1007/s12571-019-00968-1

Aura, C. M., Nyamweya, C. S., Odoli, C. O., Owiti, H., Njiru, J. M., Otuo, P. W., Waithaka, E., & Malala, J. (2020). Consequences of calamities and their management: The case of COVID-19 pandemic and flooding on inland capture fisheries in Kenya. *Journal of Great Lakes Research, 46*(6), 1767–1775.

Barrett, C. B. (2005). Rural poverty dynamics: Development policy implications. *Agricultural Economics, 32*, 45–60. https://doi.org/10.1111/j.0169-5150.2004.00013.x

Barrett, C. B. (2008). Smallholder market participation: Concepts and evidence from eastern and southern Africa. *Food Policy, 33*(4), 299–317. https://doi.org/10.1016/j.foodpol.2007.10.005

Barrett, C. B., Benton, T. G., Fanzo, J., Herrero, M., Nelson, R., Bageant, E., Buckler, E., Cooper, K. A., Culotta, I., Fan, S., Gandhi, R., James, S., Kahn, M., Lawson-Lartego, L., Liu, J., Marshall, Q., Mason-D'Croz, D., Mathys, A., Mathys, C., … Wood, S. (2021). *Socio-technical innovation bundles for agri-food systems transformation*. Palgrave Macmillan.

Barrett, C. B., & Constas, M. A. (2014). Toward a theory of resilience for international development applications. *Proceedings of the National Academy of Sciences of the United States of America, 111*(40), 14625–14630. https://doi.org/10.1073/pnas.1320880111

Barrett, C. B., Ghezzi-Kopel, K., Hoddinott, J., Homami, N., Tennant, E., Upton, J., & Wu, T. (2021). A scoping review of the development resilience literature: Theory methods and evidence. *World Development, 146*, 105612.

Barrett, C. B., & Swallow, B. M. (2006). Fractal poverty traps. *World Development, 34*(1), 1–15. https://doi.org/10.1016/j.worlddev.2005.06.008

Béné, C. (2020). Resilience of local food systems and links to food security—A review of some important concepts in the context of COVID-19 and other shocks. *Food Security, 12*(4), 805–822. https://doi.org/10.1007/s12571-020-01076-1

Béné, C., Al-Hassan, R. M., Amarasinghe, O., Fong, P., Ocran, J., Onumah, E., Ratuniata, R., Tuyen, T. V., McGregor, J. A., & Mills, D. J. (2016). Is resilience socially constructed? Empirical evidence from Fiji, Ghana, Sri Lanka, and Vietnam. *Global Environmental Change, 38*, 153–170. https://doi.org/10.1016/j.gloenvcha.2016.03.005

Ballard, T., Coates, J., Swindale, A., & Deitchler, M. (2011). Household hunger scale: indicator definition and measurement guide. *Washington, DC: Food and nutrition technical assistance II project, FHI, 360,* 23.

Béné, C., Bakker, D., Rodriguez, M. C., Even, B., Melo, J., & Sonneveld, A. (2021). *Impacts of COVID-19 on people's food security: Foundations for a more resilient food system*. International Food Policy Research Institute.

Benhamo, S. (2021). *The Big Fish Seminar: Impacts of COVID-19 on Seafood Value Chains*. University of Stirling. Online Webinar: https://www.youtube.com/watch?v=uIa7R26wQgU&t=1085s

Coates, J., Swindale, A., & Bilinsky, P. (2007). Household Food Insecurity Access Scale (HFIAS) for measurement of food access: Indicator guide: Version 3.

Constas, M., Frankenberger, T., & Hoddinott, J. (2014). Resilience measurement principles: Toward an agenda for measurement design. *Food Security Information Network, Resilience Measurement Technical Working Group, Technical Series, 1.*

Devereux, S. (2007). The impact of droughts and floods on food security and policy options to alleviate negative effects. *Agricultural Economics, 37*(s1), 47–58. https://doi.org/10.1111/j.1574-0862.2007.00234.x

de Waal, A., & Whiteside, A. (2003). New variant famine: AIDS and food crisis in southern Africa. *The Lancet, 362*(9391), 1234–1237. https://doi.org/10.1016/S0140-6736(03)14548-5

Dillon, B. M., & Barrett, C. B. (2016). Global oil prices and local food prices: Evidence from east Africa. *American Journal of Agricultural Economics, 98*(1), 154–171. https://doi.org/10.1093/ajae/aav040

Dong, E., Du, H., & Gardner, L. (2020). An interactive web-based dashboard to track COVID-19 in real time. *The Lancet Infectious Diseases, 20*(5), 533–534.

Egger, D., Miguel, E., Warren, S. S., Shenoy, A., Collins, E., Karlan, D., Parkerson, D., Mobarak, A. M., Fink, G., Udry, C., Walker, M., Haushofer, J., Larreboure, M., Athey, S., Lopez-Pena, P., Benhachmi, S., Humphreys, M., Lowe, L., Meriggi, N. F., ... Vernot, C. (2021). Falling living standards during the COVID-19 crisis: Quantitative evidence from nine developing countries. *Science Advances, 7*(6), eabe0997. https://doi.org/10.1126/sciadv.abe0997

Elston, J. W. T., Cartwright, C., Ndumbi, P., & Wright, J. (2017). The health impact of the 2014–15 Ebola outbreak. *Public Health, 143*, 60–70. https://doi.org/10.1016/j.puhe.2016.10.020

FAO. (1996). *World Food Summit: Rome Declaration on World Food Security and World Food Summit Plan of Action*. Rome.

FAO. (2016). *State of the world's fisheries and aquaculture*. Food and Agriculture Organization of the United Nations.

FAO, IFAD, UNICEF, WFP and WHO. (2021). *The State of Food Security and Nutrition in the World 2021. Transforming food systems for food security, improved nutrition and affordable healthy diets for all*. FAO. https://doi.org/10.4060/cb4474en

Fiorella, K. J., Bageant, E. R., Mojica, L., Obuya, J. A., Ochieng, J., Olela, P., Otuo, P. W., Onyango, H. O., Aura, C. M., & Okronipa, H. (2021). Small-scale fishing households facing COVID-19: The case of Lake Victoria, Kenya. *Fisheries Research, 237*, 105856. https://doi.org/10.1016/j.fishres.2020.105856

Fisher, M., & Lewin, P. A. (2013). Household, community, and policy determinants of food insecurity in rural Malawi. *Development Southern Africa, 30*(4–05), 451–467. https://doi.org/10.1080/0376835X.2013.830966

Folke, C. (2016). Resilience (Republished). *Ecology and Society, 21*(4). https://doi.org/10.5751/ES-09088-210444

Furbush, A., Josephson, A., Kilic, T., & Michler, J. D. (2021). *The evolving socioeconomic impacts of COVID-19 in four African countries* (Policy Research Working Paper No. 9556). World Bank. https://openknowledge.worldbank.org/handle/10986/35205

Gashu, D., Nalivata, P. C., Amede, T., Ander, E. L., Bailey, E. H., Botoman, L., Chagumaira, C., Gameda, S., Haefele, S. M., Hailu, K., Joy, E. J. M., Kalimbira, A. A., Kumssa, D. B., Lark, R. M., Ligowe, I. S., McGrath, S. P., Milne, A. E., Mossa, A. W., Munthali, M., … Broadley, M. R. (2021). The nutritional quality of cereals varies geospatially in Ethiopia and Malawi. *Nature, 594*(7861), 71–76. https://doi.org/10.1038/s41586-021-03559-3

Gates, S., Hegre, H., Nygård, H. M., & Strand, H. (2012). Development consequences of armed conflict. *World Development, 40*(9), 1713–1722. https://doi.org/10.1016/j.worlddev.2012.04.031

Gentilini, U., Almenfi, M., Dale, P., Palacios, R., Natarajan, H., Rabadan, G. A. G., Okamura, Y., Blomquist, J., Abels, M., Demarco, G., & Santo, I. (2021). *Social protection and jobs responses to COVID-19: A real-time review of country measures "Living paper" version 15*. https://documents1.worldbank.org/curated/en/281531621024684216/pdf/Social-Protection-and-Jobs-Responses-to-COVID-19-A-Real-Time-Review-of-Country-Measures-May-14-2021.pdf

Gómez, M. I., Barrett, C. B., Raney, T., Pinstrup-Andersen, P., Meerman, J., Croppenstedt, A., Carisma, B., & Thompson, B. (2013). Post-green revolution food systems and the triple burden of malnutrition. *Food Policy, 42*, 129–138. https://doi.org/10.1016/j.foodpol.2013.06.009

Government of Madagascar and Integrated Food Security Phase Classification (IPC). (2020). *Madagascar: IPC Acute Food Insecurity and Acute Malnutrition Analysis, October 2020–April 2021, Issued December 2020—Madagascar.* Retrieved November 15, 2021, from https://reliefweb.int/report/madagascar/madagascar-ipc-acute-food-insecurity-and-acute-malnutrition-analysis-october-2020

Hale, T., Angrist, N., Goldszmidt, R., Kira, B., Petherick, A., Phillips, T., Webster, S., Cameron-Blake, E., Hallas, L., Majumdar, S., & Tatlow, H. (2021). A global panel database of pandemic policies (Oxford COVID-19 Government Response Tracker). *Nature Human Behaviour, 5*(4), 529–538.

Hallegatte, S., Rentschler, J., & Rozenberg, J. (2019). *Lifelines: The resilient infrastructure opportunity.* The World Bank. https://doi.org/10.1596/978-1-4648-1430-3

Hallegatte, S., Vogt-Schilb, A., Bangalore, M., & Rozenberg, J. (2017). *Building the resilience of the poor in the face of natural disasters.* The World Bank. https://doi.org/10.1596/978-1-4648-1003-9

Headey, D., & Barrett, C. B. (2015). Measuring development resilience in the world's poorest countries. *Proceedings of the National Academy of Sciences, 112*(37), 11423–11425. https://doi.org/10.1073/pnas.1512215112

Headey, D., & Fan, S. (2008). Anatomy of a crisis: The causes and consequences of surging food prices. *Agricultural Economics, 39*(s1), 375–391. https://doi.org/10.1111/j.1574-0862.2008.00345.x

Hull, V., & Liu, J. (2018). Telecoupling: A new frontier for global sustainability. *Ecology and Society, 23*(4). https://doi.org/10.5751/ES-10494-230441

Integrated Food Security Phase Classification (IPC). (2021). *Malawi: Acute Food Security Situation.* Retrieved November 18, 2021, from http://www.ipcinfo.org/ipc-country-analysis/details-map/en/c/1155085/?iso3=MWI

Jain, R., & Dupas, P. (2020). The effects of India's COVID-19 lockdown on critical non-COVID health care and outcomes. *MedRxiv*, 2020.09.19.20196915. https://doi.org/10.1101/2020.09.19.20196915

Johns Hopkins University. (2021). *COVID-19 Dashboard by the Center for Systems Science and Engineering (CSSE) at Johns Hopkins University (JHU).* Retrieved September 6, 2021, from https://coronavirus.jhu.edu/map.html

Kenya National Bureau of Statistics (KNBS). (2018). *Basic Report on Well-Being in Kenya Based on the 2015/16 Kenya Integrated Household Budget Survey.*

Knippenberg, E., Jensen, N., & Constas, C. (2019). Quantifying household resilience with high frequency data: Temporal dynamics and methodological options. *World Development, 121*, 1–15. https://doi.org/10.1016/j.worlddev.2019.04.010

Knippenberg, E., Michelson, H., Lentz, E., & Lallement, T. (2020). *Early warning; better data, better algorithms, improved food insecurity predictions using machine learning and high-frequency data from Malawi* (Working Paper). https://www.semanticscholar.org/paper/Early-Warning%3B-Better-Data%2C-Better-Algorithms-Food-Knippenberg-Michelson/21d1f6e92c413b5140d22da52f0583c5be69b052

Kodish, S. R., Bio, F., Oemcke, R., Conteh, J., Beauliere, J. M., Pyne-Bailey, S., Rohner, F., Ngnie-Teta, I., Jalloh, M. B., & Wirth, J. P. (2019). A qualitative study to understand how Ebola Virus Disease affected nutrition in Sierra Leone—A food value-chain framework for improving future response strategies. *PLoS Neglected Tropical Diseases, 13*(9), 1–19. https://doi.org/10.1371/journal.pntd.0007645

Liu, J., Dietz, T., Carpenter, S. R., Folke, C., Alberti, M., Redman, C. L., Schneider, S. H., Ostrom, E., Pell, A. N., Lubchenco, J., & Taylor, W. W. (2007). Coupled human and natural systems. *Ambio, 36*(8), 639–649. https://doi.org/10.1007/s13280-020-01488-5

Manyena, S. B. (2006). The concept of resilience revisited. *Disasters, 30*(4), 434–450. https://doi.org/10.1111/j.0361-3666.2006.00331.x

Maxwell, D., Caldwell, R., & Langworthy, M. (2008). Measuring food insecurity: Can an indicator based on localized coping behaviors be used to compare across contexts? *Food Policy, 33*(6), 533–540.

Pringle, R. M. (2005). The origins of the Nile Perch. *BioScience, 55*(9), 780–787. https://doi.org/10.1641/0006-3568(2005)055[0780:TOOTNP]2.0.CO;2

Radosavljevic, S., Haider, L. J., Lade, S. J., & Schlüter, M. (2021). Implications of poverty traps across levels. *World Development, 144*, 105437. https://doi.org/10.1016/j.worlddev.2021.105437

Reyers, B., Folke, C., Moore, M.-L., Biggs, R., & Galaz, V. (2018). Social-ecological systems insights for navigating the dynamics of the Anthropocene. *Annual Review of Environment and Resources, 43*, 267–289. https://doi.org/10.1146/annurev-environ-110615-085349

Roberton, T., Carter, E. D., Chou, V. B., Stegmuller, A. R., Jackson, B. D., Tam, Y., Sawadogo-Lewis, T., & Walker, N. (2020). Early estimates of the indirect effects of the COVID-19 pandemic on maternal and child mortality in low-income and middle-income countries: A modelling study. *The Lancet Global Health, 8*(7), e901–e908. https://doi.org/10.1016/S2214-109X(20)30229-1

Rockmore, M. (2017). The cost of fear: The welfare effect of the risk of violence in Northern Uganda. *World Bank Economic Review, 31*(3), 650–669. https://doi.org/10.1093/wber/lhw025

Rockmore, M. (2020). Conflict-risk and agricultural portfolios: Evidence from Northern Uganda. *Journal of Development Studies, 56*(10), 1856–1876. https://doi.org/10.1080/00220388.2019.1703953

Seeley, J. A., & Allison, E. H. (2005). HIV/AIDS in fishing communities: Challenges to delivering antiretroviral therapy to vulnerable groups. *AIDS Care, 17*(6), 688–697. https://doi.org/10.1080/09540120412331336698

Smith, L. C., & Frankenberger, T. R. (2018). Does resilience capacity reduce the negative impact of shocks on household food security? Evidence from the 2014 floods in northern Bangladesh. *World Development, 102*, 358–376. https://doi.org/10.1016/j.worlddev.2017.07.003

Stoop, N., Desbureaux, S., Kaota, A., Lunanga, E., & Verpoorten, M. (2021). Covid-19 vs. Ebola: Impact on households and small businesses in North Kivu, Democratic Republic of Congo. *World Development, 140*, 105352. https://doi.org/10.1016/j.worlddev.2020.105352

Swindale, A., & Bilinsky, P. (2006). Household Dietary Diversity Score (HDDS) for measurement of household food access: Indicator guide. *Food and Nutrition Technical Assistance Project, Academy for Educational Development*.

Upton, J. B., Cissé, J. D., & Barrett, C. B. (2016). Food security as resilience: Reconciling definition and measurement. *Agricultural Economics, 47*(s1), 135–147. https://doi.org/10.1111/agec.12305

Upton, J. B., & Knippenberg, E. (2019). *Measurement Indicators for Resilience Analysis (MIRA)—Phase III: Malawi (UBALE) & Madagascar (HAVELO)*. Report for Catholic Relief Services.

Upton, J. B., Tennant, E. J., Michelson, H., & Lentz, E. C. (2021). *COVID-19 and food insecurity in rural Africa: Evidence from Malawi* (Manuscript submitted for publication).

Vaitla, B., Cissé, J. D., Upton, J., Tesfay, G., Abadi, N., & Maxwell, D. (2020). How the choice of food security indicators affects the assessment of resilience—An example from Northern Ethiopia. *Food Security, 12*(1), 137–150. https://doi.org/10.1007/s12571-019-00989-w

Vermeulen, S. J., Campbell, B. M., & Ingram, J. S. (2012). Climate change and food systems. *Annual Review of Environment and Resources, 37*, 195–222. https://doi.org/10.1146/annurev-environ-020411-130608

Willett, W., Rockström, J., Loken, B., Springmann, M., Lang, T., Vermeulen, S., Garnett, T., Tilman, D., DeClerck, F., Wood, A., & Jonell, M. (2019). Food in the Anthropocene: The EAT–Lancet Commission on healthy diets from sustainable food systems. *The Lancet, 393*(10170), 447–492. https://doi.org/10.1016/S0140-6736(18)31788-4

Wiesmann, D., Bassett, L., Benson, T., & Hoddinott, J. (2009). Validation of the world food programmes food consumption score and alternative indicators of household food security. *Intl Food Policy Res Inst.*

**Open Access** This chapter is licensed under the terms of the Creative Commons Attribution 4.0 International License (http://creativecommons.org/licenses/by/4.0/), which permits use, sharing, adaptation, distribution and reproduction in any medium or format, as long as you give appropriate credit to the original author(s) and the source, provide a link to the Creative Commons license and indicate if changes were made.

The images or other third party material in this chapter are included in the chapter's Creative Commons license, unless indicated otherwise in a credit line to the material. If material is not included in the chapter's Creative Commons license and your intended use is not permitted by statutory regulation or exceeds the permitted use, you will need to obtain permission directly from the copyright holder.

CHAPTER 10

# Place-Based Approaches to Food System Resilience: Emerging Trends and Lessons from South Africa

*Bruno Losch and Julian May*

## INTRODUCTION

The emergence of a globalized food system, characterized by the industrialization of food, long-distance food supply chains and growing economic

B. Losch (✉)
Centre de Coopération Internationale en Recherche Agronomique pour le Développement, ART-Dev, University of Montpellier, Montpellier, France
e-mail: bruno.losch@cirad.fr

B. Losch · J. May
DSI-NRF Centre of Excellence in Food Security, University of the Western Cape, Cape Town, South Africa
e-mail: jmay@uwc.ac.za

J. May
UNESCO Chair in African Food Systems, University of the Western Cape, Cape Town, South Africa

© The Author(s) 2023
C. Béné and S. Devereux (eds.), *Resilience and Food Security in a Food Systems Context*, Palgrave Studies in Agricultural Economics and Food Policy, https://doi.org/10.1007/978-3-031-23535-1_10

concentration, has been accompanied by a trend towards metropolitization (Sassen, 2001; Tefft et al., 2017). Among the consequences of growing populations concentrated in large cities are a heavy carbon footprint, rising sanitary risks, unhealthy ultra-processed food, economic and social exclusion, and unbalanced power relations that jeopardize democratic food choices (Lang, 2003; Puissant & Lacour, 1999). A place-based approach can help to address sustainability issues and contribute to relocalizing food supply chains and inclusive food environments (La Trobe & Acott, 2000).

This chapter posits that a territorial perspective that addresses place-based problems and their solutions can better reflect the reality of spatial dynamics. It will highlight how this perspective can contribute towards the development of resilient food systems. The food system paradox in South Africa, in which sophisticated food policies stand in contrast to poor food system outcomes, and recent state and civil society's responses to food system shocks illustrate how territoriality emerges as a solution when actors seek to build resilience.

The chapter will first summarize the systems approach, noting that some systems, such as those concerning food, operate in places that have unique physical attributes and social networks. We then explain what a place-based approach to food system change contributes. Following Harvey (2001), we emphasize that places become territories when specific conditions and actions take place, and suggest that these actions may not be state-led. As evidence of this, we then describe the South African policy context, noting the movement from the country's state-led 2017 National Food and Nutrition Security Plan (NFNSP) to the civil society-led responses to COVID-19 in 2020–2021 and the formation of Community Action Networks (CANs). Focusing on the Western Cape Province, and drawing on a systematic review of policy documents and research reports, we argue that these attempts to deal with shocks to the food system may constitute an embryonic place-based approach. This case study provides an example of a process that can be transferred to other contexts in which food systems present an inconsistency of sufficient food and a dense policy framework but with poor system outcomes and fragile resilience.

## Food Systems, Place and Resilience

Systems comprise a set of interacting or interdependent elements that function together as collective units, thereby forming a larger whole that has properties greater than the sum of its component parts (Von Bertalanffy, 1968). Activities may be arranged in sub-systems, each with their own networks and dynamics, with multiple levels, actors and boundaries. Importantly, the geographic scale of systems governance is of relevance (McGinnis & Ostrom, 2014).

A system has the capacity to adapt, change or transform in response to internal or external stimuli (Baser & Morgan, 2008), and to generate both feedback and feedforward loops (Casti & Fath, 2008).[1] Adopting a systems approach reveals the trade-offs, synergies and unintended consequences of such change (Ostrom, 2000). Sustainability and resilience questions can then be identified, as well as leverage points for action. To be resilient, a system needs to maintain diversity, manage connectivity, respond to feedback, and have the capacities to self-organize and to learn. In doing so, a resilient system becomes polycentric: "*a governance system in which there are multiple interacting governing bodies with autonomy to make and enforce rules within a specific policy arena and geography*" (Schoon et al., 2015, p. 236).[2] For public policy, a systems approach can reveal problems of coordination and cooperation, as well as opportunities to break down barriers arising from segmented approaches where departments operate in silos (Sarapuu et al., 2014, pp. 263–264).

The production, distribution and consumption of food is one such complex adaptive system (Ingram, 2019). A food system comprises "*all the elements (environment, people, inputs, processes, infrastructures, institutions, etc.) and activities that relate to the production, processing, distribution, preparation and consumption of food, and the outputs of these activities, including socio-economic and environmental outcomes*" (HLPE, 2017, p. 11). These actors and activities are embedded in "*broader economic, societal and natural environments*" (FAO, 2018a, p. 1). Regulatory institutions within the food system provide the context and rules within which the food system outcomes are produced (Ingram, 2019). Constitutional rights, international declarations, policies, strategies and local government by-laws are examples of these rules.

---

[1] Feedback loops permit responses to system disturbance to modify the system and so maintain the system's health and function. Feedforward loops anticipate system disturbance, and allow for mitigating responses prior to the disturbance taking effect.

[2] The seven principles of building resilience are well summarised in Biggs et al. (2015).

Food systems have outcomes related to three goals essential for human existence and society resilience: food and nutrition security (FNS)[3]; livelihoods and economic inclusion; and environmental sustainability (OECD, 2021). This is presented as a 'triple challenge' since pursuing the goals can result in multipliers, trade-offs and externalities. Of significance for this chapter, addressing territorial balance has been identified as an additional goal (FAO, European Union and CIRAD, 2021).

All of these goals have place-based attributes. For example, food supply chains ensure both the availability and accessibility of FNS but with different spatial spans which may be short (low-input subsistence farmer to her/his family or smallholder supplying the local market) or long (agrochemical company to urban consumer via industrialized farms and multinational retail corporations). These have differing implications for nutrition, livelihoods, inclusion and sustainability (Coley et al., 2009; Morgan et al., 2019). Following the notion of the European Union (EU) Green Deal, the approach can be thought of as 'farm-in-place' to a 'fork-in-place', with food waste and loss occurring in each place as well as during the movement from place to place.

At this point, the polysemy of wording related to places should be addressed. Several words such as region, place, space and territory refer to geography, boundaries and spatial scales. There are no standard definitions and they are often used interchangeably. This is exacerbated by different meanings and usages between different languages (particularly English and other languages with Latin roots).

While spaces are neutral and a basic category of human life (Harvey, 2001), places are specific locations where things (from natural or human origins), living species and people are situated. As reminded by Relph (1985, p. 7), they are also *"an origin; it is where one knows others and is known to others; it is where one comes from and it is one's own."* The concept of territory can also relate to identity, usage and belonging, but is rather a social construct. This is the approach adopted in economic geography. A space becomes a territory when leaders organize it to optimize economic production (Harvey, 2001), when a coalition of actors share goals, and when networks of stakeholders mobilize local resources and dedicate them to a project.

---

[3] Food and nutrition security is now defined as possessing six dimensions: availability, access, utilisation, stability, agency and sustainability (HLPE, 2017).

Therefore, territories are spaces of coordination and contestation between actors, where local resources can be 'activated' through collective action and become an answer to a shared challenge, such as adaptation to globalization or to climate change (Campagne & Pecqueur, 2014; Storper, 1997). A territory takes into consideration multiple levels of spatial organization, nexus interactions and the roles they play. Central to the concept are the multiple boundaries or borders that are implied, how these are distinguished and how they are created, regulated and enforced. In this respect, the border of a territory is an institution rather than a line that simply demarcates a geographic location (Sassen, 2005).

Territories thus include: the network of actors, their strategies and their connection to other places; the linkages between dense urban centres, surrounding peripheral areas, burgeoning small and medium towns, and the rural hinterland; flows of food products along short or long-distance supply chains; and ecosystems dynamics related to natural processes and human activity (Blay-Palmer et al., 2018). A further implication is that community-based actions and local social networks are also place-bound.

Conceptual frameworks of the food system recognize that there are multiple embedded food systems, each with their own spatial boundaries (Borman et al., 2022). Firstly, the macro or global system is predominately capitalist in nature, in which land and capital are privately owned, and food is a commodity to be acquired via market transactions. This can also be thought as the food regime, built on historical contradictions that generate crises, transition and transformation (Friedmann & McMichael, 1989).

Then there are meso-systems that are particular to a country or region. For example, Lapping (2004) refers to an American food system that is characterized by globalization, consolidation and industrialization, which succeeds in feeding the population of the USA (NRC, 2015) but less in nourishing it. Conflicting views exist about an African food system with opposite reporting about hunger, fragmentation and inefficiency (Pinstrup-Andersen, 2010), or about progression towards positive transformations (Tschirley et al., 2014). The very notion of a single food system for a continent is highly problematic due to the heterogeneity of contexts related to natural environments, population densities, agrarian systems and sociocultural patterns. However, in the case of Africa, the conditions of its integration in the world economy has resulted in important commonalities. Examples are the shared history of colonization; the adverse terms under which most countries have since been included into the world food system; and the common purpose of pan-African integration and governance started with decolonialization and the creation of a continental organization (today the African Union) focusing on infrastructural investment, regional trade collaboration and regulation reforms illustrated today by Agenda 2063, the 'One Africa Voice' initiative or the Comprehensive Africa Agriculture Development Programme (ECA et al., 2021; Forum for Agricultural Research in Africa [FARA], 2021). Applied more usefully at the country level, such meso-analysis is able to reveal strong patterns, like a dichotomous food system in South Africa with concentrated and highly skewed access to land, capital and technology, in which the corporate sector exerts considerable power in guiding the system (Greenberg, 2017; Pereira & Drimie, 2016).

Micro-food systems have been identified as those operating at the level of the neighbourhood, workplace or household (Finney et al., 2012). This focuses attention on the food environment and the capabilities of the household when managing its food security, and also considers intra-household distribution and power (Harris-Fry et al., 2017; Westbury et al., 2021). Individuals negotiate access to food within the micro-food system, and ultimately, their digestive system transforms that food into energy, nutrients and waste—which could be named the nano-food system.

Adopting a territorial approach can help articulate these components of the food system, while addressing the complexity of processes related to food in a specific area. It emphasizes that food system actors don't only produce, distribute and consume food in markets, they also live in places where the potential, constraints and plausible futures of that place matter, and where answers to challenges can be identified through collective action. It highlights the multi-dimensional nature of the food system, the diversity of its actors, their different levels of action and the need for coordination (Blay-Palmer et al., 2018; Caron et al., 2017, 2018; Cistulli et al., 2014).

As such, the territorial approach facilitates food system resilience through adaptation to changes and risk management. Risks directly affect the stability of food system outcomes through price fluctuations, sudden changes in policies, changing social relationships, unstable governments and armed conflicts (Freshwater, 2015). Moreover, repeated shocks gradually erode household livelihoods, as well as the food system resilience of the communities and places in which households are located (Davies, 1996).

Connection and complementarity between markets facilitated by globalization have always been and continue to be major answers to risks related to climate or socioeconomic and political events. These risks can result in shortages and price increases with huge impacts on food systems (Godfray et al., 2010). However, the adoption of a territorial perspective, with greater attention to the characteristics of places through local lenses, helps to mitigate the negative impacts of globalization (unsustainability of long-distance supply chains, growing economic and phytosanitary risks, exclusion and adverse inclusion due to concentration of wealth and power). It also helps to highlight within-country food system disparities related to unequal socioeconomic development (Giordano et al., 2019) often characterizing the most rural provinces or districts, ethnic communities or indigenous groups, as described by Cistulli et al. (2014) in South Africa, Ghana, Vietnam or the highlands of Latin America.

Nesting food systems governance at different geographical scales provides a link to the well-established literature on risk and vulnerability and the more recent focus on food system resilience (Béné et al., 2016; Bullock et al., 2017; Dury et al., 2019, and the current book). Some events, such as injury or illnesses from non-communicable disease, are specific to individuals or households and can be thought of as 'idiosyncratic risk' (Barrett, 2011). Others, such as drought or infectious disease, simultaneously affect many households in a community or region. This 'covariate' risk refers to the extent to which individuals, communities or sub-groups, structures and places, and indeed systems are likely to be damaged or disrupted by such shocks. The impact of HIV and AIDS is a well-documented example (Beegle et al., 2008).

The COVID-19 pandemic represents such a covariate risk, and an analysis of responses to the pandemic can provide information about the resilience of food systems (Bidisha et al., 2021; Gupta et al., 2020). Notably, the relevance of a place-based approach in articulating different levels of interventions was confirmed during the lockdowns that COVID-19 prompted. In many countries, local authorities have borne the brunt of any early response to disasters and are central in the enforcement of regulations concerning the use of space, including those relating to social distancing such as access to public facilities and the enforcement of restricted trading hours (Wright, 2020). But local governments are also proved to be well positioned to identify and help to implement nutrition-sensitive interventions targeting the vulnerable groups (FAO, 2020).

In a review of documents published since the start of the COVID-19 pandemic, Béné et al. (2021) found that food security was primarily threatened through both covariate and idiosyncratic disruptions to physical and economic access to food, with the former caused by lockdowns restricting mobility, and the latter by the loss in employment and or income. They found no evidence of disruptions to food availability, and limited evidence on utilization dimensions of food security. They interpreted lockdown interventions as disruptors to the stability aspect of FNS.

Béné et al. (2021) also note that the apparent resilience of the food system as a whole often came at the expense of smaller actors in the food system, while benefitting larger actor consolidations. For example, lockdowns meant that informal food vendors were unable to trade, while supermarkets that were given 'essential service' status, made large profits during these periods (Kroll et al., 2021). This points to the importance of understanding the manner in which food system governance responds to disruptions and operates at the level of a place rather than in aggregate (Kusumasari et al., 2010).

## Experiences of Territorial Approaches to Food System Management

If there is a polysemy of wording related to places, there is also a diversity of experiences related to territorial approaches to food system management. These experiences adopt different perspectives, and it is important to consider the level of government, national or local, at which the policies or programmes have been designed.

National governments have long been, and continue to be, the main producers of public policies. Sectorial policies remain the backbone of governments' actions, with sectorial growth as a core objective pushing spatial issues in the background. Indeed, widening regional inequality was once seen as a stage in economic growth that would follow an inverted U progression as the benefits of agglomeration trickled down (Williamson, 1965). As a result, the adoption of spatially blind policies was considered as a prerequisite for economic development even if its cost was growing territorial inequalities at a first stage (Varga, 2017). Influential international organizations such as the World Bank have been advocating for this approach: the World Development Report 2009 on Reshaping Economic

Geography being an example (World Bank, 2009). However, this conception has been criticized (Barca et al., 2012; Rodríguez-Pose, 2010), and there is evidence of the negative consequences of the rise in territorial inequality (Lessmann & Seidel, 2017). Other voices, like the European Union, particularly concerned with regional asymmetries and cohesion, and the OECD have been advocating for another approach recognizing the importance of the territorial level as a way to tailor strategies towards addressing local conditions (Barca, 2009; OECD, 2009; TP4D, 2018). This discussion is also emerging in Africa where structural challenges call for a paradigm shift towards territorial development (AfDB, OECD, UNDP, 2015; Losch, 2016).

Territorial approaches to food systems have emerged in this context (Forster et al., 2021). Still, it is necessary to clarify what is meant by the reference to a territorial approach, because some appellations can be misleading and the sectorial perspective often strongly remains behind a formal positioning on the territorial approach, which could be named 'territorial-washing'. The OECD-FAO-UNCDF study titled *"Adopting a Territorial Approach to Food Security and Nutrition Policy"* (OECD/FAO/UNCDF, 2016) is provocative. It advocates for the adoption of a territorial approach, taking stock of the limitations of existing food security policies through a series of case studies in Cambodia, Colombia, Côte d'Ivoire, Morocco, Peru, and quicker assessments in Mali and Niger. The main issues identified relate to the priority given to agricultural interventions, the persistence of project-based approaches, the insufficient coordination between sectorial departments and levels of government, the weaknesses of decentralization and the lack of spatial differentiation (i.e. spatially blind policies), which prevents taking on board within-country inequalities and disparities. It also talks to the specificities of different contexts—for instance metropolitan regions, peri-urban and remote rural regions.

Based on this assessment and common issues identified across countries, in spite of significant differences, what is proposed as a territorial approach is to implement multi-sectorial integrated policies, to promote multi-level governance and coordination and the role of sub-national institutions, and to improve territorial information for a better regional targeting of interventions. These recommendations correspond before anything else to an objective of improving national policies by a 'territorialization' of priorities and planning, as illustrated by the 3 N Initiative in Niger (Nigériens Nourish Nigériens)[4] or by the Green Morocco Plan (Faysse, 2015). Although space-based, the 3 N Initiative is far from a coalition of local actors identifying common challenges and adopting a shared vision of objectives and priorities.

Another example to mention is the experience of development corridors, which is sometimes presented as a way to engage in a territorial approach. Joining physical transport infrastructure with ad hoc facilities and services and bringing in the private sector through public-private partnerships, corridors are supposed to stimulate territorial development with improved connection to markets and reduced costs (Kuhlmann et al., 2011). Again, these corridors, now supported in many African countries, are examples of a top-down approach. They are in effect reproducing the extractive schemes of the colonial period, facilitating the transport of resources from the hinterland to the coast—a framework which has profoundly shaped the spatial pattern of many African countries (AfDB, OECD, UNDP, 2015). As such, corridors can create 'tunnel effects' with adverse impacts on the surrounding areas (Fau, 2019; Scholvin, 2021). If they offer better market opportunities with reduced transport costs and can stimulate activities, their attraction tends to marginalize neighbouring places (similar to a vacuum effect), thereby increasing spatial asymmetries. They also tend to benefit the better-off, who can quickly reap the benefits of the new infrastructure and can pave the way for outsiders. The Southern Agricultural Growth Corridor of Tanzania (SAGCOT) is a good illustration. Supposed to support agriculture and small farmers, it has been managed by foreign enterprises and has excluded smallholders (Byiers & Rampa, 2013).

Contrary to these policies designed at the national level, initiatives developed by local authorities correspond more effectively to a territorial approach to development and food system management. They need to be put in perspective with processes of decentralization, which have spread worldwide at different paces since the 1980s, with differences depending on the characteristics of the state and the political regime (Beard et al., 2008; Faguet, 2014). Roles attributed to local governments are heterogeneous, but food and food systems' related issues are generally not in their mandates, even if local food system planning was identified as requiring

---

[4] http://www.initiative3n.ne/.

attention two decades ago (Buchan et al., 2015; Pothukuchi & Kaufman, 2000).

In most parts of Africa, state consolidation objectives have delayed effective decentralization. Where administrative decentralization has occurred, with the relocation of administrative functions and executive responsibilities to different levels of government, fiscal decentralization (i.e. transfer of revenue-generating power) has been more limited (Cabral, 2011; Conyers, 2007; Crooks, 2003; Smoke, 2003). As a result, the lack of local fiscal resources prevents effective and numerous interventions and food system governance is rarely on the agenda of local governments.

This situation explains why the entrance of local governments in the food policy space first occurred in major cities of the richer countries, where the existence of a larger fiscal basis offered enough room for manoeuvre for independent action from central governments and their budgets, and where food appears as a critical issue. Among the main reasons supporting this new status of food are the rising awareness of consumers (and voters) with regard to the importance of healthy food, and growing concerns about sustainability questions (e.g. type of agricultural practices and origin of the food supply and the related transport costs). The foregoing explains why initial interest focused on localized production and the promotion of urban agriculture.

Yet, some cities have embraced a broader scope and adopted a food system approach, with pioneers like Toronto in Canada, or Belo Horizonte in Brazil (Blay-Palmer, 2009; Friedmann, 2007; Rocha & Lessa, 2009). Since the early 1990s, these cities have engaged in developing their own vision and food system planning, using the full potential of urban-rural linkages existing with their large periphery. In that context, the FAO started its 'Food for the Cities' programme[5] in 2001, with the objective of building more sustainable and resilient food systems in the conditions of rising urbanization and environmental challenges, and to develop dialogue and partnerships.

These experiences contributed to the preparation of the New Urban Agenda (NUA), adopted by the Habitat III conference in 2016. This recognizes the importance of urban-rural linkages, the need to break away from silo thinking, and to support integrated urban and territorial planning and development (UN-Habitat, 2017). International dynamics and

---

[5] https://www.fao.org/in-action/food-for-cities-programme/en/.

mobilization initiated by these processes have resulted in the development of a network of cities, formalized in 2015, with an international protocol calling for the development of more sustainable and resilient urban food systems: the Milan Urban Food Policy Pact, signed by 210 cities.[6] The approach is fostering decentralization and cooperation mechanisms between cities which have been active in sharing their experiences (Magarini et al., 2017) and is converging with other city initiatives focusing on resilience, like the Resilient Cities Network[7] which includes a food component. Significantly, several cities hold their food initiatives under their resilience units, as it is the case for instance with Cape Town and Johannesburg in South Africa.

In that context, a new conceptual and policy framework has emerged. The City Region Food System (CRFS) formalizes a possible territorial approach to food system management (Blay-Palmer et al., 2018; Forster & Escudero, 2014). With reference to its own experience with the Food for the Cities program, the FAO started a CRFS programme in 2014. Its objective is *"Reinforcing rural-urban linkages for resilient food systems,"* the CRFS being defined as *"all the actors, processes and relationships that are involved in food production, processing, distribution and consumption in a given city region"*.[8] The programme is today developed in 13 countries in six continents.

The CRFS toolkit, developed by the FAO in 2018, presents a method for assessing the performance and functioning of food systems for a city region, including defining and mapping the city region, collecting data on the food system in the city region and analysing the data through assessments to develop an understanding of the relationships between food systems components and multi-dimensional sustainability indicators in the city region (FAO, 2018b). These include indicators that reflect improved health and well-being and increased access to food and nutrition: access to affordable, sufficient, nutritious, safe, adequate and diversified food that meet dietary needs. There are also indicators showing improved social and economic conditions for workers; building local food culture, food heritage and sense of identity; strengthening rural-urban linkages including food production and flows of food, nutrients, energy, water,

---

[6] https://www.milanurbanfoodpolicypact.org/.

[7] https://resilientcitiesnetwork.org/?s=food.

[8] https://www.fao.org/in-action/food-for-cities-programme/en/.

and income across rural and urban areas; and finally protecting ecosystems and environmental resources and reducing vulnerability and increasing resilience to shocks and disasters.

This standardization of the approach by the FAO has led to critics pointing to a rural bias with an emphasis on rural-urban linkages and food supply from neighbouring areas, preventing an effective vision and specific intervention on urban food security (Battersby & Watson, 2019) and the development of urban food-sensitive planning practices (Haysom et al., 2021; see also Battersby & Haysom, this volume). The toolkit approach, with priority given to flows and resource stock, is also possibly driving the focus away from governance, which is central to an effective territorial approach.

In spite of these growing decentralized experiences, the dominance of the de facto monocentric governance of food systems by the states remains. Rooted in the history of public policies, this top-down approach has been criticized over recent decades (Candel, 2014). The food system is complex; it includes many different stakeholders characterized by huge asymmetries in terms of incomes, assets and economic power; and it is characterized by a strong interdependency and interconnectedness between issues and actors. As such, a consensus has progressively emerged calling for a necessary shift towards a new approach which should be integrative and inclusive (Termeer et al., 2018). This moves away from the traditional monocentric governance towards an approach that may be considered as polycentric and paves the way for adaptive and collaborative governance. The case of South Africa is illustrative of this emerging trend.

## Place-Based Food System Governance: Recent Experience in South Africa

Historically, food system governance has been a central issue for the state. Food policies were designed and implemented by central governments, and this characteristic remains deeply rooted in the policy practice of most countries (Toussaint-Samat, 2009). Because food was first a question of supply, production was the central issue and ministries of agriculture have quickly taken the leading role in food security policies: a position that they continue to play in a large majority of countries, particularly in developmental states such as that aspired to by the South African government. It generally results in effective support from 'strong' ministries, like economy and finance and trade which can be explained by the growth and employment potential of agro-food production. In the case of South Africa, this has been referred to as an implicit 'Economic Growth' coalition between these core ministries (Thow et al., 2018). Even if food security is recognized as a cross-sectoral issue, other ministries (such as health or social development) are in the 'second line'. The result is a specific framing of food system issues, giving priority to production and food supply.

Based on a systematic literature review on food system governance in South Africa since the end of apartheid in 1994, Adeniyi et al. (2021) analyse these tensions and point towards the need for a new approach to governance. The literature reveals the importance of state policies: it particularly highlights a national paradox in which, despite the concerns of the Economic Growth coalition, South Africa is considered food secure, with the dietary needs of its population consistently exceeded by the food that is available. However, household food and nutrition insecurity is high when compared to countries of similar economic development.

To illustrate the national paradox, in 2015, 25% of the population lived below the national food poverty line (StatsSA, 2017a) and 27% of children under the age of 59 months were found to be stunted in 2016, a situation that has not improved despite two decades of appropriate policy interventions (Devereux et al., 2019). At the same time, 68% of adult women and 31% of men are either overweight or obese, which has translated into high prevalence of diet-related non-communicable disease (StatsSA, 2017b).

These sobering results are in contrast to a sophisticated food policy framework that is rooted in the interventionist tradition of the state, dating back decades. Contrary to other African countries, the colonial history has resulted in the development of a deep state with an autonomy of government since 1910, characterized by high centralization during the apartheid regime, and then the adoption of a limited federalism with the 1996 Constitution. Yet, food and nutrition security are enshrined as a basic human right in South Africa's Bill of Rights, and the right to food for all people and the right to nutrition for children are set out in South Africa's Constitution in Sections 27(1)(b), 27(2) and 28(1). Indeed, one of the first acts of the newly elected democratic government was to introduce a National School Nutrition Program (NSNP) in 1994 that was feeding 9 million children at the start of 2020.

The last major overhaul of South Africa's long-term strategy for social and economic development, the National Development Plan (NPC,

2011), identifies food security and rural transformation as enabling milestones for the eradication of poverty and reduction of inequalities by 2030. The National Policy for Food and Nutrition Security (NPFNS) gazetted in 2014 builds on the National Development Plan (NDP) and seeks to establish a platform for increasing, and better targeting, public spending in social programmes that impact on food security (DSD/DAFF, 2013). The policy is framed in terms of the recognition of the right to food in the South African Bill of Rights, and commits government to increasing access to production inputs for the emerging agricultural sector; leveraging government food procurement to support community-based food production initiatives and smallholders; and strategically using market interventions and trade measures which will promote food security.

The NPFNS acknowledges the complex nature of food and nutrition security and the importance of interventions that encourage increased access to affordable healthy food. However, due to weak consultative processes, the state-led NPFNS is argued to have led to policy directives that were "deemed inadequate by a wide cross-section of people" (Pereira & Drimie, 2016, p. 24). Notably, after reviewing the NPFNS, the South African Human Rights Commission (SAHRC) recommends a rethink of the food system and concludes that the policy "does not speak to the need for an interconnected system" (SAHRC, 2017, p. 23).

Despite the establishment of an inter-ministerial National Food and Nutrition Plan (NFNSP) in 2017 coordinated by the Office of the President (PMG, 2017), there has been little evidence of action from national government. Although the policy framework has been put in place, both the NPFNS and NFSNP lack the legislative structures necessary to achieve their goals and objectives (Hendriks & Olivier, 2015). Jacobs & Nyamwanza (2020) go further and call for the establishment of national and sub-national forums and the involvement of non-state actors in the coordination and monitoring of both the policy and the plan.

Moreover, the South African government has been unwilling to take direct interventions such as the management of food prices, despite cycles of food inflation. The Integrated Growth and Development Plan (IGDP) of 2012 mentions addressing high food prices, improving smallholder access to markets and support services, and the need for an integrated approach to ensuring food security. The country's national agricultural policies also include strong support for food security, but again with no mention of nutrition. For example, the Agricultural Policy Action

Plan (APAP) 2015–2019 places emphasis on value-chain interventions to improve food security, and also notes the importance of research and innovation, climate-smart agriculture, trade, agribusiness development and support. Nutrition, affordability and safety are not given attention (May, 2021).

Adeniyi et al. (2021) show an analytical convergence in most of these policy documents towards a series of governance challenges which are related to the framing problem, already mentioned, and to the importance of fragmentation, siloization and (lack of) policy coherence. These core problems are aggravated by a weak coordination, limited institutional capacity, and a partial and inadequate stakeholder engagement.

Among the proposed identified solutions to these shortcomings are the need for a legislative framework, necessary to actualize existing rights and particularly the right to food, the improvement of stakeholders' participation and stronger institutions, as well as priority to be given to local food system governance (Haysom, 2015). If multiple levels of governance are necessary to address the food system complexity, a place-based approach is an opportunity for an effective understanding of food systems' issues, facilitated and improved stakeholder engagement, improved connection with local networks and grassroot movements, and the progressive adoption of a shared vision of the main challenges on which to focus action. However, the way local processes can develop and strengthen remains an open question. Experiences of food system governance in the Western Cape province of South Africa during the COVID-19 pandemic offers a case study of how place-based food system governance may be emerging.

Due to the limited federalism, South Africa's provinces have restricted competency for food system governance. Trade, industrial policy, health, social development, education, agriculture, environment and rural development are managed by the central government, although in some instances, provincial government holds the mandate for delivery. Local governments do not have any specific mandate related to FNS and food system management.

However, there is room for action and local governments do have relevant competencies, notably for zoning and trading regulations, markets and street trading (De Visser, 2019). They are often responsible for the delivery of electricity and potable water, both central to food safety. Therefore, they can, or could, influence food system outcomes through the protection of agricultural land and food trade regulation, supporting activities in the informal economy, balancing the role of large retailers and supporting local food producers and traders. They can, or could, also improve access to healthy and nutritious food through advertising and support to farmers' markets. Budgets and human resources remain a major limitation to implementation and impact, with the exception of the metropoles.

Despite the Western Cape's prosperity when compared to other provinces and its well-established food system, the largest food economy in South Africa, the prevalence of the indicators of malnutrition among its population of 6.5 million is similar to national trends (StatsSA, 2019). The Western Cape Government (WCG) recognized the urgency for action and in 2014, set out to develop a food security strategy to complement its Provincial Strategic Plan. This strategy had to align with a National Policy on Food and Nutrition Security gazetted in 2014, as well as with other national policies such as those concerning land, water, health and sanitation. The process required dialogue and co-design between multiple agents and bodies of knowledge, as well as between multiple rationalities and multiple levels, a feature identified elsewhere to co-design policy (Himanen et al., 2016).

Termed 'Nourish to Flourish', this strategy commits to a wide range of interventions that address all food system outcomes, although the focus is on FNS. These include providing food and nutrition literacy interventions targeting diverse age groups; food-sensitive economic and spatial planning; influencing municipal planning for food and nutrition; promoting a climate-resilient low-carbon agricultural sector; and building an inclusive food economy that recognizes the role of informal traders as an important source of affordable food for low-income households (WCG, 2016). The strategy also addresses food security governance and the establishment of multi-stakeholder processes. Although the implementation of the strategy appears to have stalled prior to the pandemic, the response to COVID-19 in the Western Cape largely followed its 'whole of society' approach and appears to have had a positive impact (WCEDP, 2020).

Both the provincial government and the metropolitan government of the City of Cape Town have recognized their roles in regard to improving food security. Making use of an existing agreement with the Cape Higher Education Consortium (CHEC), a network for collaboration between the four universities of the province, research reports were commissioned to provide an information base (Adelle et al., 2020). These in turn recognized the complex nature of the food system and the nature of food security as a common good: an important improvement when compared to existing national policy frameworks. The research base informed stakeholder workshops conducted during 2015. These provided opportunities to discuss the conventional collective action related to concerns of coordination, cooperation, and finding and keeping agreements at the local level (Poteete & Ostrom, 2008). The deliberations of these workshops fed into a draft strategy document approved by the provincial cabinet in 2016 for public comment, which has subsequently evolved into a programme of work that includes civil society (WCEDP, 2020).

Parallel processes took place in other forums related to food security including health, governance and agriculture (Adelle et al., 2020). Following public comment, a non-government organization, the Southern African Food Lab (SAFL) specializing in partnering convened additional groups of stakeholders including 'Transformation Laboratories'. The purpose of these meetings was to develop projects to be implemented as partnerships between government and other actors in the local food system (Drimie et al., 2018).

These provincial dynamics of consultation and local debate have provided a fertile context which allowed further engagement and the implementation of a transdisciplinary community of practice on food governance in 2018. Based on an iterative process of shared knowledge and knowledge co-production (Adelle et al., 2021), this community of practice, still active in 2022, includes: decision-makers from provincial and metropolitan governments as duty bearers for the provision of food security as a public good; the private sector as the producers and suppliers of the food itself (from farmers to processors and retailers, including from the informal sector); and civil society organizations, including consumers, as final decision-makers as well as rights-holders.

In 2017/18, extreme drought put provincial and local governments to the test and revealed the potential of community mobilization (Robins, 2019). This was followed in March 2020 by the implementation by national government of a hard lockdown in response to the first COVID-19 cases. Among the first food system impacts in 2020 was the suspension of the National School Nutrition Program (NSNP). Restrictions on informal sector food traders, the prohibition of mobility across municipal boundaries and a strict curfew also affected food systems' actors and consumers. Mitigation measures in the form of direct food assistance were slower to be implemented, and in the case of school feeding required litigation before being addressed.[9] Even then, the reintroduction of the programme was slow and uneven (Section 27, 2020).

Responding to this in the Western Cape, multiple community-led initiatives facilitated the development of a vibrant and localized food debate and action. The continuation of school feeding though the Peninsula School Feeding Association was particularly important, but this also included coordination and advocacy such as Food Dialogues, organized in 2020 and 2021 (SAUFFT, 2020) and the Food Forum hosted by the Western Cape Economic Development Partnership (Trialogue, 2021). Altogether they improved the local capacity to react to the COVID-19 crisis with the rapid development of community action networks (CANs), such as those in an umbrella organization 'Cape Town Together', which linked almost 200 such organizations.[10] These local networks, connected through information and communications technologies such as social media and WhatsApp® applications, contributed towards a localized and coordinated response to the crisis by civil society (Adelle & Haywood, 2021; Odendaal, 2021). In addition to addressing the immediate food and health crisis, a stated goal of the CANs was to "put the public back into the public sector" (Bust et al., 2021).

The Western Cape is a region characterized by extreme resource inequalities, resulting in communities that, although spatially proximate, are socially distant (Mears & Bhati, 2006). This has produced 'cities of islands' (Writers Community Action Network, 2020). To overcome this, partnering has been key, involving different forms of collaboration according to the specific issues being addressed Although partnering strategies have been widely used by local governments (Greve, 2015), they may be demanding in the context of food system governance in which there are fundamental differences in priorities, substantial material stakes and low levels of trust. In the case of the Western Cape, achieving successful partnerships required "moving at the speed of trust" as well as compromises, incentives and the enforcement of duties and rights. Despite this, the CANs still experienced push back by both local political leaders and the national government (Bust et al., 2021).

---

[9] On 17 July 2020, a consortium of NGO successfully litigated against the Minister of Basic Education to reinstate the NSNP (Section 27, 2020).

[10] See https://capetowntogether.net/ for more details.

Although still work in progress, the interventions and activities in the Western Cape since 2017 have shown the potential to produce polycentric forms of system governance and to engage in the pathway of a possible collaborative governance. They have required a dialogue between multiple agents and bodies of knowledge, as well as between multiple rationalities and multiple levels. As was shown, such dialogue is not necessarily initiated by the public sector and is unlikely to be proposed by the private sector even if some stakeholders, such as farmers, have strong incentives for collective action but generally lack adequate information. Key to the notion that these interventions may be nascent territoriality is the claim that they emerged when communities realized that "When you know that it is your neighbours who have empty cupboards, it is a political act to start cooking. As we cook, food becomes a vehicle for the sharing of social and cultural practices, as well as politicizing the hunger in the first place – generating learning, consciousness and human connection" (Writers Community Action Network, 2020).

In the case of the Western Cape, food system resilience during the COVID-19 pandemic has required assertive action on the part of consumers and civil society with the realization that they indeed hold rights for which others bear duties to fulfil. It is likely that some issues will still need to be resolved by the enforcement of duties and rights as contained in the national laws of South Africa, including the international treaties it has endorsed. As in other contexts, ensuring resilience has required litigation and civil society actions including consumer boycotts, social media campaigns and protests in order to put pressure on government and the main food corporate businesses (Huang et al., 2015). A possible outcome of these responses, and a possible objective, is the progressive adoption of a collaborative governance of the food system, where collaboration and consensus building are the rule.

The experiences since the start of the pandemic in the Western Cape may not be unique as similar responses have been documented elsewhere (Nemes et al., 2021; Zhan & Chen, 2021; Zollet et al., 2021). Within South Africa, CANs have spread to other provinces and have adopted similar modes of operation.[11] Alongside this, some umbrella associations of CANs are identifying a more ambitious agenda. One such association, Gauteng Together, states that it intends re-orientating the work of the

---

[11] For example, Eastern Cape (https://easterncapetogether.co.za/) and Gauteng (https://www.gautengtogether.org/).

CANs into a *"critical mass disruption agenda"* through the integrated zones to set up a CAN in every neighbourhood to *"save the soul of the nation"* (Nortier, 2021).

Taken together, local responses to the food system challenges arising from COVID-19 highlight how embryonic territorial approaches to addressing such challenges may contribute towards:

- rebalancing power, rebuilding capability to act, invest and influence at the local level: local governments, districts or ad hoc local cooperation bodies such as CANs help to identify the effective challenges and possible solutions through the agency and mobilization of the diversity of their stakeholders and constituents;
- building better and more resilient connections between institutions and resources, especially ecological resources: shared diagnoses and co-elaborative scenarios about plausible futures can help to design adequate strategies;
- strengthening the connection between food and innovation by building on local food cultures: local knowledge and local debate can help to build on the specific resources of a place (to be differentiated from generic resources which can be found everywhere);
- re-establishing and strengthening local flows of food and information between rural and urban areas, reducing unnecessary long-distance trade, promoting livelihoods and local multipliers.

This is not to imply a return to the localization approach ably critiqued by Born & Purcell (2006), but rather the recognition of the potential role of local governments and local actors to guide place-based food systems towards goals of economic inclusion, environmental sustainability, equity and social justice. Certainly, the long value chains, which define most present-day food systems, necessitate territorial approaches to be embedded in global contexts so that the interplay between local and global causes and effects can be clarified and understood. Furthermore, territorial approaches to food system research and development must align with national policies and commitments, including social protection measures, trade policy and corporate regulation, all of which influence food system activities and outcomes.

## Conclusion: The Spirit of Place

Despite the usefulness of toolkits of indicators and activities, like those developed in the City Region Food System (CRFS) programme presented above, local government policy development concerned with food security, and the manner of its implementation, will depend both upon the actions, negotiations and relative powers of those engaging in the process, as well as those that do not. Changing the food system is likely to produce individual and collective costs, some of which will be borne by actors who are not duty holders, nor hold rights that can be realized. A changing food system also produces new beneficiaries, some of whom are free riders able to benefit without making any contribution. Building food systems that are resilient to disruptions requires addressing the distribution of costs/benefits and how/whether these are to be managed.

Re-localizing food systems and food systems governance can significantly contribute to resilience, reduce ecological footprint and costs of transportation and transaction, foster rural-urban linkages and local activities, strengthen the social fabric, promote local food and the related cultural heritage, and enhance natural resources management and the development of new uses and services. In the context of system shocks such as COVID-19, the unintended consequences of mitigation interventions can be more quickly identified and addressed. This included expanding social protection coverage through the provision of distress grants or disaster relief, providing food parcels and permitting essential economic activities such as informal food trading.

Implementing a territorial approach to food systems in the context and the aftermath of COVID-19 in South Africa will require more than community action networks. There is a need to address funding shortages, donor and practitioner fatigue and the pull towards returning to pre-pandemic practices in the food system. Already by November 2020, 33% of the CANs in Gauteng had stopped their operations (Mahwai, 2020). To move forward, it may be necessary to draw on the existing mandate and tools of local governments and to combine the relative strengths of both state-led and community-led approaches. New forms of innovative governance will be needed to do this. This could include establishing communities of practice that build on knowledge co-production techniques, developing food charters and implementing food policy councils. Buchan et al. (2015) detail some of additional actions that are already undertaken by local governments.

To do so will require interactions with the full array of food system actors (consumers, farmers, processors, distributors, retailers, government and so forth) who are pivotal in determining food system outcomes. Although there may be failures to reach agreement, an approach to identify and provide answers to local problems can gain traction. Furthermore, a process that involves local stakeholders will facilitate territory formation, an important outcome in its own right. In this respect, the territorial approach has shown that it is well suited to addressing collective action problems concerned with the resilience of food systems, and the urban-rural interface on which they often rest.

## References

Adelle, C., & Haywood, A. (2021). *Engaging Civil Society Organisations in Food Governance in the Western Cape: Reflections from emergency food relief during Covid* (Food Security SA Working Paper Series, Working Paper 8).

Adelle, C., Kroll, F., Losch, B., & Görgens, T. (2021). Fostering communities of practice for improved food democracy: Experiences and learning from South Africa. *Urban Agriculture & Regional Food Systems*. https://doi.org/10.1002/uar2.20007

Adelle, C., Pereira, L., Görgens, T., & Losch, B. (2020). Making sense together: The role of scientists in the coproduction of knowledge for policy making. *Science and Public Policy, 47*(1), 56–66.

Adeniyi, D., Losch, B., & Adelle, C. (2021). *Investigating the South African food insecurity paradox: A systematic review of food system governance in South Africa* (Food Security SA Working Paper Series, Working Paper 9).

AfDB, OECD, UNDP. (2015). *African Economic Outlook 2015. Regional development and spatial inclusion*. OECD Publishing.

Barca, F. (2009). *An agenda for a reformed cohesion policy: A place-based approach to meeting European Union challenges and expectations*. European Commission.

Barca, F., McCann, P., & Rodríguez-Pose, A. (2012). The case for regional development intervention: Place-based versus place-neutral approaches. *Journal of Regional Science, 52*(1), 134–152.

Barrett, C. B. (2011). Covariate catastrophic risk management in the developing world: Discussion. *American Journal of Agricultural Economics, 93*(2), 512–513.

Baser, H., & Morgan, P. (2008). *Capacity, change and performance: Study report*. European Centre for Development Policy Management.

Battersby, J., & Watson, V. (2019). The planned 'city-region' in the New Urban Agenda: An appropriate framing for urban food security? *Town Planning Review*, *90*(5), 497–519.

Beard, V., Miraftab, F., & Silver, C. (2008). *Planning and decentralization: Contested spaces for public action in the global South*. Routledge.

Beegle, K., De Weerdt, J., & Dercon, S. (2008). Adult mortality and consumption growth in the age of HIV/AIDS. *Economic Development and Cultural Change*, *56*(2), 299–326.

Béné, C., Bakker, D., Rodriguez, M. C., Even, B., Melo, J., & Sonneveld, A. (2021). *Impacts of COVID-19 on people's food security: Foundations for a more resilient food system*. International Food Policy Research Institute.

Béné, C., Headey, D., Haddad, L., & von Grebmer, K. (2016). Is resilience a useful concept in the context of food security and nutrition programmes? Some conceptual and practical considerations. *Food Security*, *8*(1), 123–138.

Bidisha, S. H., Mahmood, T., & Hossain, M. B. (2021). Assessing food poverty, vulnerability and food consumption inequality in the context of COVID-19: A case of Bangladesh. *Social Indicators Research*, *155*(1), 187–210.

Biggs, R., Schlüter, M., & Schoon, M. L. (Eds.). (2015). *Principles for building resilience: Sustaining ecosystem services in social-ecological systems*. Cambridge University Press.

Blay-Palmer, A. (2009). The Canadian pioneer: The genesis of urban food policy in Toronto. *International Planning Studies*, *14*, 401–416.

Blay-Palmer, A., Santini, G., Dubbeling, M., Renting, H., Taguchi, M., & Giordano, T. (2018). Validating the City Region Food System approach: Enacting inclusive, transformational city region food systems. *Sustainability*, *10*(5), 1680. https://doi.org/10.3390/su10051680

Borman, G. D., de Boef, W. S., Dirks, F., Gonzalez, Y. S., Subedi, A., Thijssen, M. H., Jacobs, J., Schrader, T., Boyd, S., Hermine, J., & van der Maden, E. (2022). Putting food systems thinking into practice: Integrating agricultural sectors into a multi-level analytical framework. *Global Food Security*, *32*, 100591.

Born, B., & Purcell, M. (2006). Avoiding the local trap: Scale and food systems in planning research. *Journal of Planning Education and Research*, *26*(2), 195–207.

Buchan, R., Cloutier, D., Friedman, A., & Ostry, A. (2015). Local food system planning: The problem, conceptual issues, and policy tools for local government planners. *Canadian Journal of Urban Research*, *24*(1), 1–23.

Bullock, J. M., Dhanjal-Adams, K. L., Milne, A., Oliver, T. H., Todman, L. C., Whitmore, A. P., & Pywell, R. F. (2017). Resilience and food security: Rethinking an ecological concept. *Journal of Ecology*, *105*(4), 880–884.

Bust, L., Dambisya, P., Davies, B., & Patientia, R. (2021, April 22). Bottom-up responses: Lessons from Cape Town's Community Networks that inspired collective action in a time of crisis. *International Health Policies*. Retrieved 19 December 2021, from https://www.internationalhealthpolicies.org/featured-article/bottom-up-responses-lessons-from-cape-towns-community-networks-that-inspired-collective-action-in-a-time-of-crisis/

Byiers, B., & Rampa, R. (2013). *Corridors of power or plenty? Lessons from Tanzania and Mozambique and implications for CAADP* (ECDPM Discussion Paper, No. 138). European Centre for Development Policy Management.

Cabral, L. (2011). *Decentralisation in Africa: Scope, motivations and impact on service delivery and poverty* (FAC Working Paper, 20). Future Agricultures Consortium.

Campagne, P., & Pecqueur, B. (2014). *Le développement territorial: une réponse émergente à la mondialisation*. Charles Léopold Mayer.

Candel, J. J. L. (2014). Food security governance: A systematic literature review. *Food Security, 6*(4), 585–601. https://doi.org/10.1007/s12571-014-0364-2

Caron, P., Ferrero y de Loma-Osorio, G., Nabarro, D., Hainzelin, E., Guillou, M., Andersen, I., Arnold, T., Astralaga, M., Beukeboom, M., Bickersteth, S., Bwalya, M., Caballero, P., Campbell, B. M., Divine, N., Fan, S., Frick, M., Friis, A., Gallagher, M., ... Verburg, G. (2018). Food systems for sustainable development: Proposals for a profound four-part transformation. *Agronomy for Sustainable Development, 38*, 41. https://doi.org/10.1007/s13593-018-0519-1

Caron, P., Valette, E., Wassenaar, T., Coppens d'Eeckenbrugge, G., & Papazian, V. (Coord.). (2017). *Living territories to transform the world*. Quae Editions.

Casti, J. L., & Fath, B. D. (2008). Ecological complexity. In S. E. Jorgensen & B. D. Fath (Eds.), *Encyclopaedia of ecology* (Vol. 1). Amsterdam.

Cistulli, V., Rodríguez-Pose, A., Escobar, G., Marta, S., & Schejtman, A. (2014). Addressing food security and nutrition by means of a territorial approach. *Food Security, 6*(6), 879–894.

Coley, D., Howard, M., & Winter, M. (2009). Local food, food miles and carbon emissions: A comparison of farm shop and mass distribution approaches. *Food Policy, 34*, 150–155.

Conyers, D. (2007). Decentralisation and service delivery: Lessons from Sub-Saharan Africa. *IDS Bulletin, 38*(1), 18–32.

Crook, R. C. (2003). Decentralisation and poverty reduction in Africa: The politics of local-central relations. *Public Administration and Development, 23*(1), 77–88.

Davies, S. (1996). *Adaptable livelihoods: Coping with food insecurity in the Malian Sahel*. Macmillan Press.

Devereux, S., Jonah, C., & May. J. (2019). How many malnourished children are there in South Africa? What can be done? In K. Roelen, R. Morgan, & Y. Tafere (Eds.), *Putting children first: New Frontiers in the fight against child poverty in Africa*. Ibidem Press.

De Visser, J. (2019). *Multilevel government, municipalities and food security* (Food Security SA Working Paper Series No. 005). DST-NRF Centre of Excellence in Food Security.

Drimie, S., Hamann, R., Manderson, A. P., & Mlondobozi, N. (2018). Creating transformative spaces for dialogue and action. *Ecology and Society, 23*(3).

DSD/DAFF. (2013). *National Policy on Food and Nutrition Security*. Department of Social Development, Department of Agriculture, Foresty and Fisheries.

Dury, S., Bendjebbar, P., Hainzelin, E., Giordano, T., & Bricas, N. (Eds.). (2019). *Food systems at risk. New trends and challenges*. CIRAD; FAO. https://doi.org/10.19182/agritrop/00080

ECA, AUC, FAO, AUDA-NEPAD, WFP, UNICEF, IFAD, AfDB, Akademiya2063, and RUFORUM. (2021, March 4). *Background paper. Regional Dialogue: African Food Systems. Seventh Session of the Africa Regional Forum on Sustainable Development*. Retrieved 7 June 2021, from https://www.uneca.org/sites/default/files/TCND/ARFSD2021/Docume nts/Regional%20Dialogue%20-%20African%20Food%20Systems%20Backgro und%20Paper%20-%20EN.pdf

Faguet, J. P. (2014). Decentralization and governance. *World Development, 53*, 2–13.

FAO. (2018a). *Sustainable food systems: Concept and framework*. Food and Agricultural Organization.

FAO. (2018b). *City Region Food System Toolkit*. Food and Agricultural Organization.

FAO. (2020). *Cities and local governments at the forefront in building inclusive and resilient food systems: Key results from the FAO survey "Urban Food Systems and COVID-19."* Food and Agricultural Organization.

FAO, European Union and CIRAD. (2021). *Food Systems Assessment—Working towards the SDGs: Interim Synthesis Brief—September 2021*. https://doi.org/10.4060.cb6887en

FARA. (2021). *Strengthening African agricultural research and development towards an improved Africa food system: One Africa voice towards the 2021 UN Food Systems Summit* (Policy Brief). Retrieved 12 July 2021, from https://www.interacademies.org/event/strengthening-african-agricultu ral-research-and-development-towards-improved-africa-food

Fau, N. (2019). Development corridors. *EchoGéo* [Online], 49. http://jou rnals.openedition.org/echogeo/18170. https://doi.org/10.4000/echogeo. 18170

Faysse, N. (2015). The rationale of the Green Morocco Plan: Missing links between goals and implementation. *Journal of North African Studies, 20*(4), 622–634.

Finney Rutten, L., Yaroch, A. L., Patrick, H., & Story, M. (2012). Obesity prevention and national food security: A food systems approach. *International Scholarly Research Notices, 2012*.

Forster, T., & Escudero, A. G. (2014). *City regions as landscapes for people, food and nature*. EcoAgriculture Partners.

Forster, T., Penagos, A., Scherr, S., Buck, L., & Ramirez, E. (2021). *Territorial approaches for sustainable development. Stocktaking on territorial approaches—Experiences and lessons*. GIZ.

Freshwater, D. (2015). Vulnerability and resilience: Two dimensions of rurality. *Sociologia Ruralis, 55*(4), 497–515.

Friedmann, H. (2007). Scaling up: Bringing public institutions and food service corporations into the project for a local, sustainable food system in Ontario. *Agriculture and Human Values, 24*, 389–398.

Friedmann, H., & McMichael, P. (1989). Agriculture and the state system: The rise and decline of national agricultures, 1870 to the present. *Sociologia Ruralis, 29*(2), 93–117. https://doi.org/10.1111/j.1467-9523.1989.tb00360.x

Giordano, T., Losch, B., Sourisseau, J. M., & Valette, E. (2019). Risks of increasing territorial inequalities. In D. Sandrine, B. Pauline, H. Etienne, G. Thierry, & B. Nicolas (Eds.), *Food systems at risk. New trends and challenges* (pp. 83–86). CIRAD-FAO. https://doi.org/10.19182/agritrop/00099

Godfray, H. C. J., Crute, I. R., Haddad, L., Lawrence, D., Muir, J. F., Nisbett, N., Pretty, J., Robinson, S., Toulmin, C., & Whiteley, R. (2010). The future of the global food system. *Philosophical Transactions of the Royal Society, 365*, 1554. https://doi.org/10.1098/rstb.2010.0180

Greenberg, S. (2017). Corporate power in the agro-food system and the consumer food environment in South Africa. *The Journal of Peasant Studies, 44*(2), 467–496.

Greve, C. (2015). Ideas in public management reform for the 2010s. Digitalization, value creation and involvement. *Public Organization Review, 15*(1), 49–65.

Gupta, A., Zhu, H., Doan, M. K., Michuda, A., & Majumder, B. (2020). Economic impacts of the COVID-19 lockdown in a remittance-dependent region. *American Journal of Agricultural Economics, 103*(2), 466–485.

Harris-Fry, H., Shrestha, N., Costello, A., & Saville, N. M. (2017). Determinants of intra-household food allocation between adults in South Asia—A systematic review. *International Journal for Equity in Health, 16*(1), 1–21.

Harvey, D. (2001). *Spaces of capital: Towards a critical geography*. University of Edinburgh Press.

Haysom, G. (2015). Food and the city: Urban scale food system governance. *Urban Forum, 26*(3), 263–281. https://doi.org/10.1007/s12132-015-9255-7

Haysom, G., Battersby, J., & Park-Ross, R. (2021). *Food sensitive planning and Urban design—A blueprint for a future South African city?* (Food Security SA Working Paper Series, Working Paper 007).

Hendriks, S., & Olivier, N. (2015). Review of the South African agricultural legislative framework: Food and nutrition security implications. *Development Southern Africa, 32*(5), 555–576.

Himanen, S. J., Rikkonen, P., & Kahiluoto, H. (2016). Codesigning a resilient food system. *Ecology and Society, 21*(4), 41.

HLPE. (2017). *Nutrition and food systems.* A report by the High-Level Panel of Experts on Food Security and Nutrition of the Committee on World Food Security.

Huang, T. T., Cawley, J. H., Ashe, M., Costa, S. A., Frerichs, L. M., Zwicker, L., Rivera, J. A., Levy, D., Hammond, R. A., Lambert, E. V., & Kumanyika, S. K. (2015). Mobilisation of public support for policy actions to prevent obesity. *The Lancet, 385*(9985), 2422–2431.

Ingram, J. (2019). Food system models. In M. Lawrence & S. Friel (Eds.), *Healthy and sustainable food systems.* Routledge.

Jacobs, P., & Nyamwanza, A. (2020, August 28). Food for thought: Until policy meets action, hunger and poor nutrition will stalk the land. *Daily Maverick.* Retrieved 18 December 2021, from https://www.dailymaverick.co.za/article/2020-02-28-food-for-thought-until-policy-meets-action-hunger-and-poor-nutrition-will-stalk-the-land/

Kroll, F., Battersby, J., Haysom, G., Drimie, S., & Adelle, C. (2021). *Leveraging Informal Trade: Governance strategies to cultivate resilient, inclusive food economies* (Policy Brief 1/2021). DSI/NRF Centre of Excellence in Food Security.

Kuhlmann, K., Sechler, S., & Guinan, J. (2011, June 14). *Africa's development corridors as pathways to agricultural development, regional economic integration and food security in Africa.* Regional Economic Integration and Food Security in Africa. Retrieved 6 January 2022, from https://ssrn.com/abstract=3694895 or https://doi.org/10.2139/ssrn.3694895

Kusumasari, B., Alam, Q., & Siddiqui, K. (2010). Resource capability for local government in managing disaster. *Disaster Prevention and Management, 19*(4), 438–451.

Lang, T. (2003). Food industrialisation and food power: Implications for food governance. *Development Policy Review, 21*(5–6), 555–568.

Lapping, M. B. (2004). Toward the recovery of the local in the globalizing food system: The role of alternative agricultural and food models in the US. *Ethics, Place and Environment, 7*(3), 141–150.

La Trobe, H. L., & Acott, T. G. (2000). Localising the global food system. *International Journal of Sustainable Development & World Ecology, 7*(4), 309–320.

Lessmann, C., & Seidel, A. (2017). Regional inequality, convergence, and its determinants—A view from outer space. *European Economic Review, 92*, 110–132.

Losch, B. (2016). *The need for a paradigm shift towards territorial development in sub-Saharan Africa* (Rimisp Working Paper Series, 185). RIMISP. http://rimisp.org/publicaciones-documentos/documentos-de-trabajo/

Magarini, A., Nicolarea, Y., Dansero, E., & Bottiglieri, M. (2017). Urban food policies: Decentralized cooperation and African cities. *Revue internationale des études du développement, 232*(4), 67–93.

Mahwai, L. (2020). *Evaluation report for the Ahmed Kathrada Foundation on the 'Gauteng Together Initiative'*. Auwal Socio-Economic Research Institute. Retrieved 19 December 2021, from https://www.kathradafoundation.org/download/gauteng-together-report/

May, J. (2021). A political economy of persistent food insecurity in South Africa. In A. Oqubay, F. Tregenna, & I. Valodia, *The [Oxford] handbook of the South African economy*. Oxford University Press.

McGinnis, M. D., & Ostrom, E. (2014). Social-ecological system framework: Initial changes and continuing challenges. *Ecology and Society, 19*(2), 30.

Mears, D. P., & Bhati, A. S. (2006). No community is an Island: The effects of resource deprivation on urban violence in spatially and socially proximate communities. *Criminology, 44*(3), 509–548.

Morgan, E. H., Hawkes, C., Dangour, A. D., & Lock, K. (2019). Analyzing food value chains for nutrition goals. *Journal of Hunger & Environmental Nutrition, 14*(4), 447–465.

Nemes, G., Chiffoleau, Y., Zollet, S., Collison, M., Benedek, Z., Colantuono, F., Dulsrud, A., Fiore, M., Holtkamp, C., Kim, T. Y., Korzun, M., Mesa-Manzano, R., Reckinger, R., Ruiz-Martínez, I., Smith, K., Tamura, N., Viteri, M. L., & Orbán, É. (2021). The impact of COVID-19 on alternative and local food systems and the potential for the sustainability transition: Insights from 13 countries. *Sustainable Production and Consumption, 28*, 591–599.

Nortier, C. (2021). *Gauteng's community action networks saved lives in a crisis—Now it has to prepare for the long haul*. Retrieved 19 December 2021, from https://www.dailymaverick.co.za/article/2021-03-03-gautengs-community-action-networks-saved-lives-in-a-crisis-now-it-has-to-prepare-for-the-long-haul/

NPC (National Planning Commision). (2011). *National Development Plan 2030: Our future, make it work*. National Planning Commission.

NRC (National Research Council). (2015). *A framework for assessing effects of the food system*. The National Academies Press. https://doi.org/10.17226/18846

Odendaal, N. (2021). Recombining place: COVID-19 and community action networks in South Africa. *International Journal of E-Planning Research (IJEPR), 10*(2), 124–131.

OECD. (2009). *Regions matter: Economic recovery, innovation and sustainable growth*. OECD Publishing. https://doi.org/10.1787/9789264076525-en

OECD. (2021). *Making better policies for food systems*. OECD Publishing. https://doi.org/10.1787/ed93d80a-en

OECD/FAO/UNCDF. (2016). *Adopting a territorial approach to food security and nutrition policy*. OECD Paris Publishing.

Ostrom, E. (2000). Collective action and the evolution of social norms. *The Journal of Economic Perspectives, 14*(3), 137–158. https://doi.org/10.1257/jep.14.3.137

Pereira, L., & Drimie, S. (2016). Governance arrangements for the future food system: Addressing complexity in South Africa. *Environment: Science and Policy for Sustainable Development, 58*(4), 18–31.

Pinstrup-Andersen, P. (2010). The African food system and human health and nutrition: A conceptual and empirical overview. In P. Pinstrup-Andersen (Ed.), *The African food system and its interaction with human health and nutrition*. Cornell University Press.

PMG (Parliamentary Monitoring Group). (2017). *National food and nutrition security policy implementation plan: Operation Phakisa for Agriculture, Rural Development and Land Reform: Progress report*. Parliamentary Monitoring Group [online]. https://pmg.org.za/committee-meeting/25488/

Poteete, A. R., & Ostrom, E. (2008). Fifteen years of empirical research on collective action in natural resource management: Struggling to build large-N databases based on qualitative research. *World Development, 36*(1), 176–195.

Pothukuchi, K., & Kaufman, J. L. (2000). The food system: A stranger to the planning field. *Journal of the American Planning Association, 66*(2), 113–124.

Puissant, S., & Lacour, C. (1999). *La métropolisation: croissance, diversité, fractures*. Anthropos Research & Publications.

Relph, E. (1985). Geographical experiences and being-in-the-world: The phenomenological origins of geography. In D. Seamon & R. Mugerauer (Eds.), *Dwelling, place and environment: Towards a phenomenology of person and world* (pp. 15–31). Springer.

Robins, S. (2019). 'Day Zero', hydraulic citizenship and the defence of the commons in Cape Town: A case study of the politics of water and its infrastructures (2017–2018). *Journal of Southern African Studies, 45*(1), 5–29.

Rocha, C., & Lessa, I. (2009). Urban governance for food security: The alternative food system in Belo Horizonte Brazil. *International Planning Studies, 14*, 389–400.

Rodríguez-Pose, A. (2010). Economic geographers and the limelight: The reaction to the World Development Report 2009. *Economic Geography, 86*, 361–370.

SAHRC (South African Human Rights Commission). (2017). *The right to access to nutritious food in South Africa*. South African Human Rights Commission.

Sarapuu, K., Lægreid, P., Randma-Liiv, T., & Rykkja, L. H. (2014). Conclusion: Lessons learned and policy implications. In P. Lægreid, K. Sarapuu, L. H. Rykka, & T. Randma-Liiv (Eds.), *Organizing for coordination in the public sector: Practices and lessons from 12 European countries*. Palgrave Macmillan.

Sassen, S. (2001). Cities in the global economy. In R. Paddison (Ed.), *Handbook of Urban Studies* (pp. 256–272). Sage.

Sassen, S. (2005). The global city: Introducing a concept. *Brown Journal of World Affairs, 11*(2), 27–43.

SAUFFT (South African Urban Food and Farming Trust). (2020). *Food Dialogues. Report 2020*. SAUFFT.

Scholvin, S. (2021). Getting the territory wrong: The dark side of development corridors. *Area Development and Policy, 6*(4), 441–450.

Schoon, M. L., Robards, M. D., Meek, C. L., & Galaz, V. (2015). Principle 7: Promote polycentric governance systems. In R. Biggs, M. Schluter, & M. L. Schoon (Eds.), *Principles for building resilience: Sustaining ecosystem services in social-ecological systems* (pp. 226–250). Cambridge University Press.

Section 27. (2020). Retrieved 18 December 2021, from https://section27.org.za/national-school-nutrition-programme/

Smoke, P. (2003). Decentralisation in Africa: Goals, dimensions, myths and challenges. *Public Administration and Development, 23*, 7–16. https://doi.org/10.1002/pad.255

StatsSA. (2017a). *Poverty on the rise in South Africa*. Retrieved 19 December 2021, from https://www.statssa.gov.za/?p=10334

StatsSA. (2017b). *South Africa Demographic and Health Survey (SADHS)*. Retrieved 28 September 2021, from http://www.statssa.gov.za/?page_id=6634

StatsSA. (2019). *Towards measuring food security in South Africa: An examination of hunger and food inadequacy* (Report No. 03-00-14). Retrieved 19 December 2021, from http://www.statssa.gov.za/publications/03-00-14/03-00-142017.pdf

Storper, M. (1997). *The regional world: Territorial development in a global economy*. Guilford Press.

Tefft, J., Jonasova, M., Adjao, R., & Morgan, A. (2017). *Food systems for an urbanizing world*. World Bank/Food and Agricultural Organization.

Termeer, C. J. A. M., Drimie, S., Ingram, J., Pereira, L., & Whittingham, M. J. (2018). A diagnostic framework for food system governance arrangements:

The case of South Africa. *NJAS—Wageningen Journal of Life Sciences, 84*, 85–93. https://doi.org/10.1016/j.njas.2017.08.001

Thow, A. M., Greenberg, S., Hara, M., Friel, S., du Toit, A., & Sanders, D. (2018). Improving policy coherence for food security and nutrition in South Africa: A qualitative policy analysis. *Food Security, 10*(4), 1105–1130. https://doi.org/10.1007/s12571-018-0813-4

Toussaint-Samat, M. (2009). *A history of food*. Wiley.

TP4D. (2018). *Fostering territorial perspective for development. Towards a wider alliance* (White Paper). AFD, BMZ, Cirad, European Commission, FAO, GIZ, Nepad, OECD, UNCDF.

Trialogue. (2021). Connect, communicate, collaborate: How to mitigate a food crisis. In *The Trialogue Business in Society Handbook, 2021*. Retrieved 17 December 2021, from https://trialogue.co.za/wp-content/uploads/2021/11/BIS-2021.pdf

Tschirley, D., Haggblade, S., & Reardon, T. (Eds.). (2014). *Population growth, climate change and pressure on the land—Eastern and Southern Africa* (99 p). ISBN 978-0-9903005-2-6.

UN-Habitat. (2017). *Implementing the New Urban Agenda by strengthening rural-urban linkages*. United Nations Human Settlements Programme.

Varga, A. (2017). Place-based, spatially blind, or both? Challenges in estimating the impacts of modern development policies: The case of the GMR policy impact modeling approach. *International Regional Science Review, 40*(1), 12–37.

Von Bertalanffy, L. (1968). *General systems theory*. George Braziller.

WCEDP (Western Cape Economic Development Partnership). (2020, October 22). *Report: Western Cape Food Forum*. Retrieved 19 December 2021, from https://wcedp.co.za/wp-content/uploads/2020/12/Western-Cape-Food-Forum-Report-22-October-2020.pdf

WCG (Western Cape Government). (2016). *Western Cape Government household food and nutrition security strategic framework*. Retrieved 19 December 2021, from https://www.westerncape.gov.za/sites/www.westerncape.gov.za/files/assets/140916_wcg_household_food_and_nutrition_security_strategic_framework.pdf

Westbury, S., Ghosh, I., Jones, H. M., Mensah, D., Samuel, F., Irache, A., Azhar, N., Al-Khudairy, L., Iqbal, R., & Oyebode, O. (2021). The influence of the urban food environment on diet, nutrition and health outcomes in low-income and middle-income countries: A systematic review. *BMJ Global Health, 6*(10), e006358.

Williamson, J. G. (1965). Regional inequality and the process of national development: A description of the patterns. *Economic Development and Cultural Change, 13*(4, Part 2), 1–84.

World Bank. (2009). *World Development Report 2009: Reshaping economic geography*. World Bank.

Wright, C. (2020). Local government fighting Covid-19. *The Round Table, 109*(3), 338–339.

Writers Community Action Network. (2020). *Cape Town together: Organizing in a city of Islands*. ROAR: Foundation for Autonomous Media. Retrieved on 18 December 2021, from https://roarmag.org/essays/cape-town-together-organizing-in-a-city-of-islands/.

Zhan, Y., & Chen, K. Z. (2021). Building resilient food system amidst COVID-19: Responses and lessons from China. *Agricultural Systems, 190*, 103102.

Zollet, S., Colombo, L., De Meo, P., Marino, D., McGreevy, S. R., McKeon, N., & Tarra, S. (2021). Towards territorially embedded, equitable and resilient food systems? Insights from grassroots responses to COVID-19 in Italy and the city region of Rome. *Sustainability, 13*(5), 2425.

**Open Access** This chapter is licensed under the terms of the Creative Commons Attribution 4.0 International License (http://creativecommons.org/licenses/by/4.0/), which permits use, sharing, adaptation, distribution and reproduction in any medium or format, as long as you give appropriate credit to the original author(s) and the source, provide a link to the Creative Commons license and indicate if changes were made.

The images or other third party material in this chapter are included in the chapter's Creative Commons license, unless indicated otherwise in a credit line to the material. If material is not included in the chapter's Creative Commons license and your intended use is not permitted by statutory regulation or exceeds the permitted use, you will need to obtain permission directly from the copyright holder.

CHAPTER 11

# Urban Food Security and Resilience

*Gareth Haysom and Jane Battersby*

## INTRODUCTION

Many writers have effectively captured the emergence, progression, framings and limitations of the concepts of urban resilience (see Béné et al., 2018), crisis and disaster management and resilience, both urban and more generally (Boin et al., 2010) and how, given the challenges in many developing countries and societies, including socioeconomic inequities, "negotiated resilience" may be more desirable (Ziervogel et al., 2017). The concept of resilience within urban systems has gained significant academic and policy focus in the last 10 years (see Coaffee et al., 2018; Béné et al., 2018). We suggest here that the concept of urban food

G. Haysom (✉) · J. Battersby
African Centre for Cities, University of Cape Town, Cape Town, South Africa
e-mail: gareth.haysom@uct.ac.za

J. Battersby
e-mail: jane.battersby.lennard@gmail.com

J. Battersby
Environmental and Geographical Science Department, University of Cape Town, Cape Town, South Africa

© The Author(s) 2023
C. Béné and S. Devereux (eds.), *Resilience and Food Security in a Food Systems Context*, Palgrave Studies in Agricultural Economics and Food Policy, https://doi.org/10.1007/978-3-031-23535-1_11

systems resilience is inadequately framed and these framings miss the everyday resilience practices of African urban consumers. While emerging arguments encompass useful touchpoints and relevance, in general these lack contextualised, and specifically Southern, applicability.

The emergence of generalised urban and food systems resilience positions aligns both to the increased global awareness of the problem of food insecurity in urban areas, and to the increased focus on subnational policy for sustainable development. There has been a particular flurry of academic interest in the wake of COVID-19 which demonstrated a series of vulnerabilities in the food system and the urban system (Moseley & Battersby, 2020) which evolve at the intersection of multiple systems, including social systems, social services systems, health systems, infrastructure systems and economic systems at an urban scale.

The population of food and nutrition insecure urban residents is growing in absolute terms in many cities in South Asia and sub-Saharan Africa (Crush et al., 2012; Hawkes & Fanzo, 2017; Popkin et al., 2012; Ruel et al., 2017). This figure is higher in informal and marginalized settlements (Crush & Frayne, 2010). For many decades, food and nutrition insecurity has been framed as a food provision issue, with the belief that food shortages and unavailability were the key limiting factors of food security. Yet, this framing lacks the means to effectively conceptualize food and nutrition security, which are also affected by access to, use of and stability of food.

This chapter focuses on the aspirations and limitations associated with the resilience of African urban food systems. This focus is deliberate. Firstly, a key component of the wider argument of this chapter is the importance of contextual nuance and focus, rejecting travelling concepts that are often parachuted into distant contexts (Battersby, 2012a). A further reason is that while other regions, particularly South Asia, may face similar demographic and governance shifts, African cities are poised for dramatic change. The scale and pace of African urban transition requires specific considerations of what urban food system resilience requires and how resilience manifests in context. Added to this, the demographic shifts that have taken place and are currently accelerating in sub-Saharan Africa, mean that our current understanding of the drivers of food and nutrition insecurity, policy responses, as well as threats and related crises, need review and nuance.

Africa's demographic and emergent economic growth is primarily taking place in cities (Cirolia, 2020). When these changes are considered in the context of the COVID-19 pandemic, food, nutrition and urban health issues are of critical importance. Africa's urban transition is taking place in largely unplanned ways. Current urban development processes generally prioritize large scale infrastructure (Cirolia, 2020), decentralized governance (Pieterse et al., 2018) and elite developments (Myers, 2015; Turok, 2016), all with little or no cognizance of urban food system challenges, specifically food and nutrition insecurity (Battersby & Watson, 2018). Africa's urban transition requires far greater attention because "the urban transition of the next few decades will be formative of future developmental opportunities on the continent" (Pieterse et al., 2018, p. 151).

The urban food system allows for a nuanced and broader approach to questions of resilience. Urban food systems and the resultant outcomes are shaped by flows of multiple physical, economic, material and social goods. These are all supported, deepened and made resilient through constantly maintained relationships. The resilience of the food system depends on flows of food, money and social networks inter alia, each of which have unique points of vulnerability and resilience. For many urban food system actors, particularly the poor in rapidly growing African cities, the ability to use the food system is shaped by the form and function of the urban system. However, it is the agency, the ability of those different food system consumers to use the urban food system in ways that suit their daily, contextual and wider community needs, that contributes in a significant way to build their resilience.

Using food as a lens (per Steel, 2008), this chapter builds on critiques from other disciplines, such as urban poverty (Friend & Moench, 2013), taking a more critical view of current urban food systems' resilience approaches and conceptualizations, posing two questions about the intersections between the food system and the urban systems. First, we argue that in many framings of resilience, components such as agency, choice, even a diversity of options, are seldom effectively considered. Secondly, we present food security framings from FANTA (Swindale & Bilinsky, 2006)-aligned food security surveys highlighting the state of food insecurity in cities in Kenya, Namibia, South Africa, Zambia and Zimbabwe to question notions of resilience, particularly framings of resilience articulated as the ability to recover from disruption (Friend & Moench, 2013), stoicism (James, 2011) or revert to a pre-existing state of stability (Padgham et al., 2015). Our point is that many African urban food systems users are in a constant state of crisis or prolonged shock, a concept described by Béné et al. (2016, fig. 7) as a "continual state of incomplete recovery". Does a more traditional framing of the capacity to bounce back (James, 2011) or recover assume a measure of stability, or a desired previous state that is ideal? We suggest in this chapter that different measures of resilience are required, measures that take into account the constant state of crisis, and how the extremes of these perpetual stressors are mediated, diluted and overcome, albeit temporarily, while at the same time valorizing the agency and networks present and activated in these situations.

The chapter proceeds with a review of Africa's urban transition. It then engages some of the current trends evident at the intersections between resilience, urbanization and food security. The chapter then reflects on global processes, such as the Sustainable Development Goals (SDGs) and the UN Habitat New Urban Agenda (NUA) agreements. These global targets and indicators of sustainability and well-being aspire to greater sustainability and resilience. A review of the state of food insecurity is presented supported by responses to that food insecurity and the strategies applied in accessing food. The chapter closes, reflecting on these aspects, their implications for policy engagements around urban food and nutrition systems' planning. We also suggest a need for greater engagement in context, questioning generalizations and approaches that assume universal applicability of such concepts.

## Urban Food Resilience and the 2nd Urban Transition

Africa is urbanizing at a rapid rate, faster than any other continent (UN-DESA, 2019). Two aligned processes are driving this demographic transition. First is a move from rural to urban areas. The general depopulation of rural areas discourse requires interrogation though, with endogenous urban growth being an equal, if not greater driver in Africa (Crankshaw & Borel-Saladin, 2019; Menashe-Oren & Bocquier, 2021). In addition to this, Africa's median age is 19.7 years—the youngest in the world (Saleh, 2021). The rapid growth in African and Asian cities is a component of what has been termed the second urbanization transition (Pieterse, 2008).

The global North was the site of the first general societal urbanization process and was facilitated by, and resulted in, a number of societal shifts. Agricultural innovation and resulting increases in production lowered the price of food. Lower food prices meant reduced rural employment opportunities. Abundant labour and lower food prices were vital drivers of the industrialization process, particularly in rapidly growing urban areas (Beall & Fox, 2009, p. 47; Satterthwaite, 2007). The combination of cheap food, industrialization and subsequent specialization and new forms of urban governance enabled urban development. Cities in Europe and North America became the centres of economic growth. These foundational urbanization processes are playing out in different ways in cities in the global South, with great variations across African cities. The second urban transition is taking place within a particular geopolitical and economic moment (Satterthwaite, 2007). The lack of industrial growth in African cities reflects the sharp contrast between Africa's current urban transition and that of the first urban transition (Swilling & Annecke, 2012). Two differences have a direct bearing on the resilience of African urban populations. Firstly, the first urbanization process paid little attention to the sites of resource and wealth extraction and the longer-term consequences of that extraction, specifically the lock-ins now seen in terms of economy, resource types and ecological and climatic threats. Secondly, the general trend in the first urbanization wave was linked and mutually benefiting urbanization and industrialization. In Africa, urbanization is taking place in the context of a general post-Fordist environment (Pieterse, 2008) and in an economy where there is a restricted middle class and thin tax base (Pieterse et al., 2015), resulting in constrained economic activity. Earlier resource-based development models and systems (such as energy from coal) curtail options further as a result of issues such as climate change and the limitations placed on access to capital and resourcing to facilitate large-scale infrastructure development (Turok, 2016). One manifestation of this is informality, in housing, employment and other essential services. The largely informal profile of African cities presents a very different entry point to engage questions of resilience.

Rapidly growing African urban areas encounter multiple and, at times, mutually reinforcing development realities related to food, including food insecurity (Frayne et al., 2010), food systems transformation (Battersby & Watson, 2018), the growth of supermarkets (Peyton et al., 2015; Reardon et al., 2003) and the nutrition transition (Drewnowski & Popkin, 1997; Hawkes, 2006). When these food systems-related transitions intersect with other transitions, such as increased poverty and inequality, increased climatic variability (IPCC, 2021), global geopolitical shifts and global economic contraction, and watershed events such as pest infestations (often driven by climatic changes) and global health crises, urban areas face significant development and governance challenges.

In sum, Africa's urban transition presents a unique challenge. But it also offers opportunities. The scale of Africa's urbanization is unprecedented (Pieterse & Parnell, 2014; Satterthwaite, 2007). Given the state of under-development in African cities, the African city of 2030 and beyond is yet to be cast in concrete.

General descriptions of the African city, the slum city (Davis, 2006), or the auto-constructed city (Pieterse, 2013), while real, fail to effectively capture the processes, networks and dynamics of the African city. What the African city does reflect is an endless struggle. In this struggle, different forms of city-ness, networks and agency emerge, and with them different forms of resilience. Inequality and its intersection with the everyday nature of diverse African cities feeds directly into how the food system operates.

Despite what has at times been a hostile approach to cities on the African continent (see Pieterse et al., 2018), recognition of the importance of cities in Africa's development is slowly emerging. Importantly cities are also now a key part of wider global governance arrangements such as the sustainable development goals (SDGs). Tracking global processes offers a useful lens through which to understand the emergence of specific approaches and positions, specifically urban resilience.

## Evolving Global Governance Positions

There are many drivers framing what resilience might be. Attempts to articulate resilience in the urban food system, engaging processes that intersect between wider food system threats and challenges, and urban processes heightens complexity. In this chapter, we chart the wider sustainability and resilience-oriented processes and how these intersect with wider global and urban development processes. Here, we apply the challenge posed by Pieterse et al. (2018, p. 151) that

> Rather than bemoaning the lack of policy impact, we suggest it is important for scholars to engage global urban policy-making, probing where and how to augment and refine what is clearly a path-breaking moment in how development on the African continent is understood and how the life in the African metropolis is perceived.

To this end, we align the evolution of global urban and food systems-related development approaches to their impacts on urban food resilience framings. First, we track the evolution of global goals, then the explicitly urban processes. These are followed by a reflection and critical engagement with concepts and approaches that have evolved from, or been given meaning, as a result of these processes and their employment in urban food resilience.

Histories and evolving processes matter. Fukuda-Parr and Orr (2014) argue that the Millennium Development Goals (MDGs) have had important "governance" (policy) and "knowledge" (norms and concepts) effects through their goals, targets and indicators. They have both shifted policy and programmatic focus and redefined the concepts and norms that shape development discourse (Battersby, 2017). It has been argued that the MDG approach and framing may have distorted the development agenda and removed focus from some critical issues (Yamin & Falb, 2012) such as food security and urban food security specifically.

Food and nutrition security was included as part of the MDG 1, under Target 1C: "Halve, between 1990 and 2015, the proportion of people who suffer from hunger." Although the inclusion of food within the first MDG should have assured its prominence, World Bank Managing Director Wheeler called it the "forgotten MDG" (Fukuda-Parr & Orr, 2014, p. 152). This indicator-led development agenda has been critiqued for allowing these indicators to drive development agendas, rather than being used to measure progress towards broader development aims. And as has been argued "measurement drives diagnosis and response" (Barrett, 2010, p. 827).

The SDGs build on the foundation provided by the MDGs which they replaced in 2016. SDG Targets 2.1 and 2.2 are the only two food security targets within the SDG 2 targets, and both frame the food problem as one of scarcity (Burchi & Holzapfel, 2015, p. 17). This does not reflect the current and future realities of food insecurity in the global South. The lack of attention to urban and consumption issues within SDG 2 is exacerbated by the lack of dialogue across goals (Weitz et al., 2014).

SDG 11, which focuses on cities and communities, makes no reference to food and consumption in any way. As Jonathan Crush has argued in a public presentation, "SDG 2 imagines a world without cities, while SDG 11 imagines an urban world in which no-one eats" (Crush, 2017). The omission of food in SDG 11 was deliberate. In order to enable the inclusion of an explicitly urban goal in the SDGs, caution was required to avoid encroaching on other goals and interests. This was particularly evident with SDG2. While the inclusion of an exclusively urban goal is of great benefit, it does present certain challenges. Given scarce resources in many African cities, statistical data collection is limited (Acuto & Parnell, 2016). Programming and reporting on the SDGs are being prioritized over other data collection processes. In countries where limited data exists and data collection capacity is weak, this has consequences. Where SDG-related data collection becomes one of the only measurement processes in these countries, evidence for policy programming is constrained. If diagnosis of the realities is constrained by the limited focus of the goals, policy actions run the risk of missing key needs. Misaligned policy actions run the risk of undermining resilience. Arguably, it was left to the Habitat III and the policy position articulated in the New Urban Agenda to then enable some measure of urban food programming and policy action.

Habitat III, the United Nations Conference on Housing and Sustainable Urban Development, which took place in Quito, Ecuador, in 2016 formed part of the bi-decennial cycle (1976, 1996 and 2016) of urban conferences facilitated by the UN body, led by UN-Habitat. Habitat III was one of the first United Nations' global summits after the adoption of the Post-2015 Development Agenda. It was an opportunity to discuss important urban challenges and questions, such as how to plan and manage cities, towns and villages for sustainable development. In particular, the Habitat III conference elaborated on SDG11 to "Make cities and human settlements inclusive, safe, resilient, and sustainable" (Amann & Jurasszovich, 2017).

The Habitat III conference aimed to re-assert the global commitment to sustainable cities, through the implementation of the conference strategic outcome. The New Urban Agenda set out an urban vision encapsulated in three overarching principles and four commitments. The principles include:

(a) Leave no one behind, by ending poverty in all its forms and dimensions.
(b) Ensure sustainable and inclusive urban economies by leveraging the agglomeration benefits of well-planned urbanization,
(c) Ensure environmental sustainability by promoting clean energy and sustainable use of land and resources in urban development (UN-Habitat, 2017, p. 7).

The four mechanisms envisioned for effecting the New Urban Agenda are:

- National urban policies promoting "integrated systems of cities and human settlements" in furtherance of "sustainable integrated urban development".
- Stronger urban governance "with sound institutions and mechanisms that empower and include urban stakeholders" along with checks and balances, to promote predictability, social inclusion, economic growth, and environmental protection.
- Reinvigorated "long-term and integrated urban and territorial planning and design in order to optimize the spatial dimension of the urban form and deliver the positive outcomes of urbanization"; and
- Effective financing frameworks "to create, sustain and share the value generated by sustainable urban development in an inclusive manner" (UN-Habitat, 2017, p. 8).

Together, the MDGs, SDGs and the NUA established a platform for how sustainable urbanization and wider sustainability questions are framed and understood at the global scale. These framings intersect with other aligned more normative approaches that seek to connect different systems, envisioned as offering opportunities to enhance sustainability and resilience.

## Emerging Concepts and Positions

The notion of sustainability that developed following the 1972 United Nations Conference on the Human Environment, and the later Brundtland Commission that gave voice to the term sustainable development in their "Our Common Future" report (WCED, 1987) formalized earlier processes and embedded the concepts into global governance structures. Subsequently, multiple iterations of the concept sustainable development and sustainability have emerged (Swilling & Annecke, 2012). Alongside these multiple conceptualizations, various strategies and models have been suggested as tools to enact and enliven the concept of sustainable development (Dresner, 2012).

At the intersection between food and cities, varied positions are articulated. These approaches are often embedded in specific normative, ideological and value-oriented positions (see Haysom et al., 2019). Four such approaches are discussed below, selected because they align with the vision of sustainable urban food systems embedded within the New Urban Agenda, each suggesting tools through which urban food systems resilience could be enacted.

### *Urban Agriculture as a Source of Food System Resilience*

There is a body of evidence questioning the importance of urban and peri-urban agriculture (UPA), suggesting that its role in relation to food security has been overstated (Ellis & Sumberg, 1998; Zezza & Tasciotti, 2010). Others have presented concerns over the assumptions about the UPA growers (see Battersby, 2012b). Padgham et al. (2015) have identified a variety of factors that undermine the potential of UPA. Despite this, UPA is still actively advocated and promoted by some as a powerful urban food systems' resilience tool. This is best seen in a recent peer reviewed piece arguing in favour of UPA as a "necessary pathway towards urban resilience and global sustainability" (Langemeyer et al., 2021, p. 1). Here, it is argued that UPA reduces and offers resilience against social and ecological vulnerabilities and risk-related inequalities of urban inhabitants, including food shortages and different scenarios of global change, including climate change or pandemic events such as COVID-19. UPA is also posited by these authors as an effective tool to reduce the "intensified negative environmental (and related social) externalities caused by distant agricultural production, as well as lacking consideration of nutrient re-cycling potentials in cities (e.g., from wastewater) to replace emission intensive mineral fertilizer use" (Langemeyer et al., 2021, p. 1). Finally, the authors' point to the multifunctionality of UPA, and with it, the "multiple benefits it provides beyond the provision of food, including social benefits and insurance values, for instance the maintenance of cultural heritage and agro-biodiversity" (Langemeyer et al., 2021, p. 1). African urban food system studies do not find empirical evidence to support such assertions, particularly for poorer urban residents (see, e.g. Crush et al., 2017; Joubert et al., 2018; Pieterse et al., 2020).

### *Localized Food Systems as a Source of Resilience*

There have been many advocates for localized systems, some embedded in local knowledge and cultures (Norberg Hodge, 2013), embeddedness within local systems (Feenstra, 2002) and economic regeneration

(Hopkins, 2011), to name a few. However, as Born and Purcell (2006, p. 196) point out "the assumption that the local is desirable does not always hold". Born and Purcell's statement is based on case study evidence, but also a wider normative and theoretical engagement in questions of scale. These critiques challenge some of the wider SDG-related assumptions and earlier sustainability processes, such as Local Agenda 21, now Agenda21. Agenda21 is a non-binding action plan promoting a local sustainability agenda. This emerged following the 1996 Rio Summit on Sustainable Development, and guides work of a number of urban bodies, including Local Governments for Sustainability or ICLEI, previously the International Council for Local Environmental Initiatives.

### City Regional Food System as a Way of Building Resilience

More recently and embedded within the overarching urban food system framing, and by implication, resilience agenda, of Habitat III, was a wider vision of cities being able to "fulfil their territorial functions across administrative boundaries and act as hubs and drivers for balanced, sustainable and integrated urban and territorial development at all levels" (UN-Habitat, 2017, p. 7). With specific reference to food and food systems governance, the City Region Food System (CRFS) aspires to "a sustainable, resilient CRFS (…) to enhance sustainability across scales and sectors" (Blay-Palmer et al., 2018, p. 3). The CRFS is argued to increase access to food generate decent jobs and income, increase the region's resilience against shocks and lessen the dependence on distant supply sources, fostering rural-urban linkages, promotes ecosystem and natural resources management, and supports participatory governance (Blay-Palmer et al., 2018).

### The Water-Energy-Food Nexus

As a concept, the wider nexus hypothesis of the water-energy-food nexus was launched at the Bonn 2011 Nexus Conference (Endo et al., 2017, p. 20). This work is also closely linked to a publication on water security, also using the nexus concept, produced by the World Economic Forum (WEF) in 2011 (WEF, 2011). The WEF work starts with a position that assists in clarifying how this concept frames the nexus: "Water security is the gossamer that links together the web of food, energy, climate, economic growth, and human security challenges that the world economy faces over the next two decades ... the increasing volatility on food prices in 2008, 2009 and again in 2010 should be treated as early warning signs of what is to come. Arguably it is water that lies at the structural heart of these agricultural challenges" (WEF, 2011, p. 1). While the WEF presented the nexus framework from a securities' perspective (water–energy–food security), later versions have taken on various facets with alternative components (Biggs et al., 2015). Subsequently, "there is no fixed concept of nexus (...) but the (...) nexus is internationally interpreted as a process to link ideas and actions of different stakeholders under different sectors and levels for achieving sustainable development" (Endo et al., 2017, p. 22).

These positions and concepts all intersect with and are given life in an array of urban food and resilience programmes and practices that are emerging. These are generally global but are increasingly grappling with the challenges evident in African cities. Many of these, thanks to their wider urban systems framings, engage the question of urban food systems resilience, while some are deliberately urban food focused, embedded within wider sustainability and resilience conceptualizations. The leading programmes are summarized in Box 11.1.

---

**Box 11.1: Leading urban food systems sustainability/resilience programmes and initiatives**

**The Resilient Cities Network**[1]: Consists of member cities and Chief Resilience Officers from the 100 Resilient Cities programme (100RC), sharing a common lens for holistic urban resilience and projects in implementation. 100RC was initiated by The Rockefeller Foundation in 2013, as part of its Global Centennial Initiative.

**C40** is a network of mayors of nearly 100 cities[2]: C40 was founded in 2005 by the then mayor of London where agreement was sought (initially from a collection of "mega-cities"—the C20) for cooperatively reducing climate pollution. Following the formation of a small group, in 2006 a further collection of mayors were invited to ensure "balance from the Global South", creating an organization of 40 cities. Also in 2006, the Clinton Climate Initiative became the implementing partner and in 2007, then New York City mayor Michael Bloomberg hosted the second C40 Summit, and in 2010 he was elected Chair of C40.

**Milan Urban Food Policy Pact (MUFPP)**[3]: Formed in 2014 by the mayor of Milan, MUFPP is an international agreement of mayors that operates as a network of cities, providing tools and guidance for cities seeking to implement urban food actions. The MUFPP uses a Framework for Action listing 37 recommended actions, clustered in six categories, as the overarching guide to urban food systems' engagement. For each recommended action, there are specific indicators designed to assist in monitoring progress for implementing the Pact. The six MUFPP categories or indicators are:

- The presence of an active municipal interdepartmental government body for advisory and decision-making of food policies and programmes
- An active multi-stakeholder food policy and planning structure
- A municipal urban food policy or strategy and/or action plans
- An inventory of local food initiatives and practices to guide development and expansion of municipal urban food policy and programmes
- A mechanism for assembling and analysing urban food system data to monitor/evaluate and inform municipal policy-making on urban food policies
- A food supply emergency/food resilience management plan for the municipality based on vulnerability assessment

**FAO—Food for the Cities initiative**[4]: In 2001 the FAO launched this multidisciplinary initiative, aimed at addressing the challenges that urbanization brings to urban and rural populations, as well as the environment, "by building more sustainable and resilient food systems". The initiative draws on the CRFS concept as the key philosophical and operational concept informing how urban food systems are conceptualized and approached. CRFS "encompasses a complex network of actors, processes and relationships to do with food production, processing, marketing, and consumption that exist in a given geographical region that includes a more or less concentrated urban centre and its surrounding peri-urban and rural hinterland; a regional landscape across which flows of people, goods and ecosystem services are managed" (FAO, n.d.).

**ICLEI—RUAF City Food Network in Africa**[5]: The ICLEI-RUAF city food network (CITYFOOD) is a global network for local and regional governments with the ambition to develop a strategic approach to their city-region food systems. It aims to accelerate action on sustainable and resilient city-region food systems by combining networking with training, policy guidance and technical expertise. CITYFOOD Network is open to all local and regional governments. Over and above hosting the African City Food Month, ICLEI CITYFOOD projects include: Investing in Urban African Food Systems (FS-Invest); Understanding Innovation in

> Food Water Energy Nexus and Green and Blue Infrastructure (IFWEN), assisting in the Cape Town Food System Vision, developing food governance frameworks in cities such as Arusha and Antananarivo (Fisher et al., 2019), City-to-city Food System Exchange (RF) and Innovative Food System Governance.

When read as a collective, the global governance actions and the emerging concepts pertaining to urban food, and the subsequent projects, programmes and actors driving efforts to enhance urban food systems resilience, a number of key trends become evident. The first is the embedding of the urban food system within a region, evident in the framing of the NUA and local food system action, UPA and the City Region Food System. An important component is the descaling of food and food security as a general concept, something that for the most part has been seen as a rural issue, to the urban and regional scales. Additionally, these concepts actively link the urban food system to questions of sustainability and resilience.

There has, however, been little reflection on the utility of a regional focus in specific urban governance questions. How, given their distinct administrative boundaries and urban mandates, can cities engage issues that extend beyond their governance purview and legal mandates? Do concepts such as the CRFS and the Water-Energy-Food Nexus and the wider framings of food within the NUA adequately engage these scalar challenges? This is not a call for local food systems, in full recognition of Born and Purcell's (2006) warning, but it is a central question of governance. Flows of food into a city and the reliability of these flows presents a specific resilience question, but these flows do not make up the urban food system. Such systems are far more complex and require very deliberate governance approaches to effectively engage scale, but also the multiple other systems that make up the urban system. Here, the

---

[1] https://resilientcitiesnetwork.org/our-story/.

[2] https://www.c40.org/about-c40/.

[3] https://www.milanurbanfoodpolicypact.org/the-milan-pact/.

[4] https://www.fao.org/fcit/fcit-home/en/ and https://www.fao.org/in-action/food-for-cities-programme/en/.

[5] https://africa.iclei.org/cityfood/.

fundamental management question of "who wakes up and worries about an issue?" needs to be asked. Can urban food resilience be achieved if multiple actors across multiple scales, working to their own mandates and principles are assumed to hold the interests of those for whom urban food systems resilience needs to be facilitated?

Such questions are compounded by the fact that traditional framings of food, aligned to agriculture, as rural (Crush & Riley, 2018) and twin track development approaches (see Frayne et al., 2010) mean that cities have limited food systems policy mandates (Battersby & Watson, 2018). Roles are often reduced to policing and controls, not strategic governance. The result is that food security and wider food system mandates are delegated to national and regional governments, along with fiscal allocations to enable programming and action. Governing urban food systems and enabling some form of urban food system resilience therefore takes place in the context of an absent mandate, undermining strategic action and deep engagement and effective urban policies that might enable resilience (see Fig. 11.1).

Additionally, given the international profile of these global governance and programmatic initiatives seeking to enhance urban resilience, is there adequate African specificity in the targets, metrics and even overarching

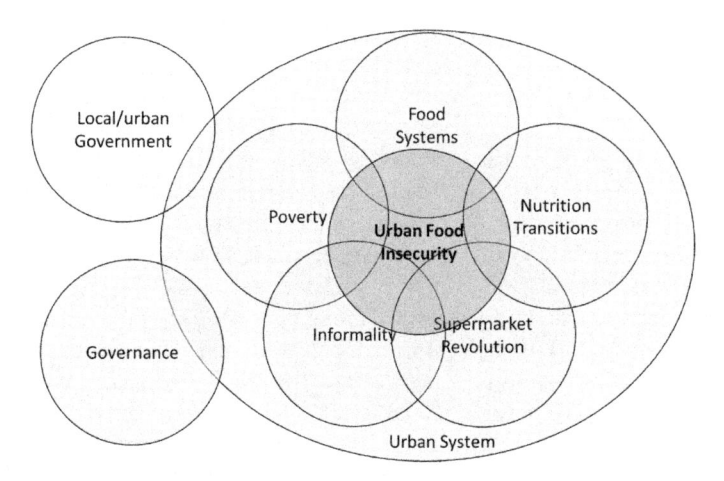

Fig. 11.1 Urban food insecurity within a wider set of intersecting urban processes and vulnerabilities despite absent governance action (*Source* Authors' own representation)

assumptions that underpin these initiatives? Measurement proxies and programming interventions are in effect generic copies of assumed pathways to sustainability and some form of resilience. Absent in these are locally specific needs and issues.

The next section draws on a selection of African food systems studies to attempt to answer these questions and challenge some of the apparent assumptions that drive such processes. These case studies also speak specifically to the challenges of food insecurity and Africa's urban transition.

## Understanding Resilience in the African Urban Food System

Urban food resilience spans different scales. While the impacts of regional disruptions to food supply, for whatever reasons, might impact the resilience of a localized food system, this chapter explicitly focuses on the strategies, and urban food systems processes evident in select African cities to attempt to answer earlier questions. This engagement works from a number of positions that require clarification, as these views inform our arguments. Firstly, the household food system status, specifically the state of food and nutrition security, is an indicator of the state of the food system in a specific urban context. Using food and nutrition security as an indicator of potential resilience is also strategic. This focus is part of a realization that households face significant resilience challenges and that the general state of food insecurity is declining globally (see, e.g. Fanzo in this volume). The focus on food security, or food insecurity, also assumes a pre-existing state of distress and as such, we avoid engaging in the ability to "bounce back" or stoicism of society, but rather, strategies to enable a measure of shorter-term stability, and longer-term sustainability as tools to secure improved or stable food systems outcomes. The primary imperative is to ensure a measure of food access, and then adequate and appropriate utilization of food. These processes are enabled through effective policy and programming. Central to ensuring food access is individual and community agency. It is for this reason that we utilize the recent FAO High Level Panel of Experts (HLPE) expanded definition of food security, one that includes availability, accessibility, utilization and stability, but also, sustainability and agency as determinants of food security (HLPE, 2020).

We draw on the HLPE definition of a food system, which makes the connection between food security, the food system and the wider set of systems in which food operates. The HLPE (2014, p. 12) defines a food system as "a system that gathers all the elements (environment, people, inputs, processes, infrastructures, institutions, etc.) and activities that relate to the production, processing, distribution, preparation and consumption of food, and the outputs of these activities, including socioeconomic and environmental outcomes".

Importantly, the articulation of food security and the food system, and how this might impact the urban scale, is necessary because "previous work on food security has conventionally focused at either the household scale or at aggregate food production, with far less focus on the food system itself and its intersection with cities" (Battersby & Watson, 2018, p. 3). These connections are important because food security and food systems are influenced by governance decisions made in the absence of consideration of their potential food security impact (Battersby & Watson, 2018, p. 1).

Using evaluations based on the food security measurement approaches developed by the United States Agency for International Development's Food and Nutrition Technical Assistance programme (FANTA) (Coates et al., 2007). Table 11.1 provides an indication of the state of food security across a collection of both primary and secondary African cities.

Table 11.1 reflects both the contextual variation across different African cities and the extreme state of food insecurity in studies targeting poorer communities only. Household dietary diversity scores (HDDS) (Swindale & Bilinsky, 2006) presented in Table 11.1 reflect a significant dietary challenge. As a general rule, HDDS scores of less than six are seen as a proxy indicator for potential risks of under nutrition. Across many cities, the average scores are less than or equal to six. Amounts reflected in Table 11.1 are from pre-pandemic research between 2015 and 2018 and as such, scores are undoubtedly worse in 2022 following the impacts of COVID-19.

The varied levels in the 12 HDDS food groups consumed from a Cape Town study provide insights into different dietary profiles across income terciles (Fig. 11.2). A similar profile was evident across the different cities assessed. Figure 11.2 also highlights an increase in grains, and less fruits, vegetables, proteins and pulses in the diets of the lowest income tercile. This raises questions about whose food system resilience requires policy support and targeting at the urban scale. Given the developmental and public health challenges associated with constrained diets, do resilience

**Table 11.1** Food security and dietary diversity scores for selected Southern cities—periods 2015–2018

| (HFIAP) categories | Cape Town (n = 1200) | Maputo (n = 2105) | Nairobi (n = 1414) | Kisumu (n = 829) | Epworth (n = 482) | Windhoek[a] (n = 875) | Kitwe[a] (n = 871) |
|---|---|---|---|---|---|---|---|
| Food secure | 45% | 29% | 29% | 20% | 8% | 6% | 6% |
| Mildly food secure | 6% | 11% | 13% | 9% | 5% | 5% | 3% |
| Moderately food insecure | 13% | 22% | 33% | 26% | 20% | 5% | 12% |
| Severely food insecure | 36% | 38% | 25% | 45% | 67% | 84% | 79% |
| Dietary diversity (HDDS/12) | 6.8 | 4.1 | 6.1 | 4.1 | 4.2 | 3.2 | 3.25 |

[a]All studies are at the city scale excepting Kitwe and Windhoek
Source Pieterse et al. (2020)

enhancing programmes need to focus on the overall food system resilience or adopt a deliberate pro-poor approach?

The results shown in Table 11.1 and Fig. 11.2 are drawn from conventional surveys and their tools, such as the FANTA. It is suggested here that these food security and food access metrics are still largely informed by the food availability and food access dimensions of the four dimensions interpretation of the earlier FAO (1996) definition of food security. Such measures have value in terms of understanding the state of food insecurity in both urban and rural households. However, they are less appropriate to offer greater detail on the profile and functioning of that food system. Second, they fail to capture some of the more multi-dimensional poverty challenges faced by those households under review. For policy and programming understanding, the state of food insecurity is essential but having greater clarity on how the specific food system is being used by different elements of society would add great utility.

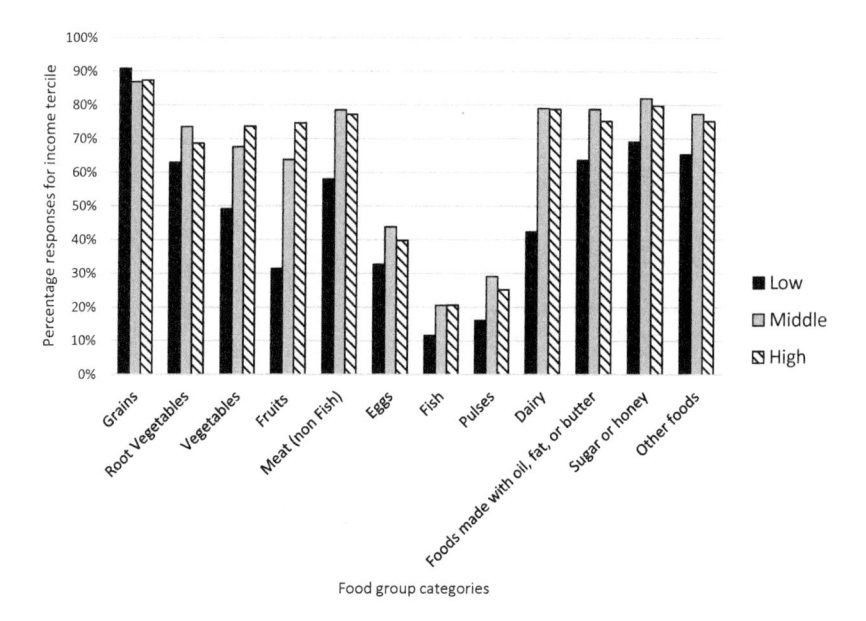

**Fig. 11.2** Household food types by HDDS food groups across all income groups in Cape Town, South Africa (High: $n = 200$; Middle: $n = 504$; Low: $n = 1800$) (*Source* Haysom et al. [2017]—from 2013 data)

Using data again drawn from Cape Town, Fig. 11.3 offers a sense of the diversity of food sources used to access food and highlights the variations across the same income terciles detailed in Fig. 11.2. Figure 11.3 suggests that while all income categories may use supermarkets, other food retail outlets are used frequently, with the informal economy being a key source of food access for low-income communities. Local *Spaza shops* (a South African term for a small informal corner-type store located in a neighbourhood) and street vendors are a key source of food access for poor households.

The frequency of use of informal food access points requires greater engagement (see Fig. 11.3). The fact that local corner stores are used as frequently as "at least five days a week" or "at least once a week" raises questions about proximity, spend profiles and other household needs, such as infrastructure. These frequency profiles are very different to the purchasing patterns of the upper income terciles, where weekly and monthly shopping dominate, indicating different food use and purchasing

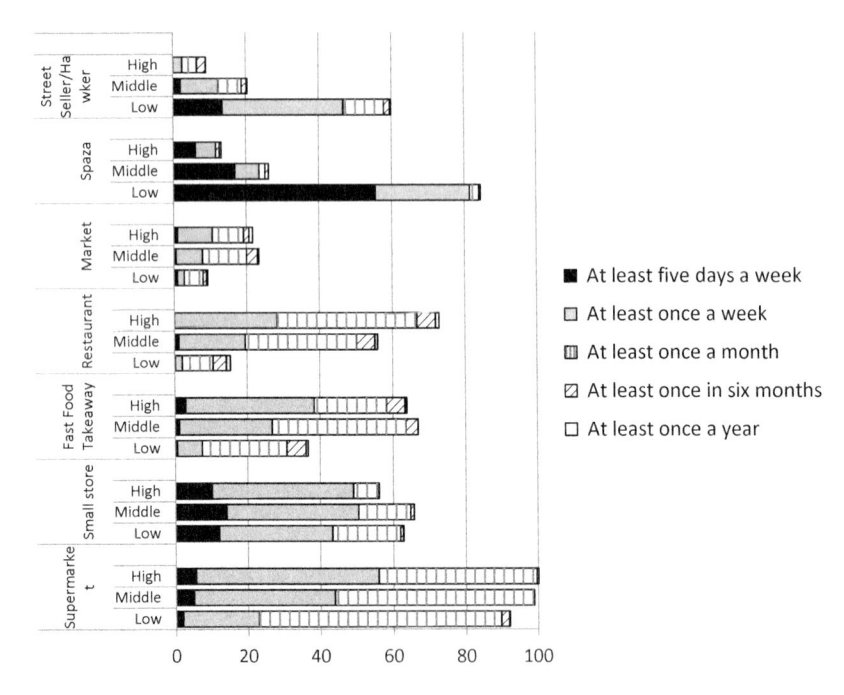

**Fig. 11.3** Food access profiles in Cape Town, South Africa ($n = 2{,}503$) (*Source* Haysom et al. [2017])

profiles. The reasons for such variations in use frequency and typology (nature of the retail outlet) of access differ in different contexts.

What the food access use and typology variation shows are varied everyday strategies applied by households in cities. This requires an engagement with the multi-dimensional nature of poverty (MDP). This is essential given that the second urban transition is not driving economic growth and not absorbing labour (Satterthwaite, 2007). While this does not directly capture food security, it is a means through which the intersections between food choice, the food system and the urban system can be better understood. The example used below in Fig. 11.4 draws on a form of MDP, the Lived Poverty Index (Mattes, 2008), which reflects a categorical scale indicating the frequency over the past year in which households went without food, fuel to cook food and clean water (Haysom & Fuseini, 2019). Here, multiple deprivations are evident.

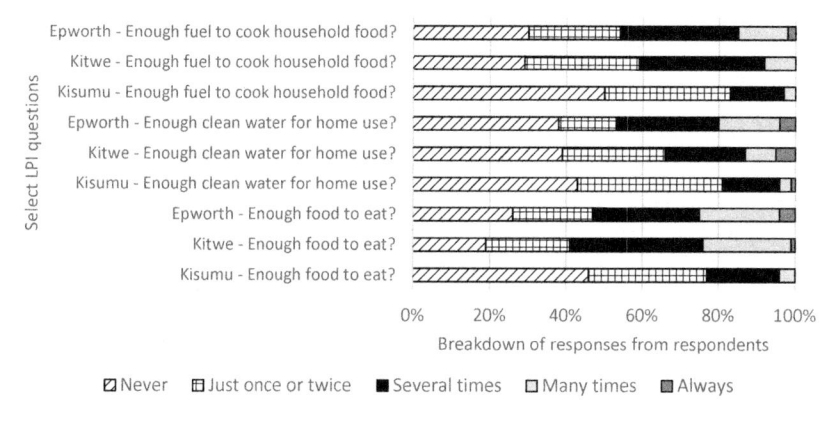

Fig. 11.4 Combined Lived Poverty Index results for key indicators in three African cities in response to the question "how often in the past year did you go without …" (Epworth, Zimbabwe $n = 483$; Kitwe, Zambia $n = 871$; Kisumu, Kenya $n = 841$) (*Source* Haysom and Fuseini [2019, p. 17])

As Kennedy et al. (2004) have pointed out, infrastructure is a key consideration to understand food and nutrition security. While the general focus on the links between food access and nutritional intake and water and sanitation has received attention (Young et al., 2021), links to the means to prepare food, household appliances and other factors that influence choice have been less considered. These questions are absent from traditional food and nutrition security assessments.

Household and community scale decisions, and the different forms of agency that inform the decisions, align with the process of iteration, projectivity and practical evaluation (Emirbayer & Mische, 1998). Such processes shift in response to wider urban disruptions and constraints, and would inform actions taken to enhance resilience, as Emirbayer and Mische point out:

> structural contexts of action are themselves temporal as well as relational fields—multiple, overlapping ways of ordering time toward which social actors can assume different simultaneous agentic orientations. Since social actors are embedded within many such temporalities at once, they can be said to be oriented toward the past, the future, and the present at any given moment, although they may be primarily oriented toward one or

another of these within any one emergent situation. (Emirbayer & Mische, 1998, pp. 963–964)

The household dynamics observed in Fig. 11.4 have an additive effect and intersect with the food retail options chosen and the nature of that food retail in many African cities. Informal vendors are able to adapt and respond to infrastructure deficiencies in ways that the formal sector cannot. In many instances, the informal sector takes on the role of the refrigerator, of the stove and of the storage cupboard for poor households living in precarious types of dwelling, across the continuum from self-built formal to largely informal dwellings. The adaptive approaches applied by informal food vendors enables a measure of food access, through bulk breaking and pre preparation of key foods.

> These days, because wood is scarce and other cooking fuel is so expensive, we cook a lot of these beans and then we place these in small packets and freeze them … otherwise no-one here would eat beans anymore; the tinned beans from the shops are also too expensive. Dry beans are not expensive, other costs [energy] are high and this takes beans from diets. Informal food trader, Epworth, Harare, 2016

The quote from the vendor in Epworth demonstrates the processes and practices that are emerging. Dry beans have traditionally been a key component of many diets across Africa, but these take time to prepare and, as a result, require energy. Infrastructure deficiencies are also faced by food vendors.

Such examples point to the fact that negative food system outcomes are at times driven by factors external to the food system. In the urban context, the functioning and distributional aspects of urban infrastructure, plays a central role in how food secure urban residents might be and influence the strategies to enhance food access. The same applies when considering urban informal economy actors.

Figure 11.5 reflects the responses from vendors in the city of Kisumu, in Kenya, where the top three costs are directly related to infrastructure. The cost of transport is linked to both the temporary nature of these stores making them difficult to secure, but also the risks associated with unreliable (but costly) energy. Here, vendors strategically choose to stock small amounts and re-stock daily. This drives up prices as bulk purchasing benefits fall away, and costs associated with transport impact food prices.

This is again reflected in the high costs associated with spoilage. Refrigeration is not listed as one of the highest cost items, but this is due to the low frequency of respondents who chose to make use of refrigeration.

The high levels of food insecurity, linked to high levels of multidimensional poverty, represent the reality in most African cities. Households make use of a diverse food system to make the best of the limited resources that they have. Equally, as the MDP analysis shows, other costs eat into food budgets. Dietary diversity is low. Staple foods dominate diets, and nutrition-providing foods are largely absent. Informal food vendors respond to the challenges faced by households but they themselves are subject to similar resource constraints. The poor state of infrastructure means that vendors change the foods that they stock to both respond to the households' resource constraints, but also to ensure minimal losses for their own businesses as a result of such infrastructure deficits. These intersecting challenges are being responded to in dynamic and resourceful ways by both vendors and households.

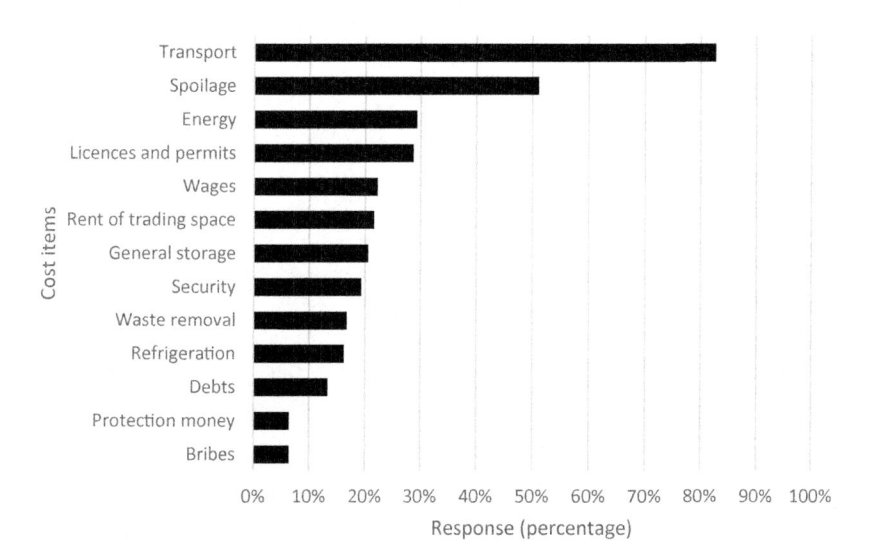

**Fig. 11.5** Kisumu food trader costs after stock purchase costs (multiple response option; $n = 1,839$) (*Source* Opiyo et al. [2018])

## Discussion and Conclusion

The vignettes used here provide insights into both the household food and nutrition challenges, as well as some wider food systems challenges. They also offer insights into the specific challenges that emerge when the urban food system intersects with the wider urban system. A resilient food system needs to be framed as a system shaped by flows of multiple physical, economic and social goods. These systems are underpinned and supported by social and economic relationships, flows of food, money and social networks, each of which having unique points of vulnerability and resilience.

Households and communities are faced with high levels of food stress (be this food insecurity, reduced diets, even hunger) and wider stressors that impact and constrain food choices, such as energy costs, time poverty, costs related to earning an income and general poverty. As a result, households strategically use the diversity of the urban food system, and the opportunities provided by the urban system, to mediate food system and food security stress. Diversity of flows exist across a series of systemic connections. Examples include economic flows (access to credit networks) and remittance flows in accessing food (both by traders and consumers) and alternative capital circulation (*stokvels*,[6] buying clubs, etc.). The urban poor navigate these multiple systems (formal and informal retail, social networks, alternative capital circulation). Households apply a diversity of access and utilization strategies. These flows are further supported by other access strategies such as social networks and material flows and materiality of flows. These terms (material and materiality) are used deliberately to elevate discussions from a single grid view, a material flow, offering ways of considering an object, service or resource and the material relations and agency, as well as the interplay between agency with the material realities of everyday life and broader political and socioeconomic structures (see Bennett et al., 2010).

We argue that it is these networks, social, material, political and institutional, coupled with the diversity of the urban food systems, and that of the urban system, spanning the economic continuum, from deeply

---

[6] *Stokvels* are invitation-only clubs of 12 or more people serving as rotating credit unions or saving schemes in South Africa, where members contribute sums of money to a central fund on a weekly, fortnightly or monthly basis. The name "stokvel" originated from the term "stock fairs", the rotating cattle auctions of English settlers in the early nineteenth century.

informal to formal, that enable urban food systems resilience. This type of resilience forms part of everyday activities and practices.

The intersections between food choices and infrastructure deficits are inadequately understood. Households are having to make critical strategic choices every day. In the absence of infrastructural justice—equitable access to reliable and affordable infrastructure—in many of Africa's urban areas, the choices that are made, we argue, are not driven by ignorance, disregard for health outcomes, as some nutrition literature might argue (Abuya et al., 2012; Igumbor et al., 2012), or laziness, even fecklessness (see Tihelková, 2015). These choices are highly strategic but do involve having to make challenging trade-offs between immediate needs and the consequences of those choices, a form of negotiated resilience (Ziervogel et al., 2017). Households' choices align directly with and counter deficits in Africa's urban systems, and the wider dynamics associated with Africa's urban transition. Here, households make use of multiple grids, physical, social, material and relational. This "griddedness", an overlay of options, as opposed to a hierarchy of options, provides a diversity of options and enables a measure of resilience which is at best not fully understood, at worst disregarded in the wider framings and programmes gathering momentum in the current urban food thematics.

Given these forms of resilience and their intricate connections to the urban and other systems, our work suggests that the existing city regional framing employed by a number of global actors needs to be cognizant of these nuanced and yet highly vulnerable systems of networks, contingency, negotiation, compromise, deep knowledge and "hustle". Principally, these programmes with a predominantly urban food-specific framing of resilience overlook the intersections with the urban system. Equally, general urban programmes miss the nuance of the urban food system as used and relied on by the urban poor in African cities.

Further, our work suggests that there is a need to situate policy recommendations within a more contextually informed framework. When food security and related policies are still viewed by most African governments as a national mandate, policy directives and fiscal flows to enable operational actions generally prohibit the establishment of municipal interdepartmental government bodies for advisory and decision-making of food policies and programmes. Where agency is constrained as a result of colonial histories and extreme inequalities, the urban poor, those actively seeking to enhance their own food system resilience, are excluded from active multi-stakeholder food policy and planning structures. The few processes that exist are not open to all, often dominated by political leaders at the urban scale and remain a collection of connected elites. As per the above governance structuring, the likelihood of local government's having an urban food policy or strategy and/or action plans is slight, but possible. For a local governance structure to effectively programme and drive real change fiscal resources are required. Having access to such resources for urban food actions is highly unlikely. Given the data paucity in African cities (see Acuto & Parnell, 2016), and the high levels of informal food system activities that are not always recognized, developing an inventory of local food initiatives and practices to guide development and expansion of municipal urban food policy and programmes would be difficult. Equally, having a mechanism for assembling and analysing urban food system data to monitor/evaluate and inform municipal policy making on urban food policies is unlikely. As a result, the general approach of city governments is to defer to case studies or studies from other cities as proxy indicators. The absence of a common urban food systems view, or political mobilization, would make the viability of any food supply emergency/food resilience management plan for the municipality based on vulnerability assessment very challenging.

Given these limitations, the concepts and arguments used to lobby for urban food systems resilience should be subjected to great critique. Urban agriculture projects offer little real food systems resilience. When an expanded view of urban resilience is taken, as argued here, adding further areas for negotiation and network building, such as facilitating access to land, securing tenure and accessing water all make resilience even more complex, thus potentially even undermining the initial objective.

The limited potential to scale out and expand such urban agriculture programmes also means that it reduces innovation, allowing city governments a sense of "doing something", but in reality, only assists a few. More problematically, the UPA focus on sites or projects constrains any attempt to engage the wider systemic issues, those that local actors are building networks to mitigate. Local is important. However, our view is that local should not be viewed as confined to a discrete bounded area. Local governance of the food system serves the discussed forms of resilience best when there is a detailed understanding of the benefits and risks in specific contexts provided by the diverse social, material, economic and wider food system flows. Local is important in terms of how resilience actions and responses can be enabled through these flows.

Importantly, given the centrality of urban systems in enabling urban food systems resilience, key enablers of urban food systems resilience are local government and local governance. While multiscale governance of the food system is essential, the current focus on territorial food systems as the primary mode of engagement serves to obscure the vitally important work of local government in addressing food systems issues that fall within their borders and within their mandates.

Based on these arguments, we see urban food systems resilience as a system in which urban households (1) can have access to and support from a diverse food system, have diverse food access options and a diversity of alternatives that align with their lived realities (accessibility, utilization); (2) can make food-related decisions and choices within an infrastructurally just urban system (utilization, stability); and (3) in which system users are seen as active agents in shaping theirs and wider food system resilience (agency and stability and sustainability).

Required is greater engagement with the urban system, approaches that engage across multiple infrastructure grids and systems. These are highly complex systems and cannot be understood in silos. Arguably, the complexity of the system offers a measure of "negotiated resilience" (as per Ziervogel et al., 2017). This resilience forms part of a collection of urban practices enlivened to secure a measure of resilience. The outcomes of such processes do not necessarily align with the framings of resilience proposed in international programmes and initiatives such as those discussed here. Existing governance structures and regimes are ill-equipped to engage in and surf the complexity of these intersecting systems, essential processes if resilience is to be facilitated. The absence of a distinct urban food mandate makes this even more challenging. Further, when urban food is equated to production and increased supply, linked to rural regions, the ability to enhance food security and urban food resilience through a focus on deliberate urban processes and mandates, such as energy, transport, water provision and economic activity is lost.

New forms of governance are required, together with new forms of measurement. Also required is the validation of local knowledge and the diversity of resilience actions. These should be integrated into new forms and understandings of African cities and African urban food systems resilience.

# References

Abuya, B. A., Ciera, J., & Kimani-Murage, E. (2012). Effect of mother's education on child's nutritional status in the slums of Nairobi. *BMC Pediatrics*, *12*(1), 1–10.

Acuto, M., & Parnell, S. (2016). Leave no city behind. *Science*, *352*(6288), 873–873.

Amann, W., & Jurasszovich, S. (2017). Habitat III-A critical review of the new urban agenda. *Housing Finance International-The Quarterly Journal of the International Union for Housing Finance*, 35–39.

Barrett, C. B. (2010). Measuring food insecurity. *Science*, *327*(5967), 825–828.

Battersby, J. (2012a). Beyond the food desert: Finding ways to speak about urban food security in South Africa. *Geografiska Annaler: Series B, Human Geography*, *94*(2), 141–159.

Battersby, J. (2012b). Urban agriculture and race. In R. Slocum & A. Saldana (Eds.), *Geographies of race and food fields, bodies, markets* (pp. 115–130). Ashgate.

Battersby, J. (2017). MDGs to SDGs—New goals, same gaps: The continued absence of urban food security in the post-2015 global development agenda. *African Geographical Review*, *36*(1), 115–129.

Battersby, J., & Watson, V. (2018). *Urban food systems governance and poverty in African cities* (p. 290). Taylor & Francis

Beall, J., & Fox, S. (2009). Urbanisation and development in historical perspective. In J. Beall & S. Fox (Eds.), *Cities and development* (pp. 34–66). Routledge.

Béné, C., Mehta, L., McGranahan, G., Cannon, T., Gupte, J., & Tanner, T. (2018). Resilience as a policy narrative: Potentials and limits in the context of urban planning. *Climate and Development*, *10*(2), 116–133.

Béné, C., Al-Hassan, R. M., Amarasinghe, O., Fong, P., Ocran, J., Onumah, E., Ratuniata, R., Van Tuyen, T., McGregor, J. A., & Mills, D. J. (2016). Is resilience socially constructed? Empirical evidence from Fiji, Ghana, Sri Lanka, and Vietnam. *Global Environmental Change*, *38*, 153–170.

Bennett, J., Cheah, P., Orlie, M. A., & Grosz, E. (2010). *New materialisms: Ontology, agency, and politics*. Duke University Press.

Biggs, E. M., Bruce, E., Boruff, B., Duncan, J. M., Horsley, J., Pauli, N., McNeill, K., Neef, A., Van Ogtrop, F., Curnow, J., & Haworth, B. (2015). Sustainable development and the water–energy–food nexus: A perspective on livelihoods. *Environmental Science & Policy*, *54*, 389–397.

Blay-Palmer, A., Santini, G., Dubbeling, M., Renting, H., Taguchi, M., & Giordano, T. (2018). Validating the city region food system approach: Enacting inclusive, transformational city region food systems. *Sustainability*, *10*(5), 1680.

Boin, A., Comfort, L. K., & Demchak, C. C. (2010). The rise of resilience. In L. K. Comfort, A. Boin, & C. C. Demchak (Eds.), *Designing resilience: Preparing for extreme events* (pp. 1–12). University of Pittsburgh Press.

Born, B., & Purcell, M. (2006). Avoiding the local trap: Scale and food systems in planning research. *Journal of Planning Education and Research, 26*(2), 195–207.

Burchi, F., & Holzapfel, S. (2015). Goal 2: End hunger, achieve food security and improved nutrition, and promote sustainable agriculture. In M. Loewe & N. Rippin (Eds.), *The Sustainable Development Goals of the Post-2015 agenda: Comments on the OWG and SDSN proposals*, revised version, 26 February 2015 (pp. 17–20). German Development Institute.

Cirolia, L. R. (2020). Fractured fiscal authority and fragmented infrastructures: Financing sustainable urban development in Sub-Saharan Africa. *Habitat International, 104*, 102233.

Coaffee, J., Therrien, M. C., Chelleri, L., Henstra, D., Aldrich, D. P., Mitchell, C. L., Tsenkova, S., Rigaud, É., & Participants. (2018). Urban resilience implementation: A policy challenge and research agenda for the 21st century. *Journal of Contingencies and Crisis Management, 26*(3), 403-410.

Coates, J., Swindale, A., & Bilinsky, P. (2007). Household Food Insecurity Access Scale (HFIAS) For Measurement of Food Access: Indicator Guide Volume.

Crankshaw, O., & Borel-Saladin, J. (2019). Causes of urbanization and counter-urbanisation in Zambia: Natural population increase or migration? *Urban Studies, 56*(10), 2005–2020.

Crush, J. (2017). The Hungry Cities Partnership and the associated implications for policy impact. Presentation at Consuming Urban Poverty meeting at Bellagio, Italy.

Crush, J., Hovorka, A., & Tevera, D. (2017). Farming the city: The broken promise of urban agriculture. In B. Frayne, J. Crush, & C. McCordic (Eds.), *Food and nutrition security in Southern African cities* (pp. 101–117). Routledge.

Crush, J., & Riley, L. (2018). Rural bias and urban food security. In J. Battersby & V. Watson (Eds.), *Urban food systems governance and poverty in African cities* (pp. 42–55). London: Routledge.

Crush, J., & Frayne, B. (2010). Feeding African cities: The growing challenge of urban food insecurity. In E. Pieterse & S. Parnell (Eds.), *Africa's urban revolution* (pp. 110–132). UCT Press.

Crush, J., Frayne, B., & Pendleton, W. (2012). The crisis of food insecurity in African cities. *Journal of Hunger & Environmental Nutrition, 7*(2–3), 271–292.

Davis, M. (2006). *Planet of slums*. Verso.

Dresner, S. (2012). *The principles of sustainability*. Routledge.

Drewnowski, A., & Popkin, B. M. (1997). The nutrition transition: New trends in the global diet. *Nutrition Reviews, 55*(2), 31–43.

Ellis, F., & Sumberg, J. (1998). Food production, urban areas and policy responses. *World Development, 26*(2), 213–225.

Emirbayer, M., & Mische, A. (1998). What is Agency? *American Journal of Sociology, 103*(4), 962–1023.

Endo, A., Tsurita, I., Burnett, K., & Orencio, P. M. (2017). A review of the current state of research on the water, energy, and food nexus. *Journal of Hydrology: Regional Studies, 11*, 20–30.

Feenstra, G. (2002). Creating space for sustainable food systems: Lessons from the field. *Agriculture and Human Values, 19*, 99–106.

Fisher, R., Currie, P., & Mongi, R. (2019). Towards a safe, nourishing, economic and inclusive food system for Arusha, based on partnering. *Urban Agriculture Magazine,* 36, 31–32. October 2019, RUAF, Netherlands.

Food and Agriculture Organisation (FAO). (1996). *World Food Summit.* Rome Declaration on World Food Security, Rome, 13 November 1996: Retrieved 19 January 2014 from http://www.fao.org/WFS/ [Online document].

Food and Agriculture Organisation (FAO). (n.d.). *City region food systems programme. Reinforcing rural-urban linkages for climate resilient food systems.* UN FAO Food for Cities Program, Rome.

Frayne, B., W. Pendleton, J. Crush, B. Acquah, J. Battersby-Lennard, E. Bras, E., & A. Chiweza. (2010). The State of Urban Food Insecurity in Southern Africa. *Urban Food Security Series No. 2.* African Food Security Urban Network (AFSUN).

Friend, R., & Moench, M. (2013). What is the purpose of urban climate resilience? Implications for addressing poverty and vulnerability. *Urban Climate, 6,* 98–113.

Fukuda-Parr, S., & Orr, A. (2014). The MDG hunger target and the competing frameworks of food security. *Journal of Human Development and Capabilities, 15*(2–3), 147–160.

Hawkes, C. (2006). Uneven dietary development: Linking the policies and processes of globalization with the nutrition transition, obesity and diet-related chronic diseases. *Globalization and Health, 2*(1), 1–18.

Hawkes, C., & Fanzo, J. (2017). Nourishing the SDGs: Global nutrition report 2017.

Haysom, G., Crush, J., & Caesar, M. (2017). *No. 3: the urban food system of Cape Town, South Africa.* The Hungry Cities Partnership, African Centre for Cities and Balsillie School of International Affairs.

Haysom, G., & Fuseini, I. (2019). Governing Food Systems in Secondary Cities in Africa, Consuming Urban Poverty Project Working Paper No. 10, African Centre for Cities, University of Cape Town.

Haysom, G., Olsson, E. G. A., Dymitrow, M., Opiyo, P., Taylor Buck, N., Oloko, M., Spring, C., Fermskog, K., Ingelhag, K., Kotze. S., & Agong, S. G. (2019). Food systems sustainability: An examination of different viewpoints on food system change. *Sustainability, 11*(12), 3337.

High Level Panel of Experts (HLPE). (2020). *Food security and nutrition: building a global narrative towards 2030.* A report by the High Level Panel of Experts on Food Security and Nutrition of the Committee on World Food Security, Rome.

High Level Panel of Experts (HLPE). (2014). *Food losses and waste in the context of sustainable food systems.* A report by the High-Level Panel of Experts on Food Security and Nutrition of the Committee on World Food Security. Rome: HLPE.

Hopkins, R. (2011). *The transition companion.* Green Books.

Igumbor, E. U., Sanders, D., Puoane, T. R., Tsolekile, L., Schwarz, C., Purdy, C., Swart, R., Durão, S., & Hawkes, C. (2012). "Big food," the consumer food environment, health, and the policy response in South Africa. *PLoS Medicine, 9*(7), e1001253.

IPCC. (2021). Climate Change 2021—The Physical Science Basis Summary for Policymakers, Working Group I Contribution to the WGI Sixth Assessment Report of the Intergovernmental Panel on Climate Change.

James, C. (2011). *The downside of the resilience discourse.* Retrieved 15 February 2022 from https://www.countercurrents.org/james211111.htm [Online document].

Joubert, L., Battersby, J., & Watson, V. (2018). *Tomatoes and taxi ranks: Running our cities to fill the food gaps.* University of Cape Town.

Kennedy, G., Nantel, G., & Shetty, P. (2004). *Globalization of food systems in developing countries: A synthesis of country case studies* (FAO Food and Nutrition Paper: 1-24).

Langemeyer, J., Madrid-Lopez, C., Beltran, A. M., & Mendez, G. V. (2021). Urban agriculture—A necessary pathway towards urban resilience and global sustainability? *Landscape and Urban Planning, 210,* 104055.

Mattes, R. (2008). The material and political bases of lived poverty in Africa: Insights from the Afrobarometer. In *Barometers of quality of life around the globe* (pp. 161–185).

Menashe-Oren, A., & Bocquier, P. (2021). Urbanization is no longer driven by migration in low-and middle-income countries (1985–2015). *Population and Development Review, 47*(3), 639–663.

Moseley, W. G., & Battersby, J. (2020). The vulnerability and resilience of African food systems, food security, and nutrition in the context of the COVID-19 pandemic. *African Studies Review, 63*(3), 449–461.

Myers, G. (2015). A world-class city-region? Envisioning the Nairobi of 2030. *American Behavioral Scientist, 59*(3), 328–346.

Norberg Hodge, H. (2013). *Ancient futures: Learning from Ladakh*. Random House.

Opiyo, P., Obange, N., Ogindo, H., & Wagah, G. (2018). *The characteristics, extent and drivers of urban food poverty in Kisumu, Kenya* (Consuming Urban Poverty Project Working Paper No. 4), African Centre for Cities, University of Cape Town.

Padgham, J., Jabbour, J., & Dietrich, K. (2015). Managing change and building resilience: A multi-stressor analysis of urban and peri-urban agriculture in Africa and Asia. *Urban Climate, 12*, 183–204.

Peyton, S., Moseley, W., & Battersby, J. (2015). Implications of supermarket expansion on urban food security in Cape Town, South Africa. *African Geographical Review, 34*(1), 36–54.

Pieterse, D. E., & Parnell, S. (2014). *Africa's urban revolution*. Zed Books Ltd.

Pieterse, E. (2008). *City futures: Confronting the crisis of urban development*. UCT Press.

Pieterse, E. (2013). City/university interplays amidst complexity. *Territorio, 66*, 26–32.

Pieterse, E., Haysom, G., & Crush, J. (2020). Hungry cities partnership: Informality, inclusive growth, and food security in cities of the global south: final project report: period May 2015–August 2020. HCP Project and IDRC, Canada.

Pieterse, E., Parnell, S., & Haysom, G. (2015). *Towards an African urban agenda*. UN-Habitat and Economic Commission for Africa, UN-Habitat.

Pieterse, E., Parnell, S., & Haysom, G. (2018). African dreams: Locating urban infrastructure in the 2030 sustainable developmental agenda. *Area Development and Policy, 3*(2), 149–169.

Popkin, B. M., Adair, L. S., & Ng, S. W. (2012). Global nutrition transition and the pandemic of obesity in developing countries. *Nutrition Reviews, 70*(1), 3–21.

Reardon, T., Timmer, C. P., Barrett, C. B., & Berdegué, J. (2003). The rise of supermarkets in Africa, Asia, and Latin America. *American Journal of Agricultural Economics, 85*(5), 1140–1146.

Ruel, M. T., Garrett, J., Yosef, S., & Olivier, M. (2017). Urbanization, food security and nutrition. In S. de Pee, D. Taren & M. Bloem (Eds,), *Nutrition and health in a developing world* (pp. 705–735).

Saleh, M. (2021). *Median age of the population of Africa 2000–2020*, April 7, 2021. Retrieved 1 October 2021 from https://www.statista.com/statistics/1226158/median-age-of-the-population-of-africa/#:~:text=In%202020%2C%20the%20median%20age,it%20was%20around%2018%20years.[Online document].

Satterthwaite, D. (2007). *The transition to a predominantly urban world and its underpinnings* (Human Settlements Discussion Paper Series). Urban Change—4. IIED, London, UK.

Steel, C. (2008). *Hungry city: How food shapes our lives.* Chatto & Windus.

Swilling, M., & Annecke, E. (2012). *Just transitions: Explorations of sustainability in an unfair world.* Juta and Company.

Swindale, A., & Bilinsky, P. (2006). Development of a universally applicable household food insecurity measurement tool: Process, current status, and outstanding issues. *The Journal of Nutrition, 136*(5), 1449S-1452S.

Tihelková, A. (2015). Framing the 'scroungers': The re-emergence of the stereotype of the undeserving poor and its reflection in the British press. *Brno Studies in English, 41*(20), 121–139.

Turok, I. (2016). Getting urbanization to work in Africa: The role of the urban land-infrastructure-finance nexus. *Area Development and Policy, 1*(1), 30–47.

UN-Habitat. (2017). New Urban Agenda, United Nations Habitat III Secretariat, United Nations, Nairobi, Kenya.

United Nations, Department of Economic and Social Affairs, Population Division (UN-DESA). (2019). *World urbanization prospects: The 2018 revision (ST/ESA/SER.A/420).* United Nations.

Weitz, N., Nilsson, M., & Davis, M. (2014). A nexus approach to the Post-2015 agenda: Formulating integrated water, energy, and food SDGs. *SAIS Review of International Affairs, 34*, 37–50.

World Commission on Environment and Development (WECD). (1987). *Our common future.* Oxford University Press.

World Economic Forum (WEF). (2011). *Water security, the water-food-energy-climate Nexus, The world economic forum water initiative.* Island Press.

Yamin, A. E., & Falb, K. L. (2012). Counting what we know; knowing what to count-sexual and reproductive rights, maternal health, and the millennium development goals. *Nordic Journal of Human Rights, 30*(3), 350–371.

Young, S. L., Frongillo, E. A., Jamaluddine, Z., Melgar-Quiñonez, H., Pérez-Escamilla, R., Ringler, C., & Rosinger, A. Y. (2021). Perspective: The importance of water security for ensuring food security, good nutrition, and well-being. *Advances in Nutrition, 12*(4), 1058–1073.

Zezza, A., & Tasciotti, L. (2010). Urban agriculture, poverty, and food security: Empirical evidence from a sample of developing countries. *Food Policy, 35*(4), 265–273.

Ziervogel, G., Pelling, M., Cartwright, A., Chu, E., Deshpande, T., Harris, L., Hyams, K., Kaunda, J., Klaus, B., Michael, K., & Pasquini, L. (2017). Inserting rights and justice into urban resilience: A focus on everyday risk. *Environment and Urbanization, 29*(1), 123–138. https://doi.org/10.1177/0956247816686905

**Open Access** This chapter is licensed under the terms of the Creative Commons Attribution 4.0 International License (http://creativecommons.org/licenses/by/4.0/), which permits use, sharing, adaptation, distribution and reproduction in any medium or format, as long as you give appropriate credit to the original author(s) and the source, provide a link to the Creative Commons license and indicate if changes were made.

The images or other third party material in this chapter are included in the chapter's Creative Commons license, unless indicated otherwise in a credit line to the material. If material is not included in the chapter's Creative Commons license and your intended use is not permitted by statutory regulation or exceeds the permitted use, you will need to obtain permission directly from the copyright holder.

CHAPTER 12

# Reflections and Conclusions

*Stephen Devereux and Christophe Béné*

The contributions to this book provoke and inspire reflections on a wide range of issues—conceptual, empirical, and policy-related. In this concluding chapter, we reflect on what we have learned, what outstanding issues remain unresolved, and the way forward for the three 'heroes' of this journey: resilience, food security, and food systems.

## THE ELUSIVENESS OF CONCEPTS

Perhaps because concepts like 'resilience', 'food system', and even 'food security' (which now has six pillars) are so abstract and elusive, they are prone to generalisation—'the global food system' (Caron et al.,

---

S. Devereux (✉)
Institute of Development Studies, University of Sussex, Brighton, UK
e-mail: s.devereux@ids.ac.uk

C. Béné
International Center for Tropical Agriculture, Cali, Colombia

Wageningen Economic Research group, Wageningen, The Netherlands
e-mail: c.bene@cgiar.org

© The Author(s) 2023
C. Béné and S. Devereux (eds.), *Resilience and Food Security in a Food Systems Context*, Palgrave Studies in Agricultural Economics and Food Policy, https://doi.org/10.1007/978-3-031-23535-1_12

Chapter 3), 'a globalized food system' (Losch & May, Chapter 10), or 'the African food system' (Pinstrup-Andersen, 2010), for example. Generally, the less precisely a concept is specified, the more likely it is to be contested and the more challenging it becomes to analyse and apply to real-world issues. But a lack of precision does not necessarily mean a lack of clarity. The contributions to this book have all added to our understanding of these three central concepts, conceptually and empirically. All the co-authors are grappling with how to make these concepts more practically useful, in terms of solving the perennial global challenges of food insecurity and hunger, not least by building the resilience of food systems against stressors like climate change and unforeseen shocks like COVID-19.

One reason why solutions are so elusive is because the questions are not clearly specified. Is there in fact a single unified global (or even African) food system—or are we edging towards this—or are there numerous interconnected food systems operating at different scales in overlapping spaces? Are we aiming to build resilient food systems, or do we look to well-functioning food systems to build resilient households and communities? Is a food system expected to deliver food, or food security? Put another way, is the persistence of food insecurity, malnutrition, and hunger the fault of the food system, or are these negative outcomes the results of failures elsewhere in local societies, national economies, and public and private policies?

While the concepts of food security and resilience both have written histories dating back several centuries, resilience has been applied to food security only very recently, probably since the turn of the century. Constas (Chapter 5) describes food security as the 'incumbent concept' and shows how the frequency of occurrence of the word 'resilience' has steadily (but erratically) increased in a key food security policy document—the FAO's annual 'State of Food Insecurity' (SOFI) report—and in two relevant peer-reviewed journals—'Global Food Security', and 'Food Policy'. De Pinto et al. (Chapter 7) find similar trends for the co-occurrence of 'resilience' and 'food security', in a wider search of publications in the Web of Science Core Collection since 1996.

As for food systems, the recent flurry of conceptual work emerged out of dissatisfaction with limited understandings of food systems that paid too much attention to individual components in isolation—notably agricultural production, food markets and food prices—while neglecting other essential components, and the linkages between them.

A food system has many actors—producers, processors, distributors, and consumers—who are all located in wider socioeconomic, environmental, and policy contexts that profoundly affect food security outcomes for individuals and households. Hoddinott (Chapter 6) points out that we know much more about producers and consumers than we do about processors and distributors, who have been relatively neglected by researchers and policy-makers. The COVID-19 pandemic drew particular attention to the processing sector, because workers handling food in close proximity were highly susceptible to transmitting the virus to each other, potentially disrupting operations due to absentee workers or temporary shutdowns of entire facilities.

This points to one advantage of broadening the lens of food security analysis—from availability to access, then adding utilisation and stability, and more recently agency and sustainability—and to analysing food security outcomes for people in terms of the food systems that deliver (or don't deliver) food to them. The complex flowcharts that capture the causal linkages between the components of and agents operating in food systems provide more 'hooks' with which to analyse food security and resilience. For instance, food resilience requires resilient producers, resilient processors, resilient distributors, and resilient consumers. Analytical tools are needed for each of these actors, and policy levers must be found and applied to strengthen resilience at each of these nodes and links in the food system.

One challenge with linking or overlaying different concepts or paradigms is that each comes with its own framings, terminologies, and unresolved debates. Constas (Chapter 5) illustrates this with his 'Integrated food security and resilience model' which has three dimensions: *food security* (World Food Summit, food sovereignty, and food systems approaches—each with their own models and pillars); *shocks and stressors* (at macro, meso, and micro scales); and *resilience capacities* (absorptive, adaptive, and transformative). While this is useful as an overarching framework and as a heuristic tool, the three-dimensional diagram (see p. 154) looks as puzzling as a Rubik's cube to 'solve', in terms of how to reshape food systems to achieve desirable outcomes such as food security and resilience.

## The Food System is Broken…

Food security, resilience, and food systems are not only academic constructs that have generated decades of peer-reviewed literature; they are all instrumentally concerned with human well-being and are central components of contemporary policy discourse (see Lindgren & Lang, Chapter 4), at global level and in countries around the world. As Fanzo shows (Chapter 2), eradicating hunger has been an explicit goal of global public policy since the Hot Springs Conference in 1943. This commitment has been repeatedly reaffirmed—by the World Food Congress (1963), World Food Conference (1974), World Food Summit (1996), Millennium Development Goals (2000), Sustainable Development Goals (2015), and, most recently, the UN Food Systems Summit (2021). However, actual achievements have lagged behind these good intentions, with 'only mixed success' (Fanzo, p. 31)—a euphemism for partial failure—to date.

One reason for this, Fanzo argues, has been a narrow framing of the problem, leading to interventions that follow either a vertically sectoral approach (e.g. Green Revolution), a technological treatment approach (e.g. GMOs), or a short-view approach (e.g. food aid). By contrast, a food systems lens offers a comprehensive framing of food insecurity problems, leading to more holistic interventions. Only by understanding better the relationships and linkages between a food system's *drivers* (population growth, urbanisation, climate change, etc.), *components* (food supply chains, food environments, consumer behaviour), and *context* (socio-cultural, political), can adverse *outcomes* (in nutrition, livelihoods, the environment, and social equity) be more effectively addressed (HLPE, 2017).

## …Or Is It?

Even if hunger has not yet been eradicated from the world, it could be argued that the key indicators are moving in the right direction, and that this is largely due to scientific advances and strengthening institutions (Pinker, 2018). As evidence of the increasing resilience of 'the global food system', Caron et al. (Chapter 3) draw attention to the fact that it withstood a historically unprecedented shock—the doubling of the world's population from three to six billion in just four decades—while continuing to produce and distribute more than enough food to feed these rapidly rising numbers. In terms of food availability, by the early 2000s the proportion of the world's population living in countries with critically low food supplies had fallen from 50+% to close to zero in just 40 years, partly driven by adoption of Green Revolution innovations in South Asia. In terms of economic access to food, international grain prices fell by 30+% over the same period (Baldos & Hertel, 2016). With hindsight, this under-appreciated achievement refuted neo-Malthusian pessimists who, even before the 1970s world food crisis, were predicting mass starvation due to unchecked population growth (Brown, 1974; Hardin, 1974; Paddock & Paddock, 1967).

The global capacity to feed a constantly growing population has been facilitated by simultaneous advances across a range of human endeavours, from yield-enhancing agricultural innovations to transport systems and the globalisation of trade, to information and communications technologies. This technical progress has been complemented by global humanitarianism: an increasingly sophisticated and responsive emergency relief system, effectively filling supply gaps that appear when food systems are disrupted by natural disasters or human conflicts—a success story for which the World Food Programme was awarded the Nobel Peace Prize in 2020.

Of course, resilience is not only about aggregate improvements over time; it is (by definition) about dealing effectively with shocks and stressors. Food systems recovered quickly after the global food crisis of 2008 (Golay, 2010), and do not appear to have been severely affected by the worldwide COVID-19 pandemic (Béné et al., 2021). The risk of famine, which killed tens of millions of people in Africa, Asia and Europe during the twentieth century, has receded and is now confined to a handful of mainly conflict-affected countries, almost all in and around the Horn of Africa (Devereux, 2020). On the other hand, there are justified concerns that climate change could threaten local, regional, and even global food security in the coming decades, through reduced or more volatile yields of staple crops and the displacement or mass migration of affected populations, many of them being farmers.

## CLIMATE CHANGES EVERYTHING

'Food resilience' and 'climate resilience' are of course intricately interconnected, since the impacts of climate change—labelled as the 'world's worst wicked problem' (Craig, 2020, p. 26)—are being felt centrally by food systems, especially by the agriculture sector, which is facing escalating shocks and stressors across the world (De Pinto et al., Chapter 7). One lesson from work on climate change for those concerned with food security is to accept that building stable food systems is an unrealistic goal in the context of a changing climate. Food systems must constantly adapt and reconfigure themselves, because the environment itself is unstable. 'Resilience theory teaches us that social-ecological systems are always changing and can act or respond in unpredictable ways, normalizing wicked problems' (Craig, 2020, p. 29). Achieving food security for all remains the ultimate goal, but the pathways to achieving that goal are shifting over time. Resilience in an era of rapid climate change is not about preserving the *status quo ante* at all costs, it is about making the necessary adjustments to continue delivering food—for instance, by adopting 'climate-smart agriculture' approaches (Lipper et al., 2018).

One shift in emphasis induced by climate change is to deprioritise perpetual increases in crop *yields* and prioritise instead reductions in yield *volatility*, with innovations that achieve both, of course, being the first prize. Beyond food production, an array of 'climate-proofing' interventions is needed in the areas of food storage, packaging, distribution, processing, marketing, and preparation—in short, at every stage of the food system. Some of these interventions require technological innovations, but others require stronger institutions and more committed governance (De Pinto et al., Chapter 7).

In other words, achieving food security and resilient food systems needs scientists and politicians, as well as activists and the private sector, to work together cooperatively towards these shared objectives (see Fanzo, Chapter 2; Caron et al., Chapter 3). Investment in research on its own is not enough. Political speeches and resolutions made at climate conferences are not enough. Defeating the unprecedented threat to global food security posed by climate change requires an alignment of scientific innovation and political commitment.

## Food System Paradoxes

A food system (or systems) that appear(s) increasingly efficient and resilient over time at the global level can conceal pockets of fragility and inability to meet food needs for vulnerable groups of people in specific places ('food deserts') at specific points in time (seasonal hunger). Even worse, efficient food systems can co-exist with high levels of persistent or chronic food insecurity over time, as reflected in indicators such as child stunting that appear to contradict the confident claims by some that the global food system is working, is fit for purpose, and 'is not broken' (Caron et al., Chapter 3).

How do we explain, for instance, the puzzling fact that 25% of children in South Africa display stunted growth, despite living in a country that not only produces enough food to meet domestic consumption needs but also exports surpluses to neighbouring countries? This paradox is easily explained by invoking the laws of supply and demand. In market-dominated economies, producers and traders respond to demand signals from consumers. In South Africa and in other countries across the world, farmers and markets deliver adequate supplies of food to local or foreign consumers who can afford to pay for it. Hunger follows not from insufficient food supplies, but from inadequate purchasing power of poor consumers— i.e. demand failure or 'entitlement failure', to use Amartya Sen's terminology (Sen, 1981).

Even a resilient food system with falling rates of undernutrition can generate other forms of food insecurity such as obesity and micronutrient deficiencies. These forms of food insecurity are increasing globally, not necessarily because food systems are not resilient but because they are producing unhealthy diets (Haddad et al., 2016). The persistent preoccupation with aggregate food availability (as quantified by the FAO's food balance sheets) has diverted policy attention away from the urgent need to shift the focus towards affordable access for all to quality diets.

It follows that well-functioning food systems are necessary but not sufficient to eradicate food insecurity (see Fanzo, Chapter 2), because food systems are themselves embedded within larger systems. Food security also depends on, *inter alia,* well-functioning health and education systems, water and sanitation facilities, as well as sound governance, economic growth, political stability, and social inclusion. Moreover, food systems resilience, defined as 'the capacity over time of a food system to sustainably provide sufficient, appropriate, and accessible food to all, in the face of shocks and stressors' (Hoddinott, Chapter 6), might not guarantee the same outcome indefinitely. Another paradox is that the trajectories taken by evolving food systems, despite generating adequate food supplies in the recent past and the present, could contain the 'seeds of their own destruction', potentially undermining their performance and resilience in the future. Examples include the harmful effects of agricultural intensification on soil quality, biodiversity, environmental sustainability, and certain rural livelihoods (on this point see Fanzo, also Caron et al., in this volume).

## Action for Resilience

Contributors to this book draw attention to the roles of various actors in constructing food resilience—governments, development agencies, the private sector, and local communities—operating at different scales, from global to local, and requiring coordination at all these levels.

Resilient households require resilient food systems. As we noted earlier (in Chapter 1), a wealthy household that is food secure and apparently resilient might nonetheless be "plunged into hunger" (Lindgren & Lang, Chapter 4) or even "plunged into starvation" (Sen, 1981, p. 47) if its access to food is severely disrupted by the closure of local markets and restrictions on mobility, due to a pandemic or military conflict. It follows that strengthening resilience has two discrete but complementary aspects: building the capacities and resources of individuals, households, groups, and communities, on the one hand, and 'shock-proofing' each component of the food system(s) with which they interact, on the other.

Hoddinott (Chapter 6) considers the role of *governments* in building resilient food systems, following the United Nations (2020) guidance on public action to support improved anticipation, prevention, absorption, adaptation, and transformation. Examples drawn from related areas include famine early warning systems (improved anticipation and prevention of food crises), social protection (improved ability to absorb shocks such as COVID-19 lockdowns), or conservation agriculture practices (better adaptation to climate shocks) as supported by the Adaptation Fund (see De Pinto et al., Chapter 7). Progress can best be described as uneven, probably because these efforts to build resilience are occurring in some of the world's most challenging contexts, in low-income food-deficit countries (LIFDCs) and in areas that are ecologically marginal, with infrastructure deficits and weak governance.

In parallel, bilateral and multilateral *donors* play important roles in conceptualising food security and food system resilience, and in designing, financing, and implementing projects in low- and middle-income countries to strengthen food security and resilience. Lindgren and Lang (Chapter 4) describe a 'fractured consensus' in terms of what development agencies mean by these concepts, and how to achieve globally endorsed objectives for ending hunger. One reason for divergence across agencies is weak global governance and institutional architecture on food, while another explanation is ideological. Some agencies favour right-of-centre neoliberal approaches driven by economic analysis and focusing on food production plus economic growth; while others are left-of-centre, advocate for a rights-based approach, and focus on sustainable and equitable access to food for all.

Moving to the *private sector*, a concentration of actors in any component of a food system generally increases risk and undermines resilience. There is a long history of literature on the risks of oligopolies or cartels in the global grain trade (Morgan, 1979) and, more recently, at national and subnational levels, of 'supermarketisation' in food wholesale and retail sectors (Crush & Frayne, 2017; Reardon et al., 2003). Too few firms in any sector raises the risk of market distortions such as price-fixing, to the detriment of poor consumers. Conversely, having many actors in the food sector, or access to many markets (including integration into global food markets), increases competition and spreads the risk of depending on a single actor or market. An open food system is likely to be more resilient than a closed system (Hoddinott, Chapter 6). On the other hand, some adaptations trade off maximising output or revenue (e.g. through specialisation) *versus* stabilising output or revenue (e.g. through diversification, or by purchasing insurance), so resilience can come at a cost.

An important implication of the recent addition of 'agency' as a pillar of food security (HLPE, 2020) is that it empowers *people* to strengthen their own resilience—as individuals, households, communities, or identity-based groups—rather than reducing them to passive recipients of government policies or donor assistance. As Haysom and Battersby argue (Chapter 11): 'Central to ensuring food access is individual and community agency', but the agency of the poor in their interactions with food systems is often constrained by social and economic inequalities, and the domination of political structures and decision-making processes by local elites. We conclude from this that food resilience can be enhanced if the agency of all food system actors and users is enhanced. However, agency has not yet been fully incorporated into mainstream analysis of food security and food systems.

## Food Security and COVID-19

COVID-19 provided an unprecedented test for the resilience of food systems across the world. In one of the most empirical contributions to this volume, Upton et al. (Chapter 9) analyse COVID-19 as a shock that compounded many other shocks and stressors faced by vulnerable people in low-income countries. As a pandemic, COVID-19 differed from shocks that are more usually studied—such as natural disasters (droughts), economic shocks (food price spikes), and sociopolitical shocks (conflicts)—but its secondary impacts, especially on livelihoods through restrictions on economic activity imposed by lockdown regulations, seem to be remarkably similar.

In general, food systems were relatively protected, as food was declared a priority sector by most governments, allowing food supply chains to continue performing their essential functions, mainly to protect consumers during lockdowns. Nonetheless, drawing on panel survey data from rural communities in three African countries—Kenya, Madagascar, and Malawi—Upton et al. find that COVID-19 impacted adversely on all pillars of food security. Barrett and Constas (2014) define resilience as the capacity for 'attaining and maintaining' food security, implying that chronically hungry households are neither food secure nor resilient. Using this definition, Upton et al. find that many households they surveyed were food insecure and therefore not resilient pre-pandemic, and that COVID-19 exacerbated both their food insecurity and their lack of resilience. In effect, COVID-19 was absorbed by these households as one more shock on top of others they were forced to deal with at the same time, notably a severe drought in Madagascar and flooding in parts of Kenya.

On the other hand, the situation did not deteriorate for all people everywhere. In Malawi, the negative shock of COVID-19 was more than compensated in many households by the positive shock of a bumper maize harvest. Upton et al. conclude that resilience must be understood and addressed not in relation to any one specific shock or stressor, but in terms of the full range of context-specific risks and vulnerabilities that households and communities face at each point in time. This brings us onto the contentious and seemingly insoluble challenge of how to measure resilience, which is discussed next.

## Measuring Resilience

Whatever their ideological differences, one commonality among almost all development agencies is a preoccupation with measurement, which is understandable given their need to demonstrate positive impacts from their investments in initiatives to improve food security and resilience in low- and middle-income countries. While a range of tools for quantifying household food security has been devised—e.g. dietary diversity score (DDS), food insecurity experience scale (FIES), and coping strategies index (CSI)—resilience is less tangible and more challenging to measure, although the Resilience Index Measurement and Analysis (RIMA-II) tool (FAO, 2016) has gained some traction, especially among United Nations agencies (Lindgren & Lang, Chapter 4).

Of course, resilience must often be measured in volatile contexts that embody a significant risk or actual occurrence of shocks to livelihoods and food systems. Recognising this reality, Haysom and Battersby (Chapter 11) suggest that 'different measures of resilience are required, measures that take into account the constant state of crisis'. Again, this is not merely of academic interest. Once the context of vulnerability and resilience is accurately understood, appropriate interventions can be designed that strengthen absorptive, adaptive, and transformative capacities of relevant populations.

However, an empirical challenge remains. Resilience implies the capacity to withstand shocks without resorting to 'distress' behaviours, such as selling productive assets to buy food, that will compromise the integrity of the affected household's livelihood. Measuring this capacity requires assessing each household's food security status and asset-holdings before and after a shock occurs, but shocks are, by their nature, unpredictable. Development programmes are much simpler to evaluate, because 'treatment' and 'control' households can be surveyed and compared before (baseline) and after (endline) an intervention is delivered. Perhaps resilience—like 'vulnerability' and 'sustainability'—is destined to remain an elusive and contested concept that is challenging to define and even more challenging to measure, for this reason.

## Disaggregating Vulnerabilities

Risks, shocks, and stressors affect different people differently, so a disaggregated analysis is needed. Bryan et al. (Chapter 8) examine the different ways that different groups of men and women are exposed and respond to shocks and stressors, by applying a framework that integrates a food systems lens to the relationships between gender, nutrition, and resilience. This framework recognises that every man and woman has a unique resilience capacity and set of response options, leading to unique outcome pathways. Importantly, Bryan et al. avoid generalisations that homogenise men and women, noting for instance that men are often more exposed to natural disasters, but women are likely to be more vulnerable because they have fewer resources and therefore lower adaptive capacity.

Another gendered determinant of exposure and sensitivity to risks is each person's positioning in the food environment, notably their respective livelihood activities in food value chains. Where women are confined to low-paid, marginal activities such as seasonal farm work or informal work in food processing or retailing, household resilience and children's nutrition status could be enhanced simply by supporting women's livelihoods.

Beyond gender, Bryan et al. note that resilience capacities are influenced by many other intersectional identities, some of which, such as marital status determining women's access to productive resources in patriarchal societies, are also gendered. Building resilience could therefore be achieved by interventions that empower women and socially marginalised groups. It follows that resilience is a matter of social justice—as is food security, given that the right to adequate food is a fundamental human right (United Nations, 1948).

## Disaggregating Places

Losch and May (Chapter 10) make the important point that food systems are grounded in places, and they argue for a territorial approach that recognises spatial dynamics. A supply chain that delivers food from farm to fork can be very short or very long—the journey can be from a farmers' cooperative to a school in the same rural community, or from a fruit farm in South Africa to a restaurant in Norway. This makes planning for food security complex, because any national food security policy must be cognisant of the local, national, and international dimensions of the country's food system.

Losch and May advocate for subnational planning for food security, and they draw on the process of producing a provincial food and nutrition security strategy in South Africa as a case study (Western Cape Government, 2016). A localised approach is well aligned with movements such as 'slow food' and food sovereignty, which promote the rights and agency of food producers, local mid-stream actors, and consumers to control their food systems, in contrast to food regimes where corporations and institutions such as supermarkets dominate (McMichael, 2005). It also ensures that all actors are fully engaged, from local government and the private sector (farmers, processors, and formal and informal retailers) to civil society and consumers.

But is there a risk that shortening food supply chains will undermine resilience? Hoddinott (Chapter 6) highlights the benefits of globalised food systems in terms of diversifying diets and spreading risk against localised shocks such as adverse weather events. Similarly, Haysom and Battersby (Chapter 11) argue that 'resilience is based in food system diversity'. Moreover, some variants of a territorial approach, such as investment in 'development corridors', could increase inequality and exclusion by privileging certain areas and communities while marginalising others (Losch and May, Chapter 10).

A decentralised view of food systems brings a sharper focus on rural areas *versus* urban areas, and the linkages between them. In parts of the world such as sub-Saharan Africa, food insecurity historically manifested as primarily a rural phenomenon, and urban food insecurity remains relatively under-analysed. In their contribution to this volume, Haysom and Battersby synthesise the growing body of thought on urban food systems in Africa, demonstrating that despite being intrinsically linked to rural hinterlands—most crudely as the mass consumption end of food value chains—they require a completely different framing. COVID-19, because it affected urban workers disproportionately through the restrictions imposed on economic activity, has sparked a flurry of interest globally in understanding urban food systems better. Urban agriculture has limited potential, so a more holistic approach is needed. One innovative concept is the city region food system (CRFS), which aspires to enhance the sustainability and resilience of food systems in urban centres (Blay-Palmer et al., 2018).

Haysom and Battersby present empirical evidence from several African cities, revealing how different urban residents engage very differently with their food systems, acquiring food from different sources in different quantities and frequencies, even consuming very different diets, depending on their relative affluence. Analysis of food systems must therefore be disaggregated even between categories of people within the same territory.

Particular attention must be paid to the role of urban governance. Infrastructure and services such as water, electricity, transport, sanitation, and waste disposal emerge centrally as contextual factors affecting urban food systems and urban food security. For the urban poor, household food choices are constrained not just by poverty, but also by infrastructure deficits. For this reason, Haysom and Battersby argue for infrastructure justice as one lever for levelling up inequalities and ensuring that no one is left behind. Public investment in pro-poor infrastructure might seem like a secondary consideration in relation to the central themes of this book, but it can contribute not only to social justice but also to more resilient and equitable food systems.

## Conclusion

Food security, food resilience, and food systems each have unique intellectual histories and policy trajectories, and all three remain as contested concepts, with no consensus even on their definitions. Yet, this book has tried to bring the three together, conceptually and analytically, not to complexify or theorise a terrain that is already overloaded with frameworks and flowcharts, but to highlight how the understanding and meaning of each one can illuminate and be illuminated by applying insights from the others.

Fundamentally, anyone concerned with food security and resilience is concerned with human well-being and how to protect or even to enhance it. Karl Marx's prescient observation, that the point is not only to interpret the world but to change it, must never be forgotten. Disagreements over definitions and measurement tools are never merely 'academic' and are sometimes passionate, because they have implications that lead to policy decisions affecting millions—or billions—of lives. To take one example that is rehearsed in the early chapters of this book: whether one believes that the food system is 'broken' or, to the contrary, increasingly efficient and resilient, has major implications for the public and private actions that are proposed and adopted against hunger and malnutrition.

This book has highlighted the persistence of several paradoxes, such as the coexistence of high levels of multiple forms of malnutrition even when food systems are functioning efficiently, and the rise of overweight and obesity even in countries (like South Africa) that simultaneously display high rates of chronic undernutrition. These paradoxes extend also to the choice of policy interventions. Should governments concerned with food security in low-income countries aim to keep food prices high to incentivise farmers, or low to keep food affordable for all consumers? How far should governments interfere with other food system actors' abilities and willingness to promote healthy diets, for instance by imposing a sugar tax? In a context of inexorable population growth and climate change, should research investments in agriculture focus on maximising yield growth or minimising yield volatility? Does strengthening food resilience require starting with households and communities, or with food systems and governance?

There are no easy answers to these and many related questions. Food (in)security, like climate change, has rightly been classified as a 'wicked problem' (Hamann et al., 2011), and in an increasingly complex and unpredictable world, achieving food resilience seems more elusive and challenging than ever. There is no doubt, however, that viewing these problems through a food systems lens can help.

## References

Baldos, U., & Hertel, T. (2016). Debunking the 'new normal': Why world food prices are expected to resume their long run downward trend. *Global Food Security, 8*, 27–38.

Barrett, C., & Constas, M. (2014). Toward a theory of resilience for international development applications. *Proceedings of the National Academy of Sciences of the United States of America, 111*(40), 14625–14630.

Béné, C., Bakker, D., Chavarro, M., Even, B., Melo, J., & Sonneveld, A. (2021). Global assessment of the impacts of COVID-19 on food security. *Global Food Security, 31*, 100575. https://doi.org/10.1016/j.gfs.2021.100575.

Blay-Palmer, A., Santini, G., Dubbeling, M., Renting, H., Taguchi, M., & Giordano, T. (2018). Validating the city region food system approach: Enacting inclusive, transformational city region food systems. *Sustainability, 10*(5), 1680.

Brown, L. (1974). *By bread alone*. Pergamon Press.

Craig, R. (2020). Resilience theory and wicked problems. *Vanderbilt Law Review, 73*(6), 1733–1775.

Crush, J., & Frayne, B. (2017). Supermarketization' of food supply and retail: private sector interests and household food security. In B. Frayne, J. Crush, & C. McCordic (Eds.), *Food and nutrition security in Southern African cities*. Routledge and Earthscan.

Devereux, S. (2020). Preventable famines: Response and coordination failures in twenty–first–century famines. Chapter 11 in J. Dijkman and B. van Leeuwen (editors), *An Economic History of Famine Resilience*. Abingdon: Routledge. 203–223.

FAO. (2016). *RIMA-II: Moving forward the development of the resilience index measurement and analysis model*. FAO.

Golay, C. (2010). The food crisis and food security: Towards a new world food order? *International Development Policy* (online). https://doi.org/10.4000/poldev.145. Accessed 4 February 2022.

Haddad, L., Hawkes, C., Waage, J., Webb, P., Godfray, C., & Toulmin, C. (2016). *Food systems and diets: Facing the challenges of the 21st century*. Global Panel on Agriculture and Food Systems for Nutrition.

Hamann, R., Giamporcaro, S., Johnston, D., & Yachkaschi, S. (2011). The role of business and cross-sector collaboration in addressing the 'wicked problem' of food insecurity. *Development Southern Africa, 28*(4), 579–594.

Hardin, G. (1974). Lifeboat ethics: The case against helping the poor. *Psychology Today, 8*, 38–43.

HLPE. (2017). Nutrition and food systems. A report by the High-Level Panel of Experts on Food Security and Nutrition of the Committee on World Food Security. HLPE report 12. Rome: HLPE.

HLPE. (2020). Food security and nutrition: Building a global narrative towards 2030. A report by the High-Level Panel of Experts on Food Security and Nutrition of the Committee on World Food Security. HLPE report 15. Rome: HLPE.

Lipper, L., McCarthy, B., Zilberman, D., Asfaw, S., & Branca, G. (Eds.). (2018). *Climate smart agriculture: Building resilience to climate change*. FAO.

McMichael, P. (2005). Global development and the corporate food regime, chapter 11. In F. Buttel & P. McMichael (Eds.), *New directions in the sociology of global development* (pp. 265–299). Emerald Group Publishing.

Morgan, D. (1979). *Merchants of grain*. Viking Press.

Paddock, W., & Paddock, P. (1967). *Famine 1975! America's decision: Who will survive?* Little, Brown and Co.

Pinker, S. (2018). *Enlightenment Now: The case for reason, science, humanism and progress*. Allen Lane.

Pinstrup-Andersen, P. (Ed.). (2010). *The African food system and its interaction with human health and nutrition*. Cornell University Press.

Reardon, T., Timmer, C., Barrett, C., & Berdegué, J. (2003). The rise of supermarkets in Africa, Asia, and Latin America. *American Journal of Agricultural Economics, 85*, 1140–1146.

Sen, A. (1981). *Poverty and famines: An essay on entitlement and deprivation*. Oxford University Press.

United Nations. (1948). *Universal declaration of human rights*. United Nations.

United Nations. (2020). *Common guidance on helping build resilient societies*. United Nations.

Western Cape Government. (2016). *Household food and nutrition security strategic framework*. Western Cape Government.

**Open Access** This chapter is licensed under the terms of the Creative Commons Attribution 4.0 International License (http://creativecommons.org/licenses/by/4.0/), which permits use, sharing, adaptation, distribution and reproduction in any medium or format, as long as you give appropriate credit to the original author(s) and the source, provide a link to the Creative Commons license and indicate if changes were made.

The images or other third party material in this chapter are included in the chapter's Creative Commons license, unless indicated otherwise in a credit line to the material. If material is not included in the chapter's Creative Commons license and your intended use is not permitted by statutory regulation or exceeds the permitted use, you will need to obtain permission directly from the copyright holder.

# Correction to: Resilience and Food Security in a Food Systems Context

*Christophe Béné and Stephen Devereux*

Correction to:
C. Béné and S. Devereux (eds.), *Resilience and Food Security in a Food Systems Context*, Palgrave Studies in Agricultural Economics and Food Policy, https://doi.org/10.1007/978-3-031-23535-1

The original version of the book was inadvertently published without the funding text in FM which has now been updated. The book has been updated with the changes.

---

The updated original version of this book can be found at
https://doi.org/10.1007/978-3-031-23535-1

© The Author(s) 2023
C. Béné and S. Devereux (eds.), *Resilience and Food Security in a Food Systems Context*, Palgrave Studies in Agricultural Economics and Food Policy, https://doi.org/10.1007/978-3-031-23535-1_13

# Index

## A
absorptive capacity, 156, 251
access to food, 92, 153, 253, 285, 326, 328, 332, 365, 393, 396, 397
adaptation, 5, 10, 21, 32, 54, 82, 193, 199, 200, 217, 222, 223, 253, 311, 325, 326, 396, 397
Adaptation Fund (AF), 6, 20, 217, 218, 224–226, 396
adaptive capacity (AC), 156, 197, 218, 223, 249, 251, 400
Africa, 10, 15, 34, 38, 57, 95, 97, 148, 247, 257, 263, 308, 313, 325, 329, 331, 356, 358–360, 370, 376, 378, 379, 393, 401
agency, 5, 13–15, 21, 35, 83, 84, 86, 89, 94, 95, 98, 101, 105, 112, 113, 150, 152, 154, 175, 246, 250, 254, 261, 264, 341, 357, 358, 360, 370, 375, 378, 379, 381, 391, 397, 400
agricultural policies, 210, 335
agricultural value chains, 244, 248, 251
agriculture, 2, 32, 33, 37, 41, 56, 61, 62, 64, 70, 87, 88, 92, 93, 95, 99, 101, 104, 111, 113, 188, 209, 217, 218, 256, 262, 263, 291, 292, 330, 333, 336, 338, 369, 394, 403
agroforestry, 222, 251, 264
Asia, 36, 57, 95, 97, 257, 393
assets, 14, 16, 68, 187, 197, 245, 247, 249–251, 253, 255, 259, 261, 263, 264, 311, 333, 399

## B
bilateral and multilateral agencies, 95
biodiversity, 3, 36, 38, 61, 70, 83, 84, 88, 222, 224, 395

## C
capacity building, 214
chronic poverty, 239

cities, 3, 16, 18, 21, 200, 254, 258, 287, 301, 322, 331, 332, 356–366, 368–371, 374, 376, 377, 379–381, 401
city regional food system, 365
climate change, 3, 7, 13, 18, 20, 35, 36, 40, 43, 62, 69, 70, 82–84, 88, 93, 97–100, 104, 105, 107, 109, 110, 113, 114, 147, 154, 167, 207–218, 223, 225–227, 240, 244, 246, 248, 249, 251, 254, 257, 258, 262, 282, 311, 325, 359, 364, 390, 392–394, 402, 403
climate resilience, 87, 94, 97, 99, 100, 112, 210, 217, 393
climate smart agriculture (CSA), 217
Committee on World Food Security (WFS), 98, 101, 114, 152–154, 156, 165
communities, 8, 12, 15, 18, 36, 43, 61, 62, 92, 101, 148, 150, 155, 198, 213, 215, 217, 218, 224, 227, 239–241, 244, 248, 254, 259, 264, 265, 282, 283, 285, 286, 292, 293, 297–299, 301, 304, 309, 310, 312, 313, 326, 327, 339, 340, 342, 361, 371, 373, 378, 390, 396–398, 401, 403
community action network (CANs), 322, 339–342
community of practice, 338
community resilience, 216, 256
conflict, 5, 7, 11, 12, 16, 17, 19, 34–36, 55, 59, 62, 64, 66, 83, 98, 106, 147, 150, 155, 165, 240, 246, 247, 253, 282, 286, 287, 302, 307, 311, 326, 393, 396, 398
conservation agriculture, 222, 260, 396

consumer behaviour, 392
consumers, 39, 56, 60, 70, 153, 186, 193, 196, 208, 213, 214, 241, 242, 244, 257, 258, 310, 312, 324, 331, 338, 340, 343, 356, 357, 378, 391, 395, 397, 398, 400, 401
coping strategies index (CSI), 399
covariate shock, 155, 284–287, 300, 305, 309–313
COVID-19 pandemic, 3, 10, 21, 41, 42, 65, 105, 109, 113, 114, 169, 185, 191, 199, 261, 281–287, 292–297, 299, 300, 305, 308, 310–312, 327, 328, 336, 340, 357, 391, 393

D

decentralization, 291, 329–333, 357, 401
development agencies, 2, 7, 12, 15, 19, 82–86, 88, 89, 105, 107, 108, 110, 111, 113, 114, 208, 396, 398
development corridor, 330, 401
diet, 32, 34, 39, 40, 43, 59, 60, 64, 84, 186, 208, 214, 245, 249, 258, 298, 312, 334, 371, 376–378, 395, 401
dietary diversity, 37, 249, 255, 296, 298, 372, 377
dietary preferences, 245
diet quality, 59, 395
diversification, 191, 196, 197, 200, 215, 216, 222, 249, 258, 260, 397

E

economic inclusion, 263, 324, 341
economies of scale, 60, 191, 194, 195, 200

ecosystem services (ES), 61, 222
empowerment, 36, 83, 250, 263–265
environment, 6, 8, 38–41, 70, 99, 195, 196, 241–243, 248, 257, 261, 262, 282, 336, 359, 367, 371, 392, 394
environmental sustainability, 33, 240, 255, 261, 324, 341, 362, 396
exposure, 148, 155, 187, 208, 243, 246, 248, 262, 282, 286, 287, 292, 299, 311, 312, 400

# F

famine, 33, 36, 57, 113, 193, 199–201, 209, 393, 396
food aid, 38, 40, 58, 208, 297, 392
food and nutrition security (FNS), 324, 328, 334–337, 356, 361, 370, 375, 400
food availability, 17, 55, 58, 208, 213, 257, 307, 312, 313, 328, 372, 393, 395
food desert, 61, 249, 394
food emergencies, 209
food environment, 39, 40, 64, 153, 241, 242, 244, 248, 249, 256, 260, 322, 326, 392, 400
food insecurity experience scale (FIES), 254, 399
food policy, 36, 108, 331, 334, 342, 367, 379, 380, 390
*Food Policy* (journal), 158, 163–165
food preparation, 244, 253, 293
food prices, 43, 54, 106, 185, 193–195, 247, 248, 295, 296, 307–309, 335, 358, 366, 376, 390, 398, 402
food processing, 56, 62, 154, 191–193, 209, 214, 400
food production, 13, 16, 20, 36, 38, 39, 54, 62, 70, 154, 185, 188–191, 208, 209, 212, 213, 243, 259, 287, 332, 334, 335, 367, 371, 394, 397
food regime, 325, 400
food resilience, 19, 82–84, 86–88, 93–95, 98–101, 104–106, 108–114, 148, 361, 367, 369, 370, 380, 381, 391, 393, 396, 397, 402, 403
food security, 1–6, 8, 11–20, 22, 32, 33, 35–37, 39–41, 43, 44, 55, 57, 63, 71, 81–89, 92–95, 97–101, 104–111, 113, 114, 147–160, 162, 165–170, 175, 186, 187, 191, 195, 196, 199, 209–212, 215–218, 222, 224–226, 240–242, 244, 248, 250, 256–258, 260, 262, 264, 281–285, 287–289, 291–293, 295–297, 299, 300, 308, 310, 311, 313, 326, 328, 329, 333, 335, 337, 338, 342, 356–358, 361, 364, 366, 368–372, 374, 378, 379, 381, 389–391, 393, 394, 396–402
food sovereignty, 65, 150, 152–154, 156, 165–167, 175, 391, 400
food storage, 34, 39, 257, 394
food system, 2–5, 8–10, 13–22, 32, 35, 36, 39–44, 53, 54, 59, 61–65, 67, 68, 71, 72, 88, 93, 104–106, 108, 109, 150, 152–154, 156, 165–167, 170, 175, 186–188, 196, 198–201, 208–210, 212, 215, 217, 225, 226, 241, 242, 247, 248, 257–259, 262, 265, 282, 283, 287, 296, 300, 310, 312, 313, 322, 323, 325–333, 336, 338, 340–343, 356, 359–361, 365, 367, 369–372, 374, 376–378, 380, 381, 389–403
food system paradox, 322, 394

Food Systems Summit (2021), 31, 392
food system transformation, 72
food value chain, 241, 242, 245, 248, 256–258, 310, 400, 401
food vendors, 258, 328, 376, 377
fractured consensus, 7, 8, 19, 81, 86, 108, 112, 396

## G

gender, vi, viii, x, xii, 3, 13, 18, 20, 60, 99, 155, 199, 209, 215, 239, 241, 242, 244, 246–248, 250–254, 257–259, 261–263, 399, 400
gender lens, 244, 259
gender sensitivity, 262
genetically modified organisms (GMOs), 37, 38, 40, 66, 392
*Global Food Security* (journal), 158–160, 164, 165, 168, 170
global food system, 36, 54, 57, 58, 65, 114, 185, 186, 395
globalisation, 393
governance, 5, 14, 21, 39, 83, 109, 156, 198, 215, 265, 323, 326–329, 331, 333, 334, 336–340, 342, 356, 357, 359–361, 363, 365, 368, 369, 371, 380, 381, 394–397, 401, 403
Green Revolution, 33, 36, 37, 40, 55, 64, 392, 393

## H

healthcare systems, 61, 300
healthy diet, 34, 43, 60, 255, 258, 265, 395, 402
High-Level Panel of Experts (HLPE), 2, 5, 8, 14, 15, 17, 35, 40, 59–66, 71, 153, 156, 175, 186, 188, 196, 242, 256, 323, 370, 371, 392, 397
HIV and AIDS, 246, 327
household, 2, 8, 10–12, 16, 17, 33, 57, 60, 105, 152, 155, 156, 187, 188, 197, 198, 213, 216, 217, 223, 241, 242, 244, 250, 252, 255–261, 263–265, 283–285, 290, 293–296, 300–303, 305, 311–313, 326, 334, 370, 371, 373, 375, 376, 378, 396, 399–401
humanitarian assistance, 149
hunger, 11, 31–33, 35, 38, 57–59, 82, 89, 92, 93, 101, 106, 153, 208, 209, 214, 247, 265, 285, 325, 340, 378, 390, 392, 395, 396, 402
hungry season, 11, 61

## I

idiosyncratic shock, 155, 187
inequality, 155, 247, 256, 328, 329, 359, 360, 401
informal economy, 336, 373, 376
infrastructure, 8, 9, 41, 42, 56, 99, 100, 109, 155, 196–199, 208, 209, 213, 224, 241, 248, 249, 260, 262, 299, 301, 304, 330, 356, 357, 359, 371, 373, 375–377, 379, 381, 401, 402
*integrated food security resilience model*, 156, 175
intersectionality, 250
intrahousehold, 242, 243, 259

## K

Kenya, 21, 260, 284, 289, 290, 293–295, 297, 299, 301, 302, 304, 307, 310, 312, 313, 357, 376, 398

INDEX   411

**L**

livelihoods, 5, 9, 11, 12, 33, 38, 39, 41, 42, 61, 62, 68, 83, 93, 101, 105, 156, 197, 212, 215, 216, 222, 224, 240, 243–245, 248–251, 253–257, 260–263, 287, 291–293, 295, 299, 302, 305, 312, 324, 326, 341, 392, 396, 398–400
local food system, 3, 16, 17, 149, 330, 336, 338, 368
local government, 323, 327, 330, 331, 336, 338, 339, 341, 342, 380, 381, 400

**M**

Madagascar, 21, 284, 289, 290, 292, 294–296, 298, 299, 301, 303, 306, 309, 310, 312, 313, 398
Malawi, 10, 21, 249, 284, 289–297, 299, 301, 303, 304, 306, 309, 310, 312, 313, 398
malnutrition, 17, 31–33, 35, 36, 59–61, 92, 153, 207–209, 214, 251, 337, 390, 402
market concentration, 193, 195
market integration, 193–196, 200
micronutrient deficiencies, 59, 214, 395
Millennium Development Goals (MDGs), 35, 361, 363, 392

**N**

natural disasters, 5, 83, 97, 100, 246, 282, 301, 307, 393, 398, 400
natural resource management, 243
New Urban Agenda (NUA), 331, 362–364, 368
NGOs, 2, 7, 16, 93, 111, 282, 283, 313

non-communicable diseases (NCDs), 32, 41, 59, 327, 334
nutrition, 3, 5, 11–14, 17, 32–34, 36, 37, 39–41, 43, 59, 61, 63, 64, 88, 101, 105, 106, 154, 209, 214, 215, 240–242, 245, 247, 251, 254, 256, 257, 261, 263, 324, 327, 332, 334, 335, 337, 356–358, 371, 377–379, 392, 399, 400
nutrition transition, 359

**O**

obesity, 32, 41, 59, 249, 395, 402
overweight, 32, 59, 334, 402

**P**

place-based approach, 21, 322, 327, 336
population growth, 392, 393, 402
post-harvest, 154, 257, 258, 282, 307
poverty, 13, 35, 37, 59, 60, 222, 264, 265, 284, 291, 295, 311, 334, 335, 359, 362, 372, 377, 378, 401
practitioners, 2, 5–7, 16, 208, 210, 217, 225, 227, 239, 265, 342
processing, 8, 34, 39, 56, 153, 191–193, 198, 200, 213, 223, 242, 248, 257, 258, 262, 367, 371, 391, 394
productivism, 100

**R**

regional development bank, 84, 88, 95, 97, 98, 100, 101, 104–107, 109–113
regional government, 367, 369
resilience capacities, 149, 156, 166, 170, 212, 241, 243–245, 247,

249, 250, 253, 256, 260–263, 265, 285, 391, 400
resilience capacity index (RCI), 197
Resilience Index Measurement and Analysis (RIMA-II), 104, 112, 197, 399
Resilience Rating System, 104, 105
retailers, 188, 264, 336, 338, 343, 401
right to food, 64, 150, 153, 154, 165, 167, 334–336
risk management, 21, 188, 210, 212, 222, 240, 245, 261, 326
rural, 10, 36, 42, 59, 61, 101, 108, 194, 213, 214, 218, 241, 247, 248, 252, 253, 256, 260, 282, 284, 285, 287, 305–308, 310, 312, 313, 325, 327, 329, 331, 333, 335, 336, 341, 343, 358, 367–369, 372, 381, 396, 398, 400, 401
rural food system, 281
rural-urban linkages, 332, 333, 342, 365

S
safety nets, 197–200, 263, 297, 311, 313
seasonal hunger, 394
seasonality, 34, 213, 284
sensitivity, 155, 243, 244, 246–249, 262, 400
shocks and stressors, 10, 12, 13, 15, 17, 155, 156, 162, 170, 187, 210, 212, 239–249, 251–255, 257, 259, 264, 265, 291, 309, 311, 313, 393–395, 398, 399
smallholder farmers, 9, 110, 218, 247, 308
social inclusion, 210, 363, 395
social justice, 154, 341, 400, 402

social networks, 215, 250, 322, 325, 357, 378
social protection, 156, 200, 209, 255, 263, 265, 283, 284, 300, 312, 341, 342, 396
soil and water conservation, 251
South Africa, 10, 17, 257, 303, 322, 326, 327, 332–337, 340, 342, 357, 395, 400, 402
South Asia, 36, 95, 247, 252, 263, 356, 393
sovereignty, 65, 68, 150
stability, 4, 5, 12, 15, 16, 34, 150, 152, 154, 165, 209, 210, 225, 257, 285–287, 313, 326, 328, 356–358, 370, 381, 391, 395
State of Food Insecurity in the World (SOFI), 83, 105, 106, 157–163, 165–167, 390
stunting, 59, 261, 395
supply chain, 9, 39, 41, 55, 56, 153, 154, 191, 192, 213, 215, 258, 302, 321, 322, 324, 325, 327, 392, 398, 400, 401
sustainability, 5, 17, 35, 61, 63, 67–71, 82, 88, 109, 110, 153, 322, 324, 331, 332, 358, 360, 363–366, 368, 370, 381, 391, 399, 401
sustainable development, 63, 209, 217, 356, 362, 363, 366
Sustainable Development Goals (SDGs), 6, 31, 35, 40, 43, 63, 92, 98, 110, 113, 149, 217, 358, 360–363, 392
sustainable livelihoods, 1, 196, 197

T
territorial approach, 326, 328–330, 332, 333, 341–343, 400, 401
territoriality, 322, 340
Theory of Change (TOC), 218

trade, 2, 3, 41, 43, 54–57, 60, 64, 65, 68, 70, 85, 92, 95, 99, 113, 152, 153, 191, 193–196, 201, 208, 209, 213, 215, 326, 328, 333, 335, 336, 341, 393, 397
trade policy, 85, 114, 341
transaction costs, 194
transformation, 10, 19, 53, 54, 59, 63–71, 169, 199, 225, 325, 335, 359, 396
transformative capacity, 156, 198, 251
triple burden of malnutrition, 59
2030 Agenda for sustainable development, 63, 71

## U

ultra-processed foods, 59, 60, 322
undernourishment, 31, 57, 59
urban, 9, 16, 21, 42, 59, 214, 241, 248, 252, 253, 256, 258, 260, 262, 294, 306, 308, 312, 325, 331, 333, 341, 343, 355–371, 374–376, 378–381, 401
urban agriculture, 258, 331, 380, 401
urban food environment, 258, 260
urban food system, 21, 332, 356, 357, 360, 364–370, 378–381, 401
urbanization, 358, 360, 363, 392
urban poverty, 36, 357
urban transition, 356–360, 370, 374, 379

utilization of food, 370

## V

vitamin A deficiency, 37
Voluntary Guidelines on Food Security and Nutrition, 69
vulnerability, 1, 6, 8–11, 187, 192, 209, 212, 218, 222, 240, 247, 251, 261, 300, 311, 327, 333, 357, 378, 399
vulnerability assessment, 224, 367, 380
vulnerable group, 60, 71, 215, 256, 265, 327, 394

## W

water and sanitation, 37, 42, 375, 395
water security, 88, 248, 365
World Food Conference (1974), 33, 81, 392
World Food Price Crisis (2007–08), 104
World Food Programme (WFP), 7, 12, 58, 85, 101, 104–107, 111, 158, 393
World Food Summit (1996), 33, 42, 81, 92, 150, 391, 392

## Z

Zero Hunger, 31, 101, 111, 149

Printed in Great Britain
by Amazon